Sea Buckthorn

Jinhu Tian • Haitian Fang
Shiguo Chen • Changqing Wei • Xiaobo Wei
Editors

Sea Buckthorn

A Functional Food Resource

Springer

Editors
Jinhu Tian
Department of Food Science and Nutrition
Zhejiang University
Hangzhou, China

Haitian Fang
Food Science and Technology
Ningxia University
Yinchuan, Ningxia, China

Shiguo Chen
Department of Food Science and Nutrition
Zhejiang University
Hangzhou, China

Changqing Wei
School of Food Science and Technology
Shihezi University
Shihezi, China

Xiaobo Wei
Food Science and Technology
Ningxia University
Yinchuan, Ningxia, China

ISBN 978-981-97-9864-3 ISBN 978-981-97-9865-0 (eBook)
https://doi.org/10.1007/978-981-97-9865-0

This Springer imprint is published by the registered company Springer Nature Singapore Pte Ltd.
The registered company address is: 152 Beach Road, #21-01/04 Gateway East, Singapore 189721, Singapore

Preface

Sea buckthorn (*Hippophae rhamnoides* L.) is a shrub or small tree of the Hippophae family and is widely distributed in the temperate regions of Eurasia. As commonly known, the roots, stems, leaves, flowers, and fruits of sea buckthorn contain many bioactive components including flavonoids, phenolics, polysaccharides, protein, and polypeptides. Meanwhile, sea buckthorn fruit contains a high content of oil, which is rich in highly active compounds such as unsaturated fatty acids, tocopherols, tocotrienols, and carotenoids. However, those active substances are not fully utilized in food, chemical, or medicine industry.

The book "Sea buckthorn—A functional food resource" will present the knowledge on the sea buckthorn chemistry and nutrition, including the identification, analysis, and application of bioactive chemical components of sea buckthorn, the functional characteristics of sea buckthorn, and the commercial products from Sea buckthorn. The authors will provide a detailed description of the sea buckthorn bioactive components as well as their role during processing and as a nutrition or function supplement. Particularly, a novel application of sea buckthorn protein, flavonoids, phenolics, polysaccharides, and the next generation of healthy sea buckthorn products will be summarized. The present work will provide students, researchers, and technologists working in the area of food, nutrition, engineering, and others who are interested in sea buckthorn, with the most recent information and state-of-the-art technology.

We aim for this book to be a valuable resource for a better understanding of sea buckthorn. Our goal is to deliver high-quality work for the bioactive compounds from sea buckthorn and its application in healthy food. This book encompasses a wide range of topics within chemical compounds of sea buckthorn and health benefits, including the description of polyphenols, flavonoid, polysaccharide, and its health benefits in antitumor, lose weight, cardiovascular disease, etc. Throughout the review and editing process, many suggestions, corrections, and suitable additions of new topics have been included. Despite best efforts, the first edition of a book often presents opportunities for improvement. We welcome and appreciate your valuable suggestions for further enhancing the content coverage.

Hangzhou, China	Jinhu Tian
Yinchuan, Ningxia, China	Haitian Fang
Hangzhou, China	Shiguo Chen
Shihezi, China	Changqing Wei
Yinchuan, Ningxia, China	Xiaobo Wei

Contents

1 **Origin, Distribution, Germplasm, and Cultivation of Sea Buckthorn** 1
Haitian Fang, Yanming Ren, and Jinhu Tian
1.1 Origin 1
1.2 Distribution 4
1.3 Germplasm 6
1.4 Cultivation 8
References 10

2 **Bioactive Compounds from Sea Buckthorn** 13
Yujing Sun, Yihua Wang, Jingtong Guo, and Yuan Zhao
2.1 Polyphenols 14
2.2 Phenolic Acids 14
2.3 Flavonoids 14
2.4 Carotenoids 15
2.5 Polypeptides and Amino Acids 16
2.6 Polysaccharides 16
2.7 Terpenes 17
2.8 Essential Oils 18
2.9 Organic Acids 19
2.10 Phytosterols 20
2.11 Phenylpropanoids and Alkaloids 21
References 22

3 **Current Knowledge of the Sea Buckthorn Composition and Health Benefits** 27
Haitian Fang, Shuifang Mao, Jinghan Wang, and Jinhu Tian
3.1 Nutrients 28
3.2 Bioactive Compounds 30
3.2.1 Carotenoids 30
3.2.2 Polyphenolics 30
3.2.3 Flavonoids 31

3.2.4 Fatty Acids 31
3.2.5 Phytosterols 31
3.2.6 Polysaccharides 32
3.2.7 SOD 32
3.2.8 Organic Acids 32
3.3 Health Benefits 33
3.3.1 Regulation of Blood Glucose and Lipid 33
3.3.2 Anti-depressant and Neuroprotective Effect 33
3.3.3 Antioxidant Mechanisms and Efficacy 34
3.3.4 Hepatoprotective Activity and Mechanism 35
3.3.5 Dermatological Benefits in Treating Skin Diseases 35
3.3.6 Anti-psoriatic and Anti-atopic Dermatitis Property 36
3.3.7 Cardiovascular Health Benefit 36
References 36

4 Active Substances and Antioxidant Activity of Different Solvent Extracts of Sea Buckthorn Leaves 41
Yaoran Li, Xiaobo Wei, Peiyi Zhu, and Jinhu Tian
4.1 Introduction 42
4.2 Morphology of Sea Buckthorn Leaves 42
4.3 Nutrients and Bioactive Compounds 43
4.3.1 Vitamins and Minerals 43
4.3.2 Carbohydrates 43
4.3.3 Organic Acids and Amino Acids 44
4.3.4 Carotenoids 44
4.3.5 Polyphenols 45
4.4 Antioxidant Activity 45
4.5 Extraction of Active Substances from Sea Buckthorn Leaves 46
4.5.1 Antioxidant Properties of Plant Extracts 48
4.5.2 Polyphenolic Extracts 49
4.5.3 Organic Acid Extracts 49
4.5.4 Volatile Oil Extracts 49
4.5.5 Other Compounds 50
4.6 Food Applications 50
4.7 Conclusion and Outlooks 51
References 51

5 Total Flavonoids and Total Phenolics Content in Different Parts of Sea Buckthorn 55
Huiyan Liu, Haitian Fang, Jinhu Tian, Yujun Zeng, and Shuifang Mao
5.1 The Bioactive Compounds of Sea Buckthorn 55
5.2 Analysis of Total Phenolics and Flavonoids 56
5.2.1 Polyphenols in Sea Buckthorn 56
5.2.2 Flavonoid Distribution in Sea Buckthorn 57

5.3 Variability in Phenolic and Flavonoid Content 57
5.3.1 Seasonal and Geographical Variations 57
5.3.2 Genotypic Effects and Extraction Methods 58
References 61

6 **Chemical Composition and Pharmacological Effects of Flavonoids and Phenolic from Sea Buckthorn** 65
Xizhe Fu and Jinhu Tian
6.1 General Introduction 66
6.2 Chemical Composition of Phenolic Substances in Sea Buckthorn 66
6.2.1 Chemical Composition of Flavonoids in Sea Buckthorn . . . 66
6.2.2 Chemical Composition of Other Phenolics in Sea Buckthorn 68
6.3 Methods for Extracting Phenols and Flavonoids in Sea Buckthorn 70
6.3.1 Microwave-Assisted Extraction Technology 70
6.3.2 Ultrasound-Assisted Extraction Technique 71
6.3.3 Enzyme-Assisted Extraction Technique 72
6.4 Biological Activity and Pharmacological Effects of Flavonoids and Phenolics in Sea Buckthorn 73
6.4.1 Role in Plants 73
6.4.2 Antioxidant Activity 73
6.4.3 Cardiovascular Disease 74
6.4.4 Obesity and Diabetes Mellitus 75
6.4.5 Antitumor Activity 76
6.4.6 Osteoporosis 77
6.5 Application Prospects and Challenges of Flavonoids and Phenolics in Sea Buckthorn 78
References 79

7 **In Vitro Anti-inflammatory and Antioxidant Activities of Flavonoids Isolated from Sea Buckthorn** 83
Haizhi Huang
7.1 Introduction 84
7.2 In Vitro Anti-inflammatory Activity of Flavonoids Extracted from Sea Buckthorn 85
7.2.1 Anti-inflammatory Activity of Flavonoids Extracted from the Seed 86
7.2.2 Anti-inflammatory Activity of Flavonoids Extracted from Berries 86
7.2.3 Anti-inflammatory Activity of Flavonoids Extracted from Leaves 87
7.3 In Vitro Anti-inflammatory of Main Flavonoids in Sea Buckthorn . 87
7.3.1 In Vitro Anti-inflammatory Activity of Isorhamnetin 88
7.3.2 In Vitro Anti-inflammatory Activity of Quercetin 89

7.3.3 In Vitro Anti-inflammatory Activity of Kaempferol 89
7.3.4 In Vitro Anti-inflammatory Activity of Myricetin 90
7.3.5 In Vitro Anti-inflammatory Activity of Flavone Glycosides in Sea Buckthorn 90
7.4 In Vitro Antioxidant Activity of Flavonoids Extracted from Sea Buckthorn 91
7.4.1 In Vitro Antioxidant Activity of Flavonoids Extracted from the Seeds 93
7.4.2 In Vitro Antioxidant Activity of Flavonoids Extracted from Berries 93
7.4.3 In Vitro Antioxidant Activity of Flavonoids Extracted from Leaves 94
7.5 In Vitro Antioxidant of Main Flavonoids of Sea Buckthorn 95
7.5.1 In Vitro Antioxidant Activity of Isorhamnetin 95
7.5.2 In Vitro Antioxidant Activity of Quercetin 96
7.5.3 In Vitro Antioxidant Activity of Kaempferol 97
7.5.4 In Vitro Antioxidant Activity of Myricetin 97
7.5.5 In Vitro Antioxidant of Flavone Glycosides in Sea Buckthorn 98
References 99

8 Hypolipidemic Effects of Flavonoids from Sea Buckthorn Leaves 105
Xiaobo Wei, Yaoran Li, and Boyu Huang
8.1 Flavonoids of Sea Buckthorn Leaves 105
8.2 Chemical Composition of Flavonoids in Sea Buckthorn Leaves 106
8.3 Extraction Technology of Flavonoids from Sea Buckthorn Leaves 106
8.3.1 Organic Solvent Extraction 106
8.3.2 Ultrasonic-Assisted Extraction 108
8.3.3 Microwave-Assisted Extraction 109
8.3.4 Supercritical Fluid Extraction 109
8.3.5 Subcritical-Assisted Extraction 110
8.3.6 Enzyme-Assisted Extraction 110
8.4 Purification Technology of Flavonoids from Sea Buckthorn Leaves 111
8.4.1 Decolorization of Macro-porous Resin 111
8.4.2 Decolorization of Magnetic Multi-walled Carbon Nanotubes 112
8.5 Hypolipidemic Effect of Flavonoids from Sea Buckthorn Leaves 112
8.5.1 Effects of Flavonoids from Sea Buckthorn Leaves on Body Weight in Hyperlipidemic Mice 113

8.5.2 Effects of Flavonoids from Sea Buckthorn Leaves on Serum Biochemical Indicators in Hyperlipidemic Mice 113
8.5.3 Effects of Flavonoids from Sea Buckthorn Leaves on Adipogenic and Lipogenic Gene Expression in Hyperlipidemic Mice 114
8.5.4 Possible Pathways of Flavonoids Sea Buckthorn Leaves in Alleviating Hyperlipidemia 114
References 115

9 Gut Microbe Regulation of Sea Buckthorn for the Metabolic Syndrome 117
Rina Wu, Weihe Cang, Shanshan Yang, Danli Yan, Xuwen Ma, Jingjing Li, Jing Li, and Wentao Xing
9.1 Introduction 118
9.2 Interactions Between Gut Microbes and Metabolic Diseases 119
9.2.1 Intestinal Barrier 119
9.2.2 Metabolites. 119
9.2.3 Immune System 120
9.3 Sea Buckthorn Modulates Gut Microbes to Influence Metabolic Diseases 120
9.3.1 Obesity 120
9.3.2 Non-alcoholic Fatty Liver Disease. 121
9.3.3 Diabetes 122
9.3.4 Cardiovascular Disease 123
9.4 Potential Mechanisms by Which Sea Buckthorn Components Affect Metabolic Diseases 124
9.4.1 Effect of Sea Buckthorn on Proteins Related to Lipid Metabolism. 124
9.4.2 Effect of Sea Buckthorn on Biomarkers of Oxidative Stress in Liver and Erythrocytes 125
9.5 Conclusion 126
References 126

10 Bioactivities of Sea Buckthorn Phenolics and Their Anti-cancer Mechanism Toward Human Breast Cancer Cells 129
Xiaobo Wei, Yaoran Li, Peiyi Zhu, and Jinhu Tian
10.1 Introduction 130
10.2 Antioxidant Activity of Sea Buckthorn Phenolics. 130
10.3 Antimicrobial Activity of Sea Buckthorn Phenolics 131
10.4 Anti-inflammatory Activity of Sea Buckthorn Phenolics 131
10.5 Anti-cancer Activity of Sea Buckthorn Phenolics 132
10.6 Protective Effect of Sea Buckthorn Phenolics on Cardiovascular System 133
10.7 Hypoglycemic Effect of Sea Buckthorn Phenolics 134

10.8 Anti-cancer Mechanism of Sea Buckthorn Polyphenols on Human Breast Cancer Cells In Vitro and In Vivo . . . 134
10.8.1 In Vitro . . . 135
10.8.2 In Vivo . . . 137
References . . . 137

11 Immunological Enhancement Effect of Sea Buckthorn Powder . . . 141
Fangyuan Song, Pin Chen, and Jinhu Tian
11.1 General Introduction Of Sea Buckthorn . . . 142
11.2 Processing Status of Sea Buckthorn Fruit Powder . . . 142
11.2.1 Processing Technology of Sea Buckthorn . . . 143
11.2.2 Processing Technology of Sea Buckthorn Fruit Powder . . . 143
11.2.3 Suitability of Variety for Processing Sea Buckthorn Powder . . . 144
11.2.4 Study on Quality of Sea Buckthorn Powder . . . 145
11.3 The Physical Basis of Immune Regulation in Sea Buckthorn . . . 145
11.3.1 Flavonoids . . . 146
11.3.2 Organic Acids . . . 146
11.3.3 Vitamins . . . 146
11.3.4 Polysaccharide . . . 147
11.3.5 Proteins and Amino Acids . . . 147
11.3.6 Terpenoids and Steroids . . . 148
11.4 Immune Regulation of Sea Buckthorn . . . 148
11.4.1 The Effect of Immune Enhancers on Immune Organs . . . 148
11.4.2 Effect of Total Flavonoids of Sea Buckthorn Powder on Immune Function . . . 148
11.4.3 Immune Effect of Vp of Sea Buckthorn . . . 149
11.4.4 Effect of Vp Powder of Sea Buckthorn . . . 150
11.4.5 Effects of Sea Buckthorn Polysaccharide on the Immune System . . . 150
11.5 The Effect of Sea Buckthorn on the Immune System . . . 150
11.5.1 Effects of Sea Buckthorn on Cellular Immunity . . . 150
11.5.2 The Effect of Sea Buckthorn on Humoral Immune Function . . . 151
11.6 Development of Sea Buckthorn Powder Product . . . 151
11.7 Problems and Prospects of Sea Buckthorn Fruit Powder Processing . . . 152
References . . . 153

12 Research Progress on Extraction and Functional Activity of Sea Buckthorn Proteins and Polypeptides . . . 157
Yue Zhao, Jinhu Tian, and Changqing Wei
12.1 Introduction . . . 158
12.2 Extraction Methods of Sea Buckthorn Proteins and Peptides . . . 159
12.2.1 Extraction Method of Sea Buckthorn Protein . . . 159

12.2.1.1 Alkali Acid Extraction Method 159
12.2.1.2 Alcohol Extraction 160
12.2.1.3 Solvent Extraction Method 160
12.2.1.4 Composite Extraction Method 160
12.2.2 Peptides Extraction Methods 161
12.2.2.1 Enzymatic Hydrolysis 161
12.2.2.2 Membrane Separation 161
12.2.2.3 Electrophoresis Method 161
12.3 Characterization of Sea Buckthorn Proteins and Peptides 161
12.3.1 Structure of Sea Buckthorn Proteins 162
12.3.2 Structure of Sea Buckthorn Peptides 162
12.3.3 Methods for the Characterization of Sea Buckthorn Proteins and Peptides 162
12.4 Functional Activity of Sea Buckthorn Proteins and Peptides 163
12.4.1 Antioxidant Activity 163
12.4.2 Immunomodulatory Activity 164
12.4.3 Anti-inflammatory Activity 164
12.4.4 Neuroprotective Activity 165
12.4.5 Hypoglycemic Activity 165
12.4.6 Antifreeze Proteins 166
12.5 Challenges and Potential Solutions 166
References 167

13 Protein Extraction Process and Anti-fatigue Activity of Sea Buckthorn Seed 171
Wei Wang
13.1 Protein in Sea Buckthorn 172
13.2 Extraction Process of Sea Buckthorn Seed Protein 172
13.2.1 Sample Preparation 172
13.2.2 Degreasing 173
13.2.2.1 Cold Pressing 173
13.2.2.2 Solvent Extraction 173
13.2.2.3 Supercritical Fluid Extraction 173
13.2.3 Protein Extraction 174
13.2.3.1 Solid-Liquid Extraction 174
13.3 Structural and Functional Properties of SSP 176
13.3.1 Determination of Solubility 177
13.3.2 Amino Acid Composition Analysis 177
13.3.3 In Vitro Protein Digestibility 177
13.3.4 Fourier Transformation Infrared Spectroscopy 178
13.3.5 Surface Hydrophobicity 178
13.3.6 Molecular Weight Distribution 178
13.3.7 Determination of Microstructure of Protein 179
13.4 Fatigue 179
13.4.1 Depletion Hypothesis 179

13.4.2 Accumulation Hypothesis . . . 180
13.4.3 Oxidative Stress . . . 180
13.4.4 Changes in Brain Neurotransmitters . . . 180
13.4.5 Inflammation . . . 181
13.4.6 Anemic . . . 181
13.5 Constituents with Anti-fatigue Activity in Sea Buckthorn Seed . . 182
13.5.1 Polyphenols . . . 182
13.5.2 Vitamin . . . 183
13.5.3 Lipids . . . 184
13.5.4 Amino Acids . . . 185
13.5.5 Carotenoids . . . 185
13.5.6 Minerals . . . 186
13.6 Conclusion . . . 187
References . . . 187

14 Isolation, Purification, and Biological Activity of Polysaccharides from Sea Buckthorn . . . 191
Jianle Chen, Qi Wang, Huan Cheng, and Pin Chen
14.1 Introduction of Polysaccharides from Sea Buckthorn . . . 192
14.2 Potential Isolation Method of Polysaccharides from Sea Buckthorn . . . 193
14.2.1 Chemical Extraction . . . 193
14.2.2 Hot Water Extraction . . . 193
14.2.3 Enzymatic Extraction . . . 194
14.2.4 Ultrasonic-Assisted Extraction . . . 194
14.2.5 Microwave-Assisted Extraction . . . 195
14.2.6 Supercritical Fluid Extraction . . . 195
14.3 Purification and Structural Analysis . . . 196
14.3.1 Purification of Polysaccharides . . . 196
14.3.1.1 Enrichment of Polysaccharides . . . 196
14.3.1.2 Purification of Polysaccharide Fraction . . . 197
14.3.2 Structural Analysis . . . 198
14.4 Biological Advances in Sea Buckthorn Polysaccharides . . . 199
14.4.1 Antiviral Activity . . . 199
14.4.2 Antitumor Activity . . . 199
14.4.3 Anti-inflammatory and Antioxidant Activities . . . 199
14.4.4 Immunomodulatory Activity . . . 200
14.5 Conclusions and Future Perspectives . . . 200
References . . . 201

15 Chemical Composition and Antioxidant Active Components of Sea Buckthorn Oil . . . 205
Changqing Wei, Wen Tan, Jinhu Tian, and Wenyu Liu
15.1 Introduction . . . 206
15.2 The Extraction Method of Sea Buckthorn Oil Chemical Constituents . . . 207

15.2.1 Pressing Method 207
15.2.2 Solvent Leaching Method 208
15.2.3 Aqueous Enzymatic Method 208
15.2.4 Ultrasonic-Assisted Extraction Method 208
15.2.5 Supercritical Fluid Extraction Method 209
15.2.6 Other Methods 209
15.3 Chemical Composition and Antioxidant Activity of Sea Buckthorn Oil 210
15.3.1 Fatty Acids 210
15.3.2 Vitamins 211
15.3.3 Phenolic Compounds 213
15.3.4 Phytosterols 214
15.4 Pharmacological Effects of Chemical Components of Sea Buckthorn Oil 214
15.4.1 Cardiovascular and Cerebrovascular Protective Effects . . 214
15.4.2 Effects on Skin 215
15.4.3 Effects on Liver and Kidney 215
15.4.4 Others 216
15.5 Application of Chemical Components of Sea Buckthorn Oil 217
References 217

16 Preparation Technology of Sea Buckthorn Essential Oil Microcapsules 221
Wenyu Liu, Changqing Wei, Qingqing Sun, and Jinhu Tian
16.1 Introduction 222
16.2 The Characteristics and Application of Sea Buckthorn Oil 222
16.2.1 Composition and Pharmacological Effects of Sea Buckthorn Oil 222
16.2.1.1 Composition of Sea Buckthorn Oil 222
16.2.1.2 Pharmacological Effects of Sea Buckthorn Oil 223
16.2.2 Sea Buckthorn Oil Product Applications 224
16.3 Introduction of Microcapsule Technology 225
16.3.1 Definition and Basic Principle of Microcapsules 225
16.3.2 Application of Microcapsules in Food and Pharmaceutical Preparations 226
16.4 Sea Buckthorn Oil Microcapsule Technology 226
16.4.1 Physical Methods 226
16.4.1.1 Spray-Drying Method 226
16.4.1.2 Freeze-Drying 227
16.4.1.3 Solvent Evaporation Method 227
16.4.1.4 Physical Adsorption Method 228
16.4.2 Physical and Chemical Method 228
16.4.2.1 Composite Condensation Method 228
16.4.2.2 Orifice Coagulation Bath 229

16.4.2.3 Holdings of Nano-precipitation 229
16.4.3 Chemical Methods 230
16.4.3.1 Interfacial Polymerization 230
16.4.3.2 In Situ Polymerization 230
16.5 Parameter Optimization in Sea Buckthorn Oil Microcapsule Preparation Process 231
16.5.1 Choice of Wall 231
16.5.2 Coating Efficiency 231
16.5.3 Microcapsule Particle Size 231
16.6 Evaluation of Sea Buckthorn Oil Microcapsule Effect 232
16.7 Summary and Outlook 232
References 233

17 Traditional Sea Buckthorn-Based Functional Food in China 237
Haitian Fang, Yanming Ren, and Jinhu Tian
17.1 The Nutritional Value of Sea Buckthorn 237
17.2 Application of Sea Buckthorn in Food 238
17.2.1 Sea Buckthorn Yogurt 239
17.2.2 Sea Buckthorn Tea 241
17.2.3 Sea Buckthorn Jam 242
17.2.4 Sea Buckthorn Wine 243
References 244

18 Effect of Sea Buckthorn on the Quality of Fermented Milk 249
Junrui Wu
18.1 Effect of Sea Buckthorn on the Physicochemical Properties of Fermented Milk 249
18.1.1 pH Value 250
18.1.2 Acidity 250
18.1.3 Ionic Concentration 250
18.1.4 Antioxidant Properties 250
18.2 Effect of Sea Buckthorn on Flavor and Sensory Properties of Fermented Milk 251
18.2.1 Aroma Characteristics 251
18.2.2 Taste Characteristics 251
18.2.3 Color Characteristics 252
18.2.4 Balanced Acidity with Sweetness 252
18.3 Effect of Sea Buckthorn Addition on the Nutritional Value of Fermented Milk 252
18.3.1 Effects of Vitamins 252
18.3.2 Effects of Minerals 253
18.3.3 Effect of Functional Ingredients 253
18.4 Effect of Sea Buckthorn on Lactic Acid Bacteria in Fermented Milk 254
18.4.1 Promotion of Lactic Acid Bacterial Growth 254
18.4.2 Increase in Lactic Acid Production 254

18.4.3 Enhancement of Lactic Acid Bacterial Activity 254
18.4.4 Increase in Nutritional Support for Lactic Acid Bacteria 255
18.5 Development and Application of Sea Buckthorn Polysaccharide Fermented Milk 255
18.5.1 Separation and Purification Techniques for Sea Buckthorn Polysaccharide 256
18.5.1.1 Solvent Extraction Method 256
18.5.1.2 Ultrasonic-Assisted Extraction Method 256
18.5.1.3 Acid Digestion, Enzyme Digestion, and Supercritical Fluid Extraction Method ... 257
18.5.2 Effect of Sea Buckthorn Polysaccharides on Fermented Milk 258
18.5.2.1 Effect of Sea Buckthorn Polysaccharide on Fermented Milk Viscosity 258
18.5.2.2 Effect of Sea Buckthorn Polysaccharides on pH and Acidity of Fermented Milk 258
18.5.2.3 Effect of Sea Buckthorn Polysaccharides on the Water-Holding Capacity of Fermented Milk 258
18.5.2.4 Effect of Sea Buckthorn Polysaccharides on the Color of Fermented Milk 259
18.5.2.5 Effect of Sea Buckthorn Polysaccharides on the Texture of Fermented Milk 259
References 260

19 Product Development of Sea Buckthorn Fruit Wine 263
Huiyan Liu, Haitian Fang, Yujun Zeng, Pin Chen, and Jinhu Tian
19.1 Sea Buckthorn Fruit Wine 263
19.2 The Production Process of Sea Buckthorn Fruit Wine 264
19.2.1 Fermentation Technology of Sea Buckthorn Wine 265
19.2.2 The Fermentation Strain of Sea Buckthorn Wine 266
19.3 The Flavor of Sea Buckthorn Wine 268
19.4 Fermentation Kinetics of Sea Buckthorn Wine 273
References 274

Chapter 1
Origin, Distribution, Germplasm, and Cultivation of Sea Buckthorn

Haitian Fang, Yanming Ren, and Jinhu Tian

Abstract Sea buckthorn (*Hippophae rhamnoides* L.), recognized by China's National Health Commission as a "medicine food homology" fruit, is rich in nearly 200 bioactive compounds, including polyunsaturated fatty acids, carotenoids, sugar alcohols, superoxide dismutase, phytosterols, and polyphenols. This classification underscores its dual role in nutrition and medicine. Sea buckthorn's notable bioactivity encompasses cardiovascular health, anti-diabetic properties, and anti-obesity effects.

Keywords Sea buckthorn · Origin · Distribution · Germplasm · Cultivation

Abbreviation

CVD Cardiovascular diseases

1.1 Origin

Sea buckthorn, also known as vinegar willow, blackthorn, or sour thorn, is scientifically termed *Hippophae rhamnoides* L. in English and *Rhamnus saxifraga* in Russia. Originating in East Asia, it was dispersed by birds of prey and rivers throughout Eurasia, particularly in Russia, Mongolia, and China. Sea buckthorn, a dioecious wild shrub or small tree, boasts a robust root system resistant to various environmental stresses including wind and sand, drought, water and moisture, salinity, barrenness, heat, and cold. Its coverage can reduce surface runoff by 85%

H. Fang
Food Science and Technology, Ningxia University, Yinchuan, Ningxia, China

Y. Ren · J. Tian (✉)
Department of Food Science and Nutrition, Zhejiang University, Hangzhou, China

J. Tian et al. (eds.), *Sea Buckthorn*, https://doi.org/10.1007/978-981-97-9865-0_1

topsoil erosion by 75%, and wind erosion by 85%, underlining its efficacy in soil improvement and conservation. The understanding of sea buckthorn's properties has evolved significantly over time. Initially regarded merely as an ornamental plant for parks or roadsides, it was later discovered that sea buckthorn's fresh fruit contains high levels of vitamin C (120–202 mg/100 g) than almost any other fruit and vegetable (Chinese kiwi about 130 mg/100 g, mandarin about 56 mg/100 g, spinach about 49 mg/100 g, and sea buckthorn about 100 mg/100 g). The advancement of modern testing instruments has led to a deeper understanding of the biology, chemistry, and pharmacology of the diverse bioactive substances in sea buckthorn. Containing a variety of bioactive substances, sea buckthorn fruit has shifted from being studied as food to medicinal use. This has expanded its development from the fruit to the comprehensive utilization of the whole plant (Olas et al. 2017). Sea buckthorn, known for its medicinal properties since ancient times, has been acknowledged as Chinese and Indian medicine for thousands of years. Its medicinal value has been acknowledged for centuries (Wen et al. 2020) and is documented in Ayurveda, the classic ancient Indian medicine system (circa 5000–500 BC), by Theophrastus and Dioscorides (Dhyani et al. 2010). In the eighth-century Tang Dynasty, the Tibetan medicinal classic *rGyud-bzi* (The Four Books of Pharmacopeia) recognized sea buckthorn for treating gastrointestinal diseases, cardiovascular diseases (CVD), as well as burn injuries (Stobdan et al. 2013).

The scientific designation of sea buckthorn, *Hippophaë rhamnoides* L., reflects its characteristics. The origin and meaning of this term differ among various authors. In *Sea Buckthorn: A Pharmacy in a Plant*, Brad et al. (2002) explain the name's origin, tracing it back to the use of sea buckthorn fruits in ancient times for treating intestinal worms in horses. They suggest that the genus name *Hippophaë*, as termed by botanists, is derived from the Greek words *hippos* (horse) and *phao* (to kill). Conversely, Olas (2016) proposes a different etymology: *hippo* (horse) and *phao* (light), translating to "shiny horse hair," thereby rendering the bush as "a shiny horse thorn."

Sea buckthorn is known variety of names globally, often reflecting regional characteristics such as location, thorn presence, fruit color, or perceived benefits. In Nordic countries, it is esteemed as the "oil tree of the north," and in Germany, it is known to as "Sanddorn." Common names included sand willow, sea thorn, shore thorn, and several others, each reflecting a unique characteristic of the plant. The Dutch have nicknamed it "laxative fruits" due to its acids that stimulate intestinal transit. In Russian dialects, there are as many as 23 different names for sea buckthorn. The Uzbekistani city of Dzbidda derives its name from sea buckthorn. Other names, like "thorny shrub" or "willow bush," reflect the plant's botanical features. In Russian, "Oblepicha" indicates fruits tightly clinging to the branches. The English name, sea buckthorn (the thorn of sea goat), is somewhat unique, with enthusiasts preferring it over any association with "backgeruch" (goat smell). In Romania, it is known by various names such as river white sea buckthorn, river sea buckthorn, and spiny sea buckthorn, among others. In Buzau County's locality of Cătina (sea buckthorn), it is revered as "the fruits of the Mother of God" (Raţi and Raţi 2003).

The therapeutic values of sea buckthorn were first discovered by various ancient cultures, including the Thracians, Hindus, Greeks, Chinese, Mongols, Celts, and Slavs. Sea buckthorn has been used extensively in traditional ethnobotanical and ethnopharmacological practices throughout history. Its earliest documentation is found in ancient Greek texts by Theophrastus and Dioscorides and in Ayurveda, the classical ancient Indian medicinal system of medicine written between circa 5000 and 500 BC, where sea buckthorn was recommended for feeding race horses to enhance muscle mass (Brad et al. 2002). Ancient Greeks utilized various parts of sea buckthorn for diverse purposes. The young shoots and leaves, used as animal feed, led to rapid weight gain and improved hair quality in animals, particularly in horses (Xing et al. 2002).

Classical Tibetan medical literature, such as the *rGyud-bzi* (Four Books of Pharmacopeia) from the Tang Dynasty (618–907 AD) also mentions sea buckthorn, recommending its consumption (*starbu*) for high-altitude travelers. The eighth-century work *Djud-shi* by the renowned physician Yuthog Yontan Gonpo lists over 300 medicinal preparations of sea buckthorn, used alone or combined with other plants, minerals or even foods, and processed in various forms such as juice, extracts, powders, pills, lard, liqueur, patchers, compresses, ointments, and pastes.

In the thirteenth century, Mongolian pharmacologist Losang Que-Pei authored a 120-chapter compilation based on Tibetan medicine. This work included numerous sea buckthorn-based recipes for treating lungs, stomach, intestines, liver, and gallstone ailments, as well as female-specific conditions, rheumatic pains, and joint edema. The *Dsejchar Migczan*, an ancient Tibetan medicinal plant reference book discovered in the nineteenth century, synthetically describes sea buckthorn's effects in the context of Mongolian medicine. Sea buckthorn is characterized by its hardening and velvety taste, acting through the "bagdan" to facilitate healing of the lungs and throat. Consequently, its concentrate is referred to as the "Blood of the King's Heart." Its seeds are also used to "dissolve evil," thin the blood, and heal the "bagdan." Khanda, the thick extract of sea buckthorn oil, was utilized to regulate mucus and treat blood swellings, including heamatomas and bleeding. It is particularly effective in the treatment of inflammatory and infectious processes, stimulating mucosal activity and beneficially affecting esophageal tumors, stomach issues, and appendix inflammation.

Even in modern times, the plant's virtues have been acknowledged, notably when Soviet cosmonauts used it experimentally in their diet. Moscow scientists discovered that it offers protection against cosmic radiation, in addition to its other benefits. The recognized potential of sea buckthorn prompted its cultivation as a crop in countries including the United States, Canada, Germany, Russia, China, and Ukraine. In Europe and Central Asia, every part of the plant-fruits, leaves, roots and bark-was utilized for various purposes including food, dietary supplement, firewood, fuel, animal feed, decoration and fencing.

Research on sea buckthorn in Romania, particularly focusing on the extraction of vitamin-rich concentrates and the enhanced utilization of the plant, commenced in 1953 under the leadership of the university professor Ion Brad. Prof. Brad pioneered to the study of sea buckthorn's biochemical composition in Romania, conducting extensive research on its applications in medicine, food, cosmetics, and animal feed. Emanoil Grigorescu, in 1963, became the first Romania doctorate in this field with his thesis "Contributions to the pharmacogenetic and Phytochemical Study of Indigenous *Hippophaë Rhamnoides* L." initially coordinated by Prof. Dr. Elemer Kopp and later by Prof. Dr. Teodor Goina. Prof. Dr. Eng. Victor Cireașă, a key figure in promoting sea buckthorn culture in Romania, recognized the plant's potential for cultivation beyond its natural habitat. He often referred to the sea buckthorn fruit as "the strongest fruit with divine oil" and "the medicinal berries of the Mother of God" (Rați and Rați 2003).

In 2001, the International Sea Buckthorn Association (ISA) was established, marking the first global organization dedicated to promoting and developing sea buckthorn. The inaugural Sea Buckthorn Congress, held from September 14 to 18, 2003, in Berlin, Germany, was organized by the German Sea Buckthorn Association with Humboldt University and Technical University Berlin.

1.2 Distribution

The term "sea buckthorn" (*H. rhamnoides* L.) encompasses all species within the *Hippophae* genus in the Elaeagnaceae family (Sharma et al. 2019). This thorny nitrogen-fixing deciduous shrub is native to regions extending from northwestern Europe to Central Asia, the Altai Mountains, western and northern China, and the northern Himalayas (Pundir et al. 2021).

Sea buckthorn naturally grows across Central Asia, China, Mongolia, the Caucasus and Siberia. In Poland, sea buckthorn is found exclusively along the Baltic coast, having been introduced there by Polish exiles returning from the Siberian gulags in the late of the nineteenth century. A distinct characteristic of sea buckthorn is its intensely orange berries, densely clustered along the shoots (Criste et al. 2020). Currently, Russia, having selected over 60 varieties, leads in the cultivation of sea buckthorn. H. *rhamnoides* is cultivated on an industrial scale in countries like China, Germany, Finland, and Estonia. In Poland, smaller plantations are found in the Suwalki region and Podlasie region, as well as the southwestern part of the country. Sea buckthorn's adaptability to poor soils is attributed to the soil bacteria Actinomycetes Frankia. These bacteria live in symbiosis with the plant's roots, enabling nitrogen fixation from the air. Its capability to bind atmospheric nitrogen and other compounds makes sea buckthorn a viable candidate for soil remediation.

Observations reveal that sea buckthorn plants are exclusive to the Eurasian continent, thriving primarily temperate and cold-temperate regions. While some species extend as far south as 27°N, they only flourish at high altitudes in these regions, maintaining a moderate ecological character. Additionally, our research suggests that the sea buckthorn genus likely originated from the Transverse Mountains and the eastern Himalayas, classifying it within the Old World temperate floristic geography. In Europe, only four subspecies of *Hippophae rhamnoidesare* recognized: subsp. Caucasia in the Caucasus, subsp. Carpatica in the Carpathians, and subsp. fluviatilis in the Alps and along the Baltic, North Sea, and Norwegian coast. These subspecies, characterized by a glossy-seeded pericarp that separates from the seed coat, fall into the tareless group. Asia hosts the greatest diversity and range of the *H. rhamnoides* genus, encompassing six species and six subspecies. This diversity is part of the genus's total of six species and ten subspecies. The distribution of *Hippophage rharmnoides* includes a variety of subspecies: subsp. mongolia from Altai to Siberia and Mongolia; subsp. turkestanika in Central Asia; subsp. Salicifolia. subsp. Yunnanensis, and subsp. Stellatopilosa, as well as subsp. goniocarpa litangensis in the Himalayan region and the Hengduan Mountains. Additionally, subsp. sinensis is found from the Transverse Mountains to northwestern and northern China. The Tibetan Plateau hosts *H. neurocarpa, H. goniocarpa*, and *H. gyantsensis*. Tibetan buckthorn (*Hippophae tibetana*) is prevalent on the Qinghai–Tibetan Plateau. In summary, six species and six subspecies are distributed from the Hengduan Mountains to the Qinghai–Tibetan Plateau, with China being the most abundant region for Buckthorn taxa, both in terms of distribution and resources (Zhao et al. 2020).

The *Hippophae* genus, originating from the Tibetan plateau, is widely distributed across the temperate regions of Eurasia. The range of Hippophae extends from Nepal's southern Himalayans slopes to Norway's Atlantic coast and from northeastern China to Spain in southern Europe, covering longitudes of 2°–123°E and latitudes from 27° to 69°N. Its distribution encompasses areas from the European coast to the Caucasus Mountains at elevations over 3000 m and the Green Plateau and the Himalayas at elevations surpassing 5000 m. The barkless group of European sea buckthorn includes four subspecies: the Caucasian, Carpathian, brook, and seaside sea buckthorn. The Caucasian sea buckthorn is primarily founded in the Caucasus, the Carpacian and stream-born sea buckthorn in the Alpine region, and the seaside sea buckthorn along the Baltic Sea, North Sea shores, and the Norwegian Atlantic coast. The continent hosts numerous sea buckthorn species and subspecies, with six species and eight subspecies out of the total of 12 subspecies. Mongolian sea buckthorn is prevalent in Altai, Siberia and Mongolia; Central Asian sea buckthorn in Central Asia; and willow-leaved sea buckthorn in the Himalayan region. The Yunnan sea buckthorn, dense-ribbed fruit buckthorn, and Litang sea buckthorn are founded throughout the Hengduan Mountains. Chinese sea buckthorn proliferates in the Hengduan Mountains as well as in China's "Three Norths" region. Additionally, sea buckthorn is common in these areas and the Tibetan Plateau. Furthermore, sea buckthorn inhabits the Qilian Mountains, river sea buckthorn thrives in the Transverse Range and the Himalayas, and Tibetan sea buckthorn is prevalent in the Tibetan Plateau, Himalayas, and Qilian Mountains. The region stretching from the Transverse Range to the Tibetan Plateau represents the densest distribution of the sea buckthorn genus (Kondrateva et al. 2021). Understanding the distribution of each sea buckthorn is essential.

1.3 Germplasm

Sea buckthorn demonstrates robust growth across various soil conditions, thriving in both arid and waterlogged environments. This species features an extensive root network, resilient roots, and significant vegetative development. It is notably effective at preventing wind, sand, and soil erosion. The United Nations has recognized it as economically valuable, and it plays a pivotal role in China's "Three Norths" protection forest system and forestry ecological initiatives. Sea buckthorn is multifunctional, improving soil, preventing wind erosion, stabilizing sand, conserving water, and offering medicinal benefits. Ancient Chinese medical texts and modern pharmacopoeias document sea buckthorn fruit's efficacy in clearing heat, relieving coughs, reducing dampness, dispelling phlegm, acting as a diuretic, lowering blood pressure, and alleviating asthma. Sea buckthorn's versatility is evident in the variety of products, including beverages, foods, medicines, and cosmetics made out of it. This underscores its significance as a species with extensive ecological, economic, and social benefits.

The genus Sea Buckthorn, comprising six species and 12 subspecies, finds its most diverse representations in China, which harbors four species and eight subspecies. Notably, the subspecies such as Seashore, Creek, Caucasian, and Carpathian Sea Buckthorn diverge from this trend. China holds the preeminent position in both the artificial cultivation of sea buckthorn and the richness of its natural germplasm resources. This prominence establishes China as a crucial player in the international sphere for breeding, conservation, and effective utilization of sea buckthorn germplasm resources. However, research and application have been largely confined to the Chinese subspecies of sea buckthorn, with limited exploration of other species. These species, despite their immense potential to adapt to harsh ecological conditions and their notable biological and economic attributes, are mostly confined to marginal areas, facing extinction threat and necessitating urgent research and conservation efforts.

Since 1987, the Chengdu Institute of Biology at the Chinese Academy of Sciences has focused on researching the conservation and use of sea buckthorn and Tibetan sea buckthorn. Since 1996, collaboration with the Sea Buckthorn Development and Management Centre of the Ministry of Water Resources has focused on identifying effective methods for utilizing sea buckthorn germplasm resources. These endeavors have led to the successful introduction of European breeds and stream-born sea buckthorn.

Since 1986, approximately 50–60 varieties of sea buckthorn from the former Soviet Union and Mongolia have been imported into China through various channels. Unfortunately, most of these varieties have shown suboptimal performance, exhibiting issues such as shrinkage disease and premature fruit drop. Nevertheless, a few have emerged as valuable genetic materials for breeding programs. Species such as Chuisk, Sun, and Ulangi have been domesticated and now play a significant role in the establishment of economic forests of buckthorn in northeastern China and parts of Xinjiang.

Dating back to 1933, Russia has invested significantly in both the research and conservation of sea buckthorn germplasm, systematically collecting and utilizing it across various geographical zones. This includes focusing on Mongolian sea buckthorn and other varieties like seaside and Caucasian sea buckthorn, with particular emphasis in Siberia. Eastern European countries, including Germany, Romania, Poland, the Czech Republic, and Nordic countries like Finland and Sweden, have dedicated considerable efforts toward the research, preservation, and application of sea buckthorn varieties, including brook and Caucasian sea buckthorn. In Germany, beyond prohibiting the commercial harvesting of natural sea buckthorn fruit, over 20 original species and new varieties have been preserved in the National Forest and Fruit Tree Germplasm Resource Bank in Dresden (Mokni Ghribi et al. 2015).

In Changji Prefecture, there is only one naturally occurring species of sea buckthorn, specifically the Central Asian sea buckthorn subspecies as identified by researcher Zhang Jianguo of the Chinese Academy of Forestry Sciences. This subspecies predominantly thrives in the Qianshan Valley area. The sea buckthorn cultivated in the Changji Prefecture predominantly comprises a high-yielding variety with large fruits, developed through selective breeding of wild sea buckthorn strains from both domestic and international sources. Upon investigation, 11 superior sea buckthorn germplasms with large fruits were identified, including Shouyuan Yellow, Deep Autumn Red, Thornless Fung, Allee, Liaofu No. 1, Qishen No. 1, Xiangyang, Giant, New Buckthorn No. 1, 2, and 3. Notably, Allee is the only male variety promoted in Xinjiang, typically maintained at male to female ratio of 8:1 (Li et al. 2008).

In Lanxian, the prevalent sea buckthorn belongs to the Chinese variety. However, specific geographical and climatic conditions have led to the emergence of several distinct types. Systematic surveys and analyses of fruits in the primary distribution areas of natural sea buckthorn resources revealed that the southern slopes exhibited higher fruiting rates, larger fruits, and more juice compared to the northern slopes. Similarly, the lowlands outperformed the slopes in these aspects. Several varieties with impressive community performance were identified within the natural forest and categorized into four types based on fruit size, shape, color, and taste. These types were classified as Arashi 1, 2, 3, and 4 and were further analyzed for their physiological and biochemical characteristics after their reintroduction to the nursery in Wangsi Township (Wang et al. 2021).

Sea buckthorn reserves are optimally established on well-drained, level, or gently sloping lands featuring loose, well-aerated, neutral, or slightly alkaline soils with a pH range of 7–8. Such fertile sandy soils necessitate irrigation. When constructing reservoirs on steep slopes or undulating terrains, it is imperative to undertake level cultivation or build horizontal terraces, choosing sites shielded from wind and offering partial shade (Sofi et al. 2020).

1.4 Cultivation

The Qinghai region's fragile ecological environment demands heightened attention and continuous effort in ecological restoration. Sea buckthorn, a versatile forest species, aligns with national policies of rehabilitating farmland and transforming decertified areas into green zones. Its adaptability to extreme climates, resistance to cold, drought, and barrenness, and well-developed root system make it a potent agent against soil erosion and nitrogen fixer. Beyond its environmental benefits, sea buckthorn offers edible, medicinal, and fodder values, contributing significantly to the region's ecological protection and yielding tangible economic benefits for farmers and herdsmen. However, a gap exists in advanced cultivation and management techniques among farmers and herdsmen, leading to an increase in pest and disease incidence, adversely impacting the growth and productivity sea buckthorn seedlings.

There are stringent requirements for the seedling base, including convenient access to transport, adequate water and electricity, favorable terrain, rich organic content, and a deep soil layer of loamy or sandy loam. The soil depth should exceed 50 cm, with a water table below 1 m, and a pH value maintained at 6.5–7.5. Pre-sowing preparations involve fertilizing and creating a seedbed with the soil should be plowed to a depth of 20–30 cm prior in autumn using a deep pine or tiller for fine breaking and aeration. Considering local climate conditions and seed ripeness, sand and seeds typically reach maturity between September and October. Harvesting is done in the following winter or spring. In Qinghai's high-altitude areas, averaging above 2000 m, establishing the optimal sowing date for sea buckthorn is crucial, considering local altitude and climate. The best sowing period at high altitudes is generally mid-April to early May, with a sowing rate of 3–5 kg per acre. Sowing methods include strip sowing, scattering, or hole sowing. The common practice involves mixing sea buckthorn seed with fine sand, evenly spreading it over the seedbed, covering it with a 0.5 cm layer of fine moist soil, gently patting it down, and applying appropriate pressure. Post-sowing, small arches or mulch can be constructed over the seedbed to maintain soil temperature, increase soil permeability, and prevent excessive water loss. Within 10–15 days of sowing, when germination reached 50% or more, shade nets can be placed over the seedbed to protect the seedlings from sunburn (Milán-Noris et al. 2018).

The introduction of sea buckthorn in the northwest of China primarily addresses the region's salinity, cold, drought, and barren planting conditions, making it well-suited for cultivation in this area. To boost sea buckthorn cultivation and enhance fruit yield for the economic development of West China, the adoption of more scientific planting methods is essential. A comprehensive analysis of factors such as pests and disease control, cultivar selection, field management, and planting patterns is crucial to effectively enhance the quality and production of sea buckthorn in China.

As sea buckthorn enters the fruiting stage, its water and more fertilizer requirement intensity. Inadequate watering and fertilizing during this period can weaken the tree, increasing susceptibility to pests and diseases. Precise timing of fertilization and consistent care during the fruiting period are vital for healthy growth. Applying approximately 3000 kg of organic fertilizer per mu (1 mu = 667 m^2) is recommended to ensure optimal growth (Pundir et al. 2021).

Regarding irrigation, it is important to monitor the frequency of watering during the budding, growth, and pre-wintering stages, with the growth stage typically requiring the most irrigation. Sea buckthorn's robust rootstocks can bear fruit within just 2 years. To achieve stable branch growth and higher yields, it is imperative to remove overgrown, diseased, or dead branches, which also extend the fruiting period. In the initial years after planting, single-stem plants are pruned to encourage lateral branching, leading to low, multi-stemmed plants. Subsequent years involve thinning, short pruning, and heart removal techniques to enhance ventilation, light penetration, and nutrient accumulation, thereby improving fruit set.

After 15 years, sea buckthorn enters senescence, necessitating repruning and renewal. On productive soils, pruning rejuvenates the main stem and fruiting branches, helping to restore the plant's vigor. In aging self-rooted fields with infertile soil, stubble renewal is often practiced to propagate new plants.

Sea buckthorn cultivation may be vulnerable to insect pests, potentially resulting in yield loss. The sea buckthorn fruit fly and wood-boring moth, primarily responsible for significant damage, can be effectively managed through three methods. Firstly, light trapping involves observing and studying the behaviors of these pests, identifying their phototropic characteristics, and utilizing specialized lights for effective trapping. Secondly, yellow board trapping, employed during the pest season, attracts and eliminates pests using strategically placed yellow boards coated with a sticky substance in sea buckthorn fields. Lastly, pharmacological insecticides, such as a 50% monocrotophos–malathion emulsion or a 20% permethrin emulsion, are injected into the insect holes to eradicate larvae within the branches (Feng et al. 2019).

The primary disease currently threatening sea buckthorn is dry shrinkage. This disease deforms the fruit, making it uneven, distorted, and sometimes unrecognizable, leading to substantial yield reductions and significant economic losses for growers. Shrinkage results from inadequate sterilization during planting and sea buckthorn's vulnerability to environmental pathogens. Consideration of local climate and soil conditions is crucial. Using disease-resistant seeds, enhancing daily management, ensuring clear drainage, and controlling soil moisture to prevent waterlogging are essential strategies. Regular soil loosening to improve ventilation and air permeability, along with reducing mechanical damage and pathogen infestation, will enhance sea buckthorn's growing conditions and resistance. Immediately remove diseased branches and apply treatments such as 800-fold diluted methomyl, 400-fold diluted 77% dornin, 500-fold diluted 40% carbendazim, or 500-fold diluted 70% red sun fungicide every 30 days, 4–5 times (Yang et al. 2020).

Enhancing the survival rate and quality of sea buckthorn requires scientifically guided planting to foster its growth and meet production goals. Sea buckthorn, a vital economic crop, necessitates scientific cultivation, including adequate fertilization and pest control, to ensure its healthy growth. This approach underpins the sustainable development of the sea buckthorn industry and boosts local economies.

References

Brad I, Brad IL, Radu F (2002) Sea buckthorn: a pharmacy in a plant. Technical Publishing House, Bucharest

Criste A, Urcan AC, Bunea A, Pripon Furtuna FR, Olah NK, Madden RH, Corcionivoschi N (2020) Phytochemical composition and biological activity of berries and leaves from four Romanian sea buckthorn (*Hippophae rhamnoides* L.) varieties. Molecules 25(5):5

Dhyani D, Maikhuri RK, Misra S, Rao KS (2010) Endorsing the declining indigenous ethnobotanical knowledge system of sea buckthorn in Central Himalaya, India. J Ethnopharmacol 127(2):329–334

Feng L, Peng F, Wang X, Li M, Lei H, Xu H (2019) Identification and characterization of antioxidative peptides derived from simulated in vitro gastrointestinal digestion of walnut meal proteins. Food Res Int 116:518–526

Kondrateva IA, Borodulina ID, Zubarev YA, Zemtsova AY (2021) Integrated biochemical estimation of seabuckthorn berries in conditions of the forest-steppe area of the Altai Krai. IOP Conf Ser Earth Environ Sci 670(1):012009

Li Y, Jiang B, Zhang T, Mu W, Liu J (2008) Antioxidant and free radical-scavenging activities of chickpea protein hydrolysate (CPH). Food Chem 106(2):444–450

Milán-Noris AK, Gutiérrez-Uribe JA, Santacruz A, Serna-Saldívar SO, Martínez-Villaluenga C (2018) Peptides and isoflavones in gastrointestinal digests contribute to the anti-inflammatory potential of cooked or germinated desi and kabuli chickpea (*Cicer arietinum* L.). Food Chem 268:66–76

Mokni Ghribi A, Maklouf Gafsi I, Sila A, Blecker C, Danthine S, Attia H, Bougatef A, Besbes S (2015) Effects of enzymatic hydrolysis on conformational and functional properties of chickpea protein isolate. Food Chem 187:322–330

Olas B (2016) Sea buckthorn as a source of important bioactive compounds in cardiovascular diseases. Food Chem Toxicol 97:199–204

Olas B, Kontek B, Szczesna M, Grabarczyk L, Stochmal A, Zuchowski J (2017) Inhibition of blood platelet adhesion by phenolics' rich fraction of *Hippophae rhamnoides* L. fruits. J Physiol Pharmacol 68(2):223–229

Pundir S, Garg P, Dviwedi A, Ali A, Kapoor VK, Kapoor D, Kulshrestha S, Lal UR, Negi P (2021) Ethnomedicinal uses, phytochemistry and dermatological effects of *Hippophae rhamnoides L.*: a review. J Ethnopharmacol 266:113434

Raţi IV, Raţi L (2003) Sea buckthorn in agricultural holdings. Ministry of Agriculture, Forests, Waters and Environment, National Agricultural Consulting Agency, Bucharest

Sharma A, Singh V, Sharma A, Negi N (2019) Seabuckthorn a new approach in ecological restoration of Himalayan ecosystem: a review. Int J Chem Stud 7:1219–1226

Sofi SA, Singh J, Mir SA, Dar BN (2020) *In vitro* starch digestibility, cooking quality, rheology and sensory properties of gluten-free pregelatinized rice noodle enriched with germinated chickpea flour. LWT-Food Sci Technol 133:110090

Stobdan T, Korekar G, Srivastava R (2013) Nutritional attributes and health application of seabuckthorn (*Hippophae rhamnoides* L.)- a review. Curr Nutr Food Sci 9:151–165

Wang K, Xu Z, Liao X (2021) Bioactive compounds, health benefits and functional food products of sea buckthorn: a review. Crit Rev Food Sci Nutr 5:1–22
Wen P, Zhao P, Qin G, Tang S, Li B, Zhang J, Peng L (2020) Genotoxicity and teratogenicity of sea buckthorn (*Hippophae rhamnoides* L.) berry oil. Drug Chem Toxicol 43(4):391–397
Xing J, Yang B, Dong Y, Wang B, Wang J, Kallio HP (2002) Effects of sea buckthorn (*Hippophaë rhamnoides L.*) seed and pulp oils on experimental models of gastric ulcer in rats. Fitoterapia 73(7–8):644–650
Yang J, Huang J, Dong X, Zhang Y, Zhou X, Huang M, Zhou G (2020) Purification and identification of antioxidant peptides from duck plasma proteins. Food Chem 319:126534
Zhao L, Geng T, Sun K et al (2020) Proteomic analysis reveals the molecular mechanism of Hippophae rhamnoides polysaccharide intervention in LPS-induced inflammation of IPEC-J2 cells in piglets. Int J Biol Macromol 164:3294–3304

Chapter 2
Bioactive Compounds from Sea Buckthorn

Yujing Sun, Yihua Wang, Jingtong Guo, and Yuan Zhao

Abstract Sea buckthorn is a rich source of bioactive compounds including carotenoids, polyphenols, polysaccharides, terpenes, organic acids, and phytosterols. These compounds contribute to its various health benefits, such as antioxidant, anticancer, anti-hyperlipidemic, anti-obesity, anti-inflammatory, antimicrobial, antiviral, dermatological, neuroprotective, and hepatoprotective properties. The bioactive composition of sea buckthorn is influenced by several factors including genetic variation, plant part analyzed, climatic and growth conditions, year and degree of harvest, storage conditions, harvest time, and processing and analytic methods. This chapter focuses on key substances such as polyphenols, carotenoids, polypeptides, polysaccharides, terpenes, essential oil, organic acids, phytosterols, phenylpropanoids, and alkaloids.

Keywords Detergent Fiber · Gallic Acid Equivalents · Carotenoids · Polyphenols · Polysaccharides · Terpenes · Organic acids · Phytosterols

Abbreviations

GAE	gallic acid equivalents
FW	fresh weight
DW	dry weight
DF	detergent fiber
NDF	neutral detergent fiber
ADF	acid detergent fiber
HDL-C	high-density lipoprotein cholesterol
LDL-C	low-density lipoprotein cholesterol

Y. Sun (✉) · Y. Wang · J. Guo · Y. Zhao
College of Food Science and Technology, Zhejiang University of Technology, Hangzhou, China
e-mail: yjsun01@zjut.edu.cn

J. Tian et al. (eds.), *Sea Buckthorn*, https://doi.org/10.1007/978-981-97-9865-0_2

2.1 Polyphenols

Polyphenols represent the most abundant and widely distributed class of specialized metabolites in plants, typically classified based on the number of phenolic rings and other structural elements (Chen et al. 2023). They are the main antioxidants in sea buckthorn, with their content in the fruit ranging from 12.36 to 34.6 mg GAE/g; this concentration surpasses that in oranges (1.27 mg GAE/g), mandarins (1.16 mg GAE/g), blueberries (2.19 mg GAE/g), sour cherries (2.56 mg GAE/g), and strawberries (1.12 mg GAE/g) (Kuhkheil et al. 2017; Verica et al. 2009). All parts of the sea buckthorn plant, including berries, roots, leaves, stems, and branches, contain polyphenols. However, the polyphenol concentrations in the fruits and leaves vary depending on the variety, geographical location, and physiological maturity of a plant (Chen et al. 2023). A recent review has identified that nearly 100 polyphenolic compounds have been isolated and identified from sea buckthorn, primarily phenolic acids and flavonoids (Ji et al. 2020). The total phenol content, calculated in gallic acid equivalent (GAE), ranges from 32.93 to 1417 mg/100 g of tissue, varying mainly among different sea buckthorn varieties (Guo et al. 2017a; Rop et al. 2014).

2.2 Phenolic Acids

Phenolic acids, simple non-flavonoid compounds, form a significant group of plant phenolics. They exhibit a broad spectrum of pharmacological activities, including antioxidant, antimutagenic, antitumor, and anticarcinogenic properties (Kampa et al. 2004). Comprising mainly hydroxybenzoic and hydroxycinnamic acid derivatives, these acids vary in the number and position of hydroxylation and methoxylation on the aromatic ring. Found in both free and bound forms in nature, phenolic acids occur more commonly as esters and glycosides (Arimboor et al. 2008). Gallic acid, predominantly in its free form, is the most abundant phenolic acid in sea buckthorn, with differing concentrations in leaves (79 mg/kg) and berries (16.9 mg/kg). Other phenolic acids, such as caffeic acid, *p*-coumaric acid, and ferulic acid, are present at much lower concentrations in sea buckthorn (Bittova et al. 2014).

2.3 Flavonoids

Flavonoids are implicated in the prevention of chronic diseases, including diabetes, cardiovascular disease, and cancer (Ciesarova et al. 2020). Guo et al. (2017b) revealed that the total phenols and flavonoid aglycones in sea buckthorn extracts exhibit antioxidant and anti-proliferative properties. Sea buckthorn berries contain flavonoids in concentrations several times higher than other high-flavonoid plants, such as hawthorn, dogwood cherry, and blueberry (Ji et al. 2020). Predominant phenolic compounds in sea buckthorn include flavonoid glycosides, such as isorhamnetin, quercetin, myricetin, and kaempferol (Guo et al. 2017b). To date, 95 flavonoids have been identified in sea buckthorn, encompassing 75 flavonols, 2 dihydroflavones, 6 catechins, 1 leucocyanidin, 9 anthocyanidins, 1 proanthocyanidin, and 1 chalcone (Liu et al. 2021a).

The content of flavonoids varies significantly across different parts of the sea buckthorn plant. For instance, sea buckthorn leaves 2.5 to 64 times more total flavonoids than the fruit, making them an important flavonoid source. Key flavonoids in the leaves include rutin, Q-3-Ogalactoside, I-3-O-glucoside, quercetin, K-3-O-glucoside, and kaempferol (Morgenstern et al. 2014). In contrast, berries are rich in flavonoid glycosides like I-3-O-rutinoside, I-3-O-glucoside, Q-3-rutinoside, Q-3-O-glucoside, I-3-glucoside-7-rhamnoside, K-3-sorphoroside-7-O-rhamnoside, I-3-O-sorphoroside-7-O-rhamnoside, and rutin. The major aglycons present in berries include kaempferol (32–72 mg/kg FW), quercetin (67–175 mg/kg FW), myricetin (36–172 mg/kg FW), and isorhamnetin (45–106 mg/kg FW) (Fatima et al. 2015; Guo et al. 2017a). In sea buckthorn seeds, prominent flavonoid glycosides are Q-3-rutinoside and I-3-O-rutinoside, along with I-3-O-glucosides, 3-O-sophroside-7-O--rhamnosides of quercetin and kaempferol, and 3-O-glucoside-7-O-rhamnosides of quercetin and isorhamnetin, with significant quantities of isorhamnetin, quercetin, and kaempferol also present (Arimboor and Arumughan 2012).

2.4 Carotenoids

Carotenoids play crucial physiological roles in human health, notably β-carotene, a potent antioxidant. β-Carotene efficiently neutralizes singlet oxygen and counteracts the adverse effects of free radicals in the body, thereby bolstering the immune response system. Sea buckthorn fruits contain high levels of carotenoids, giving them their distinctive orange-yellow hue. However, the content of carotenoids varies significantly among different species and plant parts of sea buckthorn (Wang et al. 2022). β-Carotene predominates in sea buckthorn, constituting 15–55% in berries, and 26–34% in peel, pulp, and seed oil (Seglina et al. 2021; Teleszko et al. 2015). In addition, carotenoids include γ-carotene, cis-lycopene, lycopene, cis-γ-carotene, β-cryptoxanthin, α-carotene, and so on. Levels of β-carotene range from 0.2 to 17 mg/100 g, with total carotenoids (calculated as β-carotene) varying from 1 to 120 mg/100 g in fresh berries (Yang and Kallio 2002).

Previous studies have identified variability in the composition and concentration of carotenoids in sea buckthorn berries (Beveridge et al. 1999; Yang and Kallio 2002). This variation is attributed to factors like genetic diversity, berry parts analyzed, climate and growth conditions, interannual variability, harvest maturity, storage conditions, and analysis methods. Optimal conditions for carotenoid accumulation in sea buckthorn berries include warm, sunny weather with moderate precipitation (Bekker and Glushenkova 2001).

2.5 Polypeptides and Amino Acids

Sea buckthorn is rich in amino acids, essential for human health. These amino acids, which are the basic units that make up proteins, play a vital role in life processes. Seventeen amino acids, including seven essential ones (threonine, valine, methionine, isoleucine, leucine, phenylalanine, and lysine), have been identified in sea buckthorn fruits, leaves, branches, and seeds of sea buckthorn. The amino acid content in sea buckthorn seeds is 18.63%, leaves 15.41%, branches 11.62%, and fruits 6.89%. Notably, aspartic acid and glutamic acid are most abundant in fruits, leaves, and branches, with 1.11% and 1.24% in fruits, 2.42% and 1.60% in leaves, and 3.71% and 0.97% in branches. The highest content of tyrosine and glutamic acid can be found in sea buckthorn seeds, at 4.72% and 3.42%, respectively (Wang et al. 2022).

Sea buckthorn polypeptide, protein hydrolysis products, have molecular weights ranging from 50 to 5000 Da, classified as peptides. Macropeptides, with molecular weights between 5000 and 10,000 Da, and small peptides (50–1000 Da), also known as oligopeptides, are observed. These low molecular weight peptides often exhibit significant biological activity and are referred to as bioactive peptides. Typically, bioactive peptides consist of 2–50 amino acid residues, forming specific structures that play active biological roles.

2.6 Polysaccharides

Carbohydrate distribution in sea buckthorn berries varies across different morphological parts. The berries primarily contain non-starch-type polysaccharides like cellulose, hemicelluloses, pectin, and hydrocolloids, major dietary fiber components (Ciesarova et al. 2020). Starch, found only in seeds, was measured at 49 g/kg DW. The raw fiber content in seeds, peel, and pulp is 130 g/kg DW, 66 g/kg DW, and 47 g/kg DW, respectively, with small quantities of pectin in peel (5 g/kg DW) and pulp (15 g/kg DW) (Jaroszewska et al. 2018).

The high fiber content of sea buckthorn contributes significantly to its health benefits, influenced by weather conditions and maturity of the berries. The levels of crude fiber in sea buckthorn range from 62 to 100 g/kg DW. The dietary fiber fractions in the berries are distributed as follows: neutral detergent fiber160–200 g/kg DW, acid detergent fiber 120–145 g/kg DW, acid detergent lignin 50–70 g/kg DW, hemicellulose 45–55 g/kg DW, and cellulose 60–75 g/kg DW (Jaroszewska et al. 2018). The percentage proportions of the sea buckthorn DF fractions with different solubility properties were determined as the ratio of soluble DF to insoluble DF equals 2.8% to 65.2%. The insoluble and soluble properties of DF are related to differences in their physiological functions and effects (Jimenez-Escrig and Sanchez-Muniz 2000). Compared with other fruits assayed, the sea buckthorn has an abundant content of DF. The average content of neutral detergent fiber (NDF) in sea buckthorn berries was 160–200 g/kg DW (Jaroszewska et al. 2018), which is in the range typical of vegetables considered a very good source of DF (97–219 g/kg DW) (Ciesarova et al. 2020; Marciniak et al. 2021). Acid detergent fiber (ADF) in sea buckthorn berries was similar to that in blueberry, kiwifruit, and watermelon (Ciesarova et al. 2020).

Additionally, it was found that sea buckthorn leaves are also a rich fiber source, with a high concentration of crude fiber (122 g/kg DW) and both NDF (284 g/kg DW) and ADF (181 g/kg DW) fraction (Bieł and Jaroszewska 2017). The NDF fraction includes insoluble fibers, hemicellulose, cellulose, lignocellulose, and lignin. The fraction ADF is a subset of NDF, including cellulose, lignocellulose, and lignin. ADF is the least digestible portion (Bieł and Jaroszewska 2017). Moreover, in sea buckthorn leaves, high contents of cellulose (123 g/kg DW) and hemicellulose (80 g/kg DW) were found, which can possibly increase the health-promoting value of the product (Ciesarova et al. 2020; Marciniak et al. 2021).

Sea buckthorn polysaccharides have demonstrated various health benefits, including blood sugar and lipid reduction and cancer and bacterial inhibition (Zhang et al. 2017; Zhao et al. 2020b), fatigue resistance (Ni et al. 2013), immune regulation, antioxidant (Liu et al. 2021b; Ren et al. 2020; Zhao et al. 2020a), anti-aging, lipid metabolism regulation, liver protection, and more (Suryakumar and Gupta 2011; Zhao et al. 2020c). These polysaccharides have also shown protective effects against acetaminophen-induced liver toxicity in mice by activating the Nrf-2/HO-1-SOD-2 pathway (Wang et al. 2018). Zhang et al. (2017) revealed the protective effect and mechanism of sea buckthorn polysaccharides on carbon tetrachloride (CCl_4) hepatotoxicity. In addition, the hypoglycemic and anti-inflammatory effects of sea buckthorn polysaccharides on diabetic mice and their mechanisms have also been reported (Guo et al. 2013).

2.7 Terpenes

Terpenoids, predominantly polymers and the derivatives of isoprenes, constitute essential components of sea buckthorn, known for their wide distribution, complex structures, and notable bioactivities. In sea buckthorn, various terpenoids have been identified in the fruits, leaves, and branches. To date, 30 terpenoids have been isolated from sea buckthorn, including compounds like 2α-hydroxyursolic acid, 19-α-hydroxyursolic acid, 3-*O*-trans-*p*-coumaroyl oleanolic acid, 3-*O*-caffeoyl oleanolic acid, lupeol, ursolic acid, ursolic aldehyde, oleanolic acid, α-amyrin, β-amyrin, 28-norurs-12-ene-3β, 17β-diol, pomolic acid, 2-O-trans-pcoumaroyl maslinic acid, 2-*O*-caffeoyl maslinic acid, 2-*O*-[β-D-glucopyranosyl (1→2)-β-D-glucopyranosyl-(1→3)]-[α-L-rhamnopyranosyl-(1→2)]-α-L-arabinopyranosyl-13-ene-19-one-28-oic acid 28-*O*-β-D-glucopyranosyl ester, 3-O-[β-D-glucopyranosylrhamnopyranosyl-(1→2)]-α-L-arabinopyranosyl-13-ene-19-one-30-hydroxyolean-28-oic acid 28-O-β-D-glucopyranosyl ester, 3-*O*-[β-Dglucopyranosyl(1→2)-β-D-glucopyranosyl-(1→3)]-[α-Lrhamnopyranosyl-

(1→2)]-β-D-glucopyranosyl-13-ene-19-one-28-oic acid 28-*O*-β-D-glucopyranosyl ester, 3-*O*-[β-D-glucopyranosyl(1→2)-β-D-glucopyranosyl-(1→3)]-[α-L-rhamnopyranosyl-(1→2)]-β-D-glucopyranosyl-13-ene-19-one-3-*O*-hydroxyolean-28-oic acid 28-*O*-β-D-glucopyranosylester, violaxanthin, δ-carotene, β-cryptoxanthin, zeaxanthin dipalmitate, B2 cryptoxanthin palmitate, neoxanthin, lutein, zeaxanthin, lycopene, β-carotene, α-carotene, and γ-carotene. These compounds are generally categorized into pentacyclic triterpenoid sand monocyclic triterpenoids. In pentacyclic triterpenoids, the position of C-3 typically features a hydroxyl or glycoside substituiton (Ma et al. 2023).

Both sea buckthorn fruits and leaves contain terpenoids with leaves and stems showing higher total triterpenoids content than fruits. Pentacyclic triterpenoids in sea buckthorn are divided into two groups. One group includes maslinic acid, pomolic acid, corosolic acid, betulinic acid, oleanolic acid, ursolic acid, and botulin, found in significant quantities, whereas compounds like tormentic acid, α-boswellic acid, uvaol, and erythrodiol (only in branches and leaves) were found in amounts below 1 mg/100 g DW, and their sum represented approximately 2% of all triterpenoids in each part (Tkacz et al. 2021). Oleanolic and ursolic acids are prevalent triterpenoids in various plants and have been extensively investigated in different parts of sea buckthorn, such as whole berries, seeds, leaves, roots, twigs, and extract of branch bark of sea buckthorn (Grey et al. 2010; Marciniak et al. 2021; Michel et al. 2012; Olas 2018; Olas et al. 2018; Rozalska et al. 2018; Tkacz et al. 2021; Yang et al. 2007). Leaves are particularly high in triterpenoid saponins, contrasting with the twigs and fruits rich in free triterpenoids and triterpenoids acylated with phenolic acids, mainly coumaric and ferulic acids (Yang et al. 2007).

Triterpenoids and certain derivatives exhibit antioxidant and cytotoxic properties, including inhibiting proliferation and differentiation of tumor cells and angiogenesis, as well also acting cytoprotectively (Marciniak et al. 2021). For instance, corosolic acid shows insulin-like effects and benefits in type II diabetes management (Ni et al. 2019). Oleanolic acid is known for its liver protection effect and the effect of treating diabetes (Castellano et al. 2022), while ursolic acid is noted its anti-tumor properties, including inhibiting tumor cell proliferation, inducing tumor differentiation and apoptosis, anti-invasive and anti-tumor angiogenesis, and other anti-tumor effects (Cargnin and Gnoatto 2017).

2.8 Essential Oils

Volatile oils, also known as essential oils, are characterized as oily liquids that volatilize at room temperature. These oils can be steam-distilled but are insoluble in water. They are predominantly found in the fruits of sea buckthorn but relatively rare in sea buckthorn leaves, exhibiting complex and diverse structures (Ma et al. 2023; Ran and Li 2019). At present, 64 volatile oils have been isolated from sea buckthorn. These compounds have been identified as ethyl benzoic acid ester,

2-naphthalenemethanol, benzaldehyde, 3-methyl-1-butanol benzoate, toluene,1,2-dimethylbenzene,[1R-(1R,9S)]-4,11,11-trimethyl-8-methylene-bicyclo-1,3-dimethylbenzene, 1,2,3-trimethylbenzene, phenylacetaldehyde, phenylethanol, furfural, dehydroelsholtzia ketone, γ-valerolactone, 1,4-dimethylcyclohexane, terpinene,limonene,undec-4-ene, 3-methyldiadamantane, octadecanoic acid, octadecanal, hexadecanal, 2-pentadecanone, 4,5-dimethyl-nonane,11-bromoundecanoic acid methyl ester, pentadecane, pentacosane, pentadecanoic acid, 7-hexadecenoic acid, 12-pentacosene, 9-octadecenoic acidmethyl ester, dodecanoic acid, 8-dodecen-1-ol acetate, 9,12,15-octadecatrienoic acid methyl ester, tetradecanoic acid, elaidic acid, 2,6,10,15-tetramethyl-heptadecane, 6,10,14-trimethyl-2-pentadecanone, 6,10-dimethyl-5,9-undecadien-2-one, 11-dodecenol, 2-tridecanone, docosane,14-methyl-pentadecanoic acid ester, n-decanoic acid, 6,10,14-trimethyl-5,9,13-pentadecatrien-2-one,6-octadecenoic acid, traumatic acid, octacosane, 9-tricosene, 8-heptadecene, hexadecadienal, octane, 2,3-dimethyloctane, 2,4-dimethyloctane, 4-methyloctane, nonane, heptanal, decane, nonanal, dodecane, tridecane, cetane, (E)-hexadecenoic acid ethyl ester, eicosane, and pentacosane. Volatile oils constitute key bioactive components in sea buckthorn. Structurally, these oils are categorized into two main types: aromatic small molecules and chain hydrocarbons (Cakir 2004; Ma et al. 2023; Yue et al. 2017). Research has shown that sea buckthorn essential oils from various parts possess antibacterial activities (Yue et al. 2017), In addition, some of these compounds also have been identified with antioxidant and anti-aging effects.

2.9 Organic Acids

Aliphatic and alicyclic carboxylic acids, commonly known as organic acids, play a vital role in plant innate immunity. Their content is highly dependent on growing conditions (climate, plant pathogens attack, plant variety) (Ciesarova et al. 2020). Sea buckthorn fruit contains several organic acids and their derivatives, which have been shown to promote bone differentiation and contribute to the differentiation of mesenchymal stem cells into osteoblasts (Lee et al. 2021).

Sea buckthorn is particularly high in organic acids, including malic acid, citric acid, succinic acid, quinic acid, tartaric acid, lactic acid, ascorbic acid, succinic acid, oxalic acid, gallic acid, mustard acid, gentisic acid, salicylic acid, caffeic acid, cinnamic acid, and others. Notably, malic acid ranges from 11 to 60 mg/L of juice, quinic acid from 7 to 49 mg/L, citric acid constitute up to 0.3% of all acids, tartaric acid to 0.013%, and succinic acid to 0.6% (Ciesarova et al. 2020). The total organic acid content in sea buckthorn fruits is approximately 3.86–4.52%, with malic acid, citric acid, tartaric acid, and quinic acid (e.g., oxalic acid, succinic acid) being the most prevalent. Sea buckthorn oil mainly contains malic and citric acid, while other plant parts, like leaves, roots, predominantly contain succinic acid (Zielinska and Nowak 2017).

The composition and concentrations of organic acids vary among different sea buckthorn species. Russian subspecies tend to have lower total acidity, with organic acid concentrations of 2.1–3.2 g/100 mL, while Finnish genotypes fall in the mid-range (4.2–6.5 g/100 mL), and Chinese genotypes showed the highest concentration, ranging between 3.5 and 9.1 g/100 mL (Bal et al. 2011). Studies have identified nine organic acids in sea buckthorn juice, including quinic acid, L-malic acid, D-malic acid, succinic acid, pyruvic acid, tartaric acid, acetic acid, formic acid, and citric acid (Odgerel et al. 2021). Additionally, research on Polish sea buckthorn varieties discovered the presence of oxalic acid and isocitric acid (Tkacz et al. 2019).

2.10 Phytosterols

Phytosterols, recognized for their role in preventing cardiovascular diseases, are crucial bioactive components in the lipids of sea buckthorn pulp. Specially, they are effective in combating hypercholesterolemia, a significant risk factor for cardiovascular disease (Zhou et al. 2022). A meta-analysis from 11 independent randomized controlled trials revealed that supplementation with sea buckthorn berries/extracts significantly improved total cholesterol, triglyceride, LDL-C, and HDL-C in subjects with hyperlipidemia, but not in healthy individuals (Guo et al. 2017c).

Sea buckthorn's fruit, seed, and oil are rich sources of phytosterols, notable for contributing to cardiovascular disease prevention and treatment. Phytosterols are the major constituents of the unsaponifiable fraction of sea buckthorn oils. The major phytosterol in sea buckthorn oil is sitosterol (β-sitosterol), with 5-avenasterol second in quantitative importance. Although other phytosterols are present in relatively minor quantities, the overall phytosterol content is remarkably high, surpassing soybean oil by 4–20 times. The sterol content in different varieties ranged from 1.3% to 2% (Bal et al. 2011).

To date, 20 different phytosterols have been identified in sea buckthorn (Chen et al. 2023). The fruit contains 20–30 g/kg of phytosterols, which is 4–20 times higher than the levels in soybean oil. The sterol content in sea buckthorn berries ranges from 2.2% to 8.8%, with seeds containing about 0.1–0.2% and the soft parts 0.02–0.04%. The total sterol content in seed oil, soft parts, and whole fruit is 12–23, 10–29, and 13–33 g/kg, respectively. β-Sitosterol is the most abundant, estimated at approximately 5.2–5.7 g/kg FW, making sea buckthorn a superior source of β-sitosterol compared to oils like sunflower oil or virgin olive oil (Mihalcea et al. 2021). In seeds, sitosterol comprises 57–76% of total sterols, and 61–83% of total sterols in soft parts (Yang et al. 2001). δ-5-Avenasterol and obtusifoliol are the most valuable phytosterols in seeds (about 15–17% of total sterols) and together with stigmasta-8-ene-3β-ol are about 5–6% and 8–10% of total sterols from soft parts and whole fruits, respectively. The other abundant phytosterols are campesterol, stigmasterol, and Δ^5-avenasterol (Zheng et al. 2017).

In fresh berries of sea buckthorn, the sterol content ranges from 350 to 500 mg/kg, with the majority (70–80%) existing in the soft parts. The two subspecies show no significant difference in sterol content. The levels of oils vary based on both raw materials and extraction methods, typically being 1–2% in seed oil and 1–3% in oil from the soft parts. Sitosterol constitutes 60–70% of seed sterols and up to 80% of those in soft parts, while isofucosterol is another major sterol component representing 10–20% of seed sterols and 2–5% of sterols of the soft parts. Campesterol, stigmastanol, citrostadienol, avenasterol, cycloartenol, 24-methylenecycloartanol, and obtusifoliol each represent 1–5% of total sterols of both fractions (Yang and Kallio 2001). Recent research found that the total phytosterol content of berry lipids from eight Russian sea buckthorn species ranged from 6168.24 to 13,378.22 μg/100 mL. Fourteen sterol compounds have been detected in sea buckthorn pulp lipids, namely 4-desmethyl sterols (cholestanol derivatives, including β-sitosterol, stigmasterol, campesterol, and Δ5-avenasterol), 4α-monomethyl sterols (e.g. citrostadienol), and 4,4-dimethylsterols (e.g., 24-methylenecycloartanol) (Teleszko et al. 2015).

2.11 Phenylpropanoids and Alkaloids

Phenylpropanoids, naturally occurring compounds characterized by benzene rings and three straight-chain carbons, are found in various parts of sea buckthorn, including the stems, leaves, and fruits. This group encompasses a range of compounds, such as phenylpropene, phenylpropionic acid, and coumarins (such as coumarin, methoxycoumarin, scopoletin, and isoscopoletin). Notably, 13 different phenylpropanoids have been identified in sea buckthorn, including cinnamaldehyde, p-hydroxy cinnamaldehyde, methyl eugenol, and coniferaldehyde. A specific compound, 1-feroyl-β 1-feruloyl-β-D-glucopyranoside, has been detected in sea buckthorn leaves (Kim et al. 2011).

Alkaloids are nitrogen-containing organic compounds, often with complex heterocyclic structures. Sea buckthorn contains various alkaloid compounds, such as nicotinamide, fenugreek hydrochloride, cordycepin, and glycoside alkaloids (including crotonoside, guanosine and adenosine), which are present in its fruits, stems, and leaves. The seeds of sea buckthorn seeds are notable for alkaloids like 5,11-dihydroxychromamine, 5-hydroxychromamine, Shepherdine, 4-[(E)-p-coumaroylamino] butan-1-ol, 4-[(Z)-pcoumaroylamino] butan-1-ol, and hippophamide. These seed alkaloids demonstrate significant anti-inflammatory and analgesic effects, potentially due to their heat-clearing, blood-dispersing, and swelling-reducing effects (OuYang et al. 2015).

References

Arimboor R, Arumughan C (2012) HPLC-DAD-MS/MS profiling of antioxidant flavonoid glycosides in sea buckthorn (*Hippophae rhamnoides* L.) seeds. Int J Food Sci Nutr 63(6):730–738

Arimboor R, Kumar KS, Arumughan C (2008) Simultaneous estimation of phenolic acids in sea buckthorn (Hippophae rhamnoides) using RP-HPLC with DAD. J Pharm Biomed Anal 47(1):31–38

Bal LM, Meda V, Naik SN, Satya S (2011) Sea buckthorn berries: a potential source of valuable nutrients for nutraceuticals and cosmoceuticals. Food Res Int 44(7):1718–1727

Bekker NP, Glushenkova AI (2001) Components of certain species of the Elaeagnaceae family. Chem Nat Compd 37(2):97–116

Beveridge T, Li TSC, Oomah BD, Smith A (1999) Sea buckthorn products: manufacture and composition. J Agric Food Chem 47(9):3480–3488

Bieł W, Jaroszewska A (2017) The nutritional value of leaves of selected berry species. Sci Agric 74(5):405–410

Bittova M, Krejzova E, Roblova V, Kuban P, Kuban V (2014) Monitoring of HPLC profiles of selected polyphenolic compounds in sea buckthorn (*Hippophae rhamnoides* L.) plant parts during annual growth cycle and estimation of their antioxidant potential. Cent Eur J Chem 12(11):1152–1161

Cakir A (2004) Essential oil and fatty acid composition of the fruits of Hippophae rhamnoides L. (Sea Buckthorn) and Myrtus communis L. from Turkey. Biochem Syst Ecol 32(9):809–816

Cargnin ST, Gnoatto SB (2017) Ursolic acid from apple pomace and traditional plants: a valuable triterpenoid with functional properties. Food Chem 220:477–489

Castellano JM, Ramos-Romero S, Perona JS (2022) Oleanolic acid: extraction, characterization and biological activity. Nutrients 14(3):623

Chen Y, Cai YF, Wang K, Wang YS (2023) Bioactive compounds in sea buckthorn and their efficacy in preventing and treating metabolic syndrome. Food Secur 12(10):1985

Ciesarova Z, Murkovic M, Cejpek K, Kreps F, Tobolkova B, Koplik R, Burcova Z (2020) Why is sea buckthorn (*Hippophae rhamnoides* L.) so exceptional? A review. Food Res Int 133:109170

Fatima T, Kesari V, Watt I, Wishart D, Todd JF, Schroeder WR, Krishna P (2015) Metabolite profiling and expression analysis of flavonoid, vitamin C and tocopherol biosynthesis genes in the antioxidant-rich sea buckthorn (*Hippophae rhamnoides* L.). Phytochemistry 118:181–191

Grey C, Widen C, Adlercreutz P, Rumpunen K, Duan RD (2010) Antiproliferative effects of sea buckthorn (*Hippophae rhamnoides* L.) extracts on human colon and liver cancer cell lines. Food Chem 120(4):1004–1010

Guo FX, Zeng Y, Li JP, Chen ZN, Ma JX (2013) Effects of Hippophae rhamnoides L. subsp. chinensis Rousi polysaccharide on alpha-glucosidase enzyme activity and level of blood glucose. Yaoxue Xuebao(in Chinese) 48(4):604–608

Guo RX, Chang XX, Guo XB, Brennan CS, Li T, Fu X, Liu RH (2017a) Phenolic compounds, antioxidant activity, antiproliferative activity and bioaccessibility of Sea buckthorn (*Hippophae rhamnoides* L.) berries as affected by in vitro digestion. Food Funct 8(11):4229–4240

Guo RX, Guo XB, Li T, Fu X, Liu RH (2017b) Comparative assessment of phytochemical profiles, antioxidant and antiproliferative activities of Sea buckthorn (*Hippophae rhamnoides* L.) berries. Food Chem 221:997–1003

Guo XF, Yang B, Cai W, Li D (2017c) Effect of sea buckthorn (*Hippophae rhamnoides* L.) on blood lipid profiles: a systematic review and meta-analysis from 11 independent randomized controlled trials. Trends Food Sci Technol 61:1–10

Jaroszewska A, Biel W, Telesinski A (2018) Effect of mycorrhization and variety on the chemical composition and antioxidant activity of sea buckthorn berries. J Elem 23(2):673–684

Ji MY, Gong X, Li X, Wang CC, Li MH (2020) Advanced research on the antioxidant activity and mechanism of polyphenols from hippophae species-A review. Molecules 25(4):917

Jimenez-Escrig A, Sanchez-Muniz FJ (2000) Dietary fibre from edible seaweeds: chemical structure, physicochemical properties and effects on cholesterol metabolism. Nutr Res 20(4):585–598

Kampa M, Alexaki VI, Notas G, Nifli AP, Nistikaki A, Hatzoglou A, Castanas E (2004) Antiproliferative and apoptotic effects of selective phenolic acids on T47D human breast cancer cells: potential mechanisms of action. Breast Cancer Res 6(2):R63–R74

Kim JS, Kwon YS, Sa YJ, Kim MJ (2011) Isolation and identification of sea buckthorn (*Hippophae rhamnoides*) phenolics with antioxidant activity and alpha-glucosidase inhibitory effect. J Agric Food Chem 59(1):138–144

Kuhkheil A, Badi HN, Mehrafarin A, Abdossi V (2017) Chemical constituents of sea buckthorn (*Hippophae rhamnoides* L.) fruit in populations of central Alborz Mountains in Iran. Res J Pharmacogn 4:1–12

Lee YH, Jang HJ, Park KH, Kim SH, Kim JK, Kim JC, Kim KH (2021) Phytochemical analysis of the fruits of sea buckthorn (Hippophae rhamnoides): identification of organic acid derivatives. Plants-Basel 10(5):860

Liu SY, Xiao PT, Kuang YJ, Hao JH, Huang TQ, Liu EH (2021a) Flavonoids from sea buckthorn: a review on phytochemistry, pharmacokinetics and role in metabolic diseases. J Food Biochem 45(5):e13724

Liu JF, Xu DX, Chen S, Yuan F, Mao LK, Gao YX (2021b) Superfruits in China: bioactive phytochemicals and their potential health benefits-A review. Food Sci Nutr 9(12):6892–6902

Ma QG, He NX, Huang HL, Fu XM, Zhang ZL, Shu JC, Wei RR (2023) *Hippophae rhamnoides* L.: a comprehensive review on the botany, traditional uses, phytonutrients, health benefits, quality markers, and applications. J Agric Food Chem 71(12):4769–4788

Marciniak B, Kontek R, Zuchowski J, Stochmal A (2021) Novel bioactive properties of low-polarity fractions from sea-buckthorn extracts (Elaeagnus rhamnoides (L.) A. Nelson)-(*in vitro*). Biomed Pharmacother 135:111141

Michel T, Destandau E, Le Floch G, Lucchesi ME, Elfakir C (2012) Antimicrobial, antioxidant and phytochemical investigations of sea buckthorn (*Hippophae rhamnoides* L.) leaf, stem, root and seed. Food Chem 131(3):754–760

Mihalcea L, Turturica M, Cucolea EI, Danila GM, Dumitrascu L, Coman G, Stanciuc N (2021) CO2 Supercritical fluid extraction of oleoresins from sea buckthorn pomace: evidence of advanced bioactive profile and selected functionality. Antioxidants 10(11):1681

Morgenstern A, Ekholm A, Scheewe P, Rumpunen K (2014) Changes in content of major phenolic compounds during leaf development of sea buckthorn (*Hippophae rhamnoides L.*). Agric Food Sci 23(3):207–219

Ni WH, Gao TT, Wang HL, Du YZ, Li JY, Li C, Bi HT (2013) Anti-fatigue activity of polysaccharides from the fruits of four Tibetan plateau indigenous medicinal plants. J Ethnopharmacol 150(2):529–535

Ni MT, Pan JH, Hu X, Gong DM, Zhang GW (2019) Inhibitory effect of corosolic acid on alpha-glucosidase: kinetics, interaction mechanism, and molecular simulation. J Sci Food Agric 99(13):5881–5889

Odgerel U, Islam MZ, Kitamura Y, Kokawa M, Odbayar TO (2021) Effect of micro wet milling process on particle sizes, antioxidants, organic acids, and specific phenolic compounds of whole sea buckthorn (*Hippophae rhamnoides* L.) juices. J Food Proc Preserv 45(6):e15474

Olas B (2018) The beneficial health aspects of sea buckthorn (*Elaeagnus rhamnoides* L.A.Nelson) oil. J Ethnopharmacol 213:183–190

Olas B, Zuchowski J, Lis B, Skalski B, Kontek B, Grabarczyk L, Stochmal A (2018) Comparative chemical composition, antioxidant and anticoagulant properties of phenolic fraction (a rich in non-acylated and acylated flavonoids and non-polar compounds) and non-polar fraction from Elaeagnus rhamnoides (L.) A. Nelson fruits. Food Chem 247:39–45

OuYang J, Zhou WN, Li G, Wang XY, Ding CX, Suo YR, Wang HL (2015) Three new alkaloids from Hippophae rhamnoides LINN. subsp sinensis ROUSI. Helv Chim Acta 98(9):1287–1291

Ran BB, Li WD (2019) Research progress on chemical constituents and their differences between sea buckthorn berries and leaves. China J Chinese Mater Med 44(9):1767–1773

Ren RR, Li N, Su C, Wang YL, Zhao XJ, Yang LL, Ma XQ (2020) The bioactive components as well as the nutritional and health effects of sea buckthorn. RSC Adv 10(73):44654–44671

Rop O, Ercisli S, Mlcek J, Jurikova T, Hoza I (2014) Antioxidant and radical scavenging activities in fruits of 6 sea buckthorn (*Hippophae rhamnoides* L.) cultivars. Turk J Agric For 38(2):224–232

Rozalska B, Sadowska B, Zuchowski J, Wieckowska-Szakiel M, Budzynska A, Wojcik U, Stochmal A (2018) Phenolic and nonpolar fractions of elaeagnus rhamnoides (L.) A. nelson extracts as virulence modulators-in vitro study on bacteria, fungi, and epithelial cells. Molecules 23(7):19

Seglina D, Krasnova I, Grygier A, Radziejewska-Kubzdela E, Rudzinska M, Gornas P (2021) Unique bioactive molecule composition of sea buckthorn (*Hippophae rhamnoides* L.) oils obtained from the peel, pulp, and seeds via physical "solvent-free" approaches. J Am Oil Chem Soc 98(10):1009–1020

Suryakumar G, Gupta A (2011) Medicinal and therapeutic potential of sea buckthorn (*Hippophae rhamnoides* L.). J Ethnopharmacol 138(2):268–278

Teleszko M, Wojdylo A, Rudzinska M, Oszmianski J, Golis T (2015) Analysis of lipophilic and hydrophilic bioactive compounds content in sea buckthorn (*Hippophae rhamnoides* L.) berries. J Agric Food Chem 63(16):4120–4129

Tkacz K, Wojdylo A, Turkiewicz IP, Bobak L, Nowicka P (2019) Anti-oxidant and anti-enzymatic activities of sea buckthorn (*Hippophae rhamnoides* L.) fruits modulated by chemical components. Antioxidants 8(12):618

Tkacz K, Wojdylo A, Turkiewicz IP, Nowicka P (2021) Triterpenoids, phenolic compounds, macro- and microelements in anatomical parts of sea buckthorn (*Hippophae rhamnoides* L.) berries, branches and leaves. J Food Compos Anal 103:104107

Verica DU, Bursać Kovačević D, Levaj B, Sandra P, Mirela M, Ana T (2009) Polyphenols and antioxidant capacity in fruits and vegetables common in the croatian diet. Agric Conspect Scientif:74, 175

Wang X, Liu JR, Zhang XH, Zhao SM, Zou K, Xie JM, Wang YZ (2018) Seabuckthorn berry polysaccharide extracts protect against acetaminophen induced hepatotoxicity in mice via activating the Nrf-2/HO-1-SOD-2 signaling pathway. Phytomedicine 38:90–97

Wang Z, Zhao FL, Wei PP, Chai XY, Hou GG, Meng QG (2022) Phytochemistry, health benefits, and food applications of sea buckthorn (*Hippophae rhamnoide*s L.): a comprehensive review. Front Nutr 9:1036295

Yang BR, Kallio HP (2001) Fatty acid composition of lipids in sea buckthorn (*Hippophae rhamnoides* L.) berries of different origins. J Agric Food Chem 49(4):1939–1947

Yang BR, Kallio H (2002) Composition and physiological effects of sea buckthorn (Hippophae) lipids. Trends Food Sci Technol 13(5):160–167

Yang B, Karlsson RM, Oksman PH, Kallio HP (2001) Phytosterols in sea buckthorn (*Hippophae rhamnoides* L.) berries: identification and effects of different origins and harvesting times. J Agric Food Chem 49(11):5620–5629

Yang ZG, Li HR, Wang LY, Li YH, Lu SG, Wen XF, Kitanaka S (2007) Triterpenoids from Hippophae rhamnoides L. and their nitric oxide production-inhibitory and DPPH radical-scavenging activities. Chem Pharm Bull 55(1):15–18

Yue XF, Shang X, Zhang ZJ, Zhang YN (2017) Phytochemical composition and antibacterial activity of the essential oils from different parts of sea buckthorn (*Hippophae rhamnoides* L.). J Food Drug Anal 25(2):327–332

Zhang W, Zhang XH, Zou K, Xie JM, Zhao SM, Liu JR, Wang YZ (2017) Seabuckthorn berry polysaccharide protects against carbon tetrachloride-induced hepatotoxicity in mice via anti-oxidative and anti-inflammatory activities. Food Funct 8(9):3130–3138

Zhao L, Li MY, Su S, Geng TT, Sun H (2020a) Hippophae rhamnoides L. polysaccharide enhances antioxidant enzyme activity, cytokine level, and related mRNA expression in intestinal porcine epithelial cells. Can J Anim Sci 100(1):193–204

Zhao L, Li MY, Sun KC, Su S, Geng TT, Sun H (2020b) Hippophae rhamnoides polysaccharides protect IPEC-J2 cells from LPS-induced inflammation, apoptosis and barrier dysfunction in vitro via inhibiting TLR4/NF-kappa B signaling pathway. Int J Biol Macromol 155:1202–1215

Zhao SJ, Sun HY, Liu QB, Shen Y, Jiang YJ, Li YH, Shao ML (2020c) Protective effect of sea-buckthorn berry juice against acrylamide-induced oxidative damage in rats. J Food Sci 85(7):2245–2254

Zheng L, Shi LK, Zhao CW, Jin QZ, Wang XG (2017) Fatty acid, phytochemical, oxidative stability and in vitro antioxidant property of sea buckthorn (*Hippophaë rhamnoides* L.) oils extracted by supercritical and subcritical technologies. LWT-Food Sci Technol 86:507–513

Zhou FF, Zhang JA, Zhao A, Zhang YM, Wang PY (2022) Effects of sea buckthorn puree on risk factors of cardiovascular disease in hypercholesterolemia population: a double-blind, randomized, placebo-controlled trial. Anim Biotechnol 33(5):955–963

Zielinska A, Nowak I (2017) Abundance of active ingredients in sea-buckthorn oil. Lipids Health Dis 16:95

Chapter 3
Current Knowledge of the Sea Buckthorn Composition and Health Benefits

Haitian Fang, Shuifang Mao, Jinghan Wang, and Jinhu Tian

Abstract Sea buckthorn, a versatile broad-leaved shrub, is acclaimed for its dual medicinal and nutritional applications. Modern scientific researches have illuminated its rich profile of bioactive substances and nutrients, garnering interest in various fields such as medicine, food, beverages, and cosmetics. The health benefits of sea buckthorn, including cytoprotection, stress mitigation, immunomodulation, hepatoprotection, anti-tumor activity, and promotion of tissue regeneration, can be largely attributed to its diverse range of bioactive compounds and nutrients.

Keywords Sea buckthorn · Nutrients · Composition · Health benefits

Abbreviations

FW	fresh weight
DW	dry weight
SOD	superoxide dismutase
SBT oil	sea buckthorn oil
TTAs	three triterpenoid acids
TC	triglyceride
LDL	low-density lipoprotein
HDL-c	high-density lipoprotein cholesterol
UVB	ultraviolet radiation
V_A	vitamin A
V_{B1}	vitamin B_1
V_{B2}	vitamin B_2

H. Fang
Food Science and Technology, Ningxia University, Yinchuan, Ningxia,, China

S. Mao · J. Wang · J. Tian (✉)
Department of Food Science and Nutrition, Zhejiang University, Hangzhou, China

J. Tian et al. (eds.), *Sea Buckthorn*, https://doi.org/10.1007/978-981-97-9865-0_3

V_{B6}	vitamin B_6
V_C	vitamin C
V_E	vitamin E

3.1 Nutrients

Sea buckthorn is notable for its high protein content across all parts, including a significant amount of non-protein nitrogen in the leaves and fruits (Guo et al. 2017a, b, c). Research indicates that sea buckthorn leaves contain about 15–20% protein, while the seed meal has approximately 30% protein (Ren et al. 2019). The overall protein content in sea buckthorn ranges from 0.4% to 2.5% fresh weight (FW) (Nazir and Bashir 2017), with the free amino acid content spanning from 0.77% to 2.19% FW (Zhang et al. 1989). This plant is a valuable source of vegetable protein, largely due to its composition of essential amino acids, including a predominance of asparagine (Zhou et al. 2018).

In addition to its protein content, sea buckthorn is rich in amino acids, essential for the human health and fundamental to protein structure. It contains 17 amino acids, including seven essential amino acids (threonine, valine, methionine, isoleucine, leucine, phenylalanine, and lysine), found in sea buckthorn fruits (Ma et al. 2019a, b, c), leaves, branches, and seeds (Tan et al. 2018). The amino acid content in sea buckthorn seeds is notable at 18.63%, compared to 15.41% in leaves, 11.62% in branches, and 6.89% in fruits. Moreover, the concentrations of aspartic acid and glutamic acid are highest in fruits, leaves, and branches (1.11% and 1.24% in fruits; 2.42% and 1.60% in leaves; 3.71% and 0.97% in branches), while the seeds contain the highest levels of tyrosine and glutamic acid at 4.72% and 3.42%, respectively (Tan et al. 2018).

Sea buckthorn peptides, which are the hydrolysis products of proteins, are classified based on their molecular weight. Peptides with a molecular weight of 50–5000 Da are classified as small peptides or oligopeptides. In contrast, those ranging from 5000 to 10,000 Da are classified as large peptides. The smaller peptides, especially those within the 50–1000 Da range, are known for their significant biological activities and are often termed as bioactive peptides. These peptides, typically composed of 2–50 amino acid residues, adopt specific structures that are crucial for their biological functions (Cui et al. 2018). Currently, an increasing number of studies have demonstrated that these bioactive peptides exhibit a range of beneficial properties, including hypotensive, hypoglycemic, antioxidant, antibacterial, anti-inflammatory, and immunomodulatory properties (Yuan et al. 2018; Shobako and Ohinata 2020).

Sea buckthorn is also distinguished by its rich vitamin content, encompassing vitamin C (V_C), V_E, V_A, folic acid, and B-vitamins (V_{B1}, V_{B2}, V_{B6}). V_C, the predominant vitamin in sea buckthorn fruit, is renowned for its strong antioxidant capabilities. The V_C content in sea buckthorn fruit is influenced by various factors, including the variety, origin, harvest time, fruit maturity, and storage conditions of sea buckthorn (Li et al. 2018). Fresh sea buckthorn exhibits exceptionally high V_C levels, averaging 7950 mg/kg (Hussain et al. 2014; Wu et al. 2016), nearly five times higher than the amount found in yellow kiwi. Additionally, the plant contains fat-soluble vitamins V_E and V_K.V_E is the term used to describe a group of eight fat-soluble compounds, that is, tocopherols (a, b, c, and d isomers) and tocotrienols (a, b, c, and d isomers), which is essential for its antioxidant activity. Among these compounds, a-tocopherol has the highest biological activity, with its content in sea buckthorn ranging from 43 to 223 mg/kg FW (Kenan et al. 2010; Pop et al. 2015).

Mineral elements are crucial for human tissues formation and the maintenance of regular physiological processes. Sea buckthorn berries are abundant in several minerals, such as phosphorus, iron, magnesium, boron, calcium, aluminum, and potassium (Saeidi et al. 2016; Vaitkevièienë et al. 2019). Studies have revealed significant variation in the mineral content of sea buckthorn fruits at different maturity stages. For instance, ripe fruits contain the highest levels of calcium (68.28 mg/kg), magnesium (145.67 mg/kg), and phosphorus (457.7 mg/kg) (Arif et al. 2010).

Carbohydrates, as the primary constituents of dry matter, play vital roles in living organisms, acting as primary energy sources for human metabolism and structural components (Biel and Jaroszewska 2017). Sea buckthorn's carbohydrate profile predominantly consists of monosaccharides and disaccharides, with a total carbohydrate content (excluding dietary fiber) ranging from 0.48%–2.87% FW (Tkacz et al. 2019; Wu et al. 2016). The main monosaccharides include glucose, fructose, rhamnose, and xylose, while sucrose and maltose are the most typical disaccharides. The contents of glucose and fructose are approximately 0.42%–2.70% dry weight (DW) and 0.03%–1.30% DW (dry weight), respectively (Green and Low 2013; Tkacz et al. 2019; Wu et al. 2016).

The lipid content in sea buckthorn accounts for 1.2%–7.8% FW (Nazir and Bashir 2017; Zadernowski et al. 1997), with fatty acids being the predominant lipids, comprising 8.8%–11.1% of the total lipids (Qiao et al. 1988). The fatty acid profile, especially its polyunsaturated fatty acid content, is crucial in evaluating the nutritional value of sea buckthorn (Cristina et al. 2020).

Furthermore, sea buckthorn fruit is rich in various organic acids and their derivatives, contributing to bone differentiation and the development of mesenchymal stem cells into osteoblasts (Lee et al. 2021). Different sea buckthorn species vary in their organic acid types and concentrations. For example, subspecies of Russian sea buckthorn exhibit relatively low total acidity, with organic acid concentrations of 2.1–3.2 g/100 mL, while Finnish varieties fell within the intermediate range, with concentrations, ranging from 4.2 to 6.5 g/100 mL, in contrast to Chinese varieties, which demonstrated the highest levels of organic acids, with values between 3.5 and 9.1 g/100 mL (Bal et al. 2011). Sea buckthorn juice contains a range of organic acids, including quinic acid, L-malic acid, D-malic acid, succinic acid, pyruvic acid, tartaric acid, acetic acid, formic acid, and citric acid (Odgerel et al. 2021). An additional study on six Polish varieties of sea buckthorn identified oxalic acid and isocitric acid (Tkacz et al. 2019).

3.2 Bioactive Compounds

Bioactive compounds, plant-derived non-nutrient substances, have gained recognition in the past decade for their health-promoting benefits (Cristina et al. 2020). The bioactive compounds in sea buckthorn are primarily include polyphenols, polysaccharides and sugar alcohols, superoxide dismutase (SOD), organic acids, and phytosterols (Wang et al. 2021).

3.2.1 Carotenoids

Sea buckthorn berries are rich in carotenoids, contributing to their distinctive orange-yellow hue. Carotenoids, primarily functioning as antioxidants, also play various roles in health, such as β-carotene acting as a precursor for vitamin A and lutein/zeaxanthin forming the macular pigment in the eye (Eggersdorfer and Wyss 2018). These compounds are linked to reduced risks of eye diseases and certain cancers (Johnson 2002). The content of carotenoids varies significantly among different species and parts of sea buckthorn. Teleszko et al. (2015) detected an average of 11 mg/100 g FW of total carotenoids in eight species of Russian sea buckthorn. In another study on six Romanian sea buckthorn varieties (*H. rhamnoides* ssp. *carpatica*), the total carotenoid content ranged from 53 to 97 mg/100 g DW in berries and ranged from 3.5 to 4.2 mg/100 g DW in leaves (Pop et al. 2014). β-Carotene is the predominant carotenoid in sea buckthorn, constituting 15%–55% in berries and 26%–34% in the peel, pulp, and seed oil (Teleszko et al. 2015; Segliòa et al. 2021). Other carotenoids include γ-carotene, *cis*-lycopene, lycopene, *cis*-γ-carotene, β-cryptoxanthin, α-carotene.

3.2.2 Polyphenolics

Polyphenolics, a vast and widespread group of secondary metabolites, are abundant in sea buckthorn (de la Rosa et al. 2019). The polyphenol content in the fruit ranges from 12.36 to 34.6 mg GAE/g (GAE, gallic acid equivalents), which is higher than that in oranges (1.27 mg GAE/g) mandarins (1.16 mg GAE/g), blueberries (2.19 mg GAE/g), sour cherries (2.56 mg GAE/g), and strawberries (1.12 mg GAE/g) (Kuhkheil et al. 2017; Dragović-Uzelac et al. 2009; Li et al. 2021). Nearly 100 polyphenolic compounds have been isolated and identified from sea buckthorn, primarily consisting of phenolic acids and flavonoids (Ji et al. 2020). Seventeen phenolic acids are found in sea buckthorn berries, with salicylic acid being the main phenolic acid in berries, accounting for 55%–74.3% of the total phenolic acids

(Zadernowski et al. 2005). However, other studies have identified gallic acid as the primary phenolic acid found in both sea buckthorn fruit and leaves (Ji et al. 2020).

3.2.3 *Flavonoids*

Flavonoids, known for their potential roles in the prevention of chronic diseases, such as diabetes, cardiovascular disease, and cancer, are a significant component of sea buckthorn. Research by Guo et al. (2017a, b, c) demonstrated that the total phenols and flavonoid aglycones in sea buckthorn extract exhibit notable antioxidant and anti-proliferative properties. A comprehensive analysis has identified 95 distinct flavonoids in sea buckthorn, including 75 flavonols, 2 dihydroflavones, six catechins, one leucocyanidin, nine anthocyanidins, one proanthocyanidin, and one chalcone (Liu et al. 2021). A study by Raudonis et al. (2014) reported that the flavonoid content in 11 kinds of sea buckthorn varieties in Lithuanian ranged from 385 to 616 μg/g FW, with flavonols likequercetin, isorhamnetin, and kaempferol being predominant (Liu et al. 2021). Specifically, these findings, combined with the identification of compounds such as kaempferol, isorhamnetin, catechin, and epicatechin (Guo et al. 2017a, b, c), as well as B-type proanthocyanins, underscore the complexity and diversity of flavonoids in sea buckthorn (Ma et al. 2019a, b, c), contributing significantly to its antioxidant capacity (Kim et al. 2011).

3.2.4 *Fatty Acids*

Sea buckthorn is a valuable source of diverse fatty acids, which are essential for human health. These fatty acids are instrumental in treating skin and mucous membrane disorders, alleviating dry eyes syndrome, and reducing the risk of cardiovascular disease (Saeidi et al. 2016). Teleszko et al. (2015) identified 11 fatty acids in sea buckthorn pulp oil, with palmitic, palmitoleic, and oleic acids being the primary components in the berries.

3.2.5 *Phytosterols*

Phytosterols in sea buckthorn have been recognized for their potential to mitigate the risk of cardiovascular diseases. A comprehensive study has shown that the combined phytosterol quantity in berry lipids derived from eight Russian sea buckthorn species ranged from 6168.24 to 13,378.22 μg/100 mL. Fourteen sterol compounds have been detected in sea buckthorn pulp lipids, namely, 4-desmethyl sterols (cholestanol derivatives, including β-sitosterol, stigmasterol, campesterol, and

Δ5-avenasterol), 4α-monomethyl sterols (e.g., citrostadienol), and 4,4-dimethylsterols (e.g., 24-methylenecycloartanol) (Teleszko et al. 2015). The highest content of phytosterol is sitosterol, approximately 160–330 mg/kg FW (Yang and Kallio 2001). The remaining phytosterols consist primarily of campestero, stigmastadienol, stigmastanol, and α-amyrin (Yang and Kallio 2001). Due to their structural similarity to cholesterol, phytosterols are effective in lowering the plasma cholesterol levels in humans (Li 2002).

3.2.6 *Polysaccharides*

Plant polysaccharides, such as those found in sea buckthorn, have gained significant interest due to their multifunctional properties, low toxicity, and minimal side effects (Cristina et al. 2020). In sea buckthorn, polysaccharides primarily consist of dietary fiber, amounting to 4.20% FW. This is comparatively lower than the highest dietary fiber content in wild jujube. Pectin, a type of soluble dietary fiber in sea buckthorn, varies between 0.21% and 0.68% FW. Furthermore, sea buckthorn contains sugar alcohols including sorbitol, mannitol, xylitol, inositol, and methyl inositol, although in low content ranging from 0.013 to 3.0 g/kg DW (Green and Low 2013; Tkacz et al. 2019).

3.2.7 *SOD*

Superoxide dismutase (SOD), a critical antioxidant enzyme present in aerobic cells, acts as a primary defense against superoxide anion radicals (O_2) (Ighodaro and Akinloye 2018). Hou et al. (2019) determined The SOD content in fresh sea buckthorn is remarkably high, ranging from 1427.5 to 3175.1 U/mg, surpassing that in fresh olive (19.29–28.01 U/mg), litchi (24–96 U/mg), "Daw" longan fruit (90–160 U/mg), and even chestnut rose (1680 U/mg), commonly referred to as the "SOD king" (Ali et al. 2018; Lopez-Huertas and Palma 2020). This highlights the exceptional richness in SOD and its potential antioxidant capacity of sea buckthorn.

3.2.8 *Organic Acids*

The organic acid profile in sea buckthorn varies among species, both in types and concentrations. It has been reported that sea buckthorn juice contains nine organic acids, including quinic acid, L-malic acid, D-malic acid, succinic acid, pyruvic acid, tartaric acid, acetic acid, formic acid, and citric acid (Odgerel et al. 2021). L-malic acid (0.82%–6.08% FW) is the predominant organic acid, followed by succinic acid (0.20%–0.64% FW) and quinic acid (0.07%–1.74% FW) (Tkacz et al. 2019).

Additionally, three triterpenoid acids (TTAs), specifically including ursolic acid, oleanolic acid, and betulonic acid, have been identified in sea buckthorn, with total TTA concentration ranging from 52.64 to 82.58 mg/kg (Hu et al. 2015). These organic acids not only contribute to the low pH of plants, enhancing its preservation capabilities, but also exhibit diverse biological effects such as anti-tumor, anti-hyperlipidemic, antioxidant, and antibacterial properties (Hu et al. 2015).

3.3 Health Benefits

3.3.1 Regulation of Blood Glucose and Lipid

Sea buckthorn berries, integral to various dietary studies, have shown efficacy in regulating obesity and managing blood sugar levels. Studies demonstrate that sea buckthorn polysaccharides stimulate the activation of brown adipocytes and enhance thermogenesis through the upregulation of PPARγ-coactivator 1α (PGC1α), uncoupling protein-1 (UCP-1), and PR domain containing 16 (PRDM 16). This activation results in reduced lipid accumulation and weight gain (Ma et al. 2022). Additionally, palmitic acid-rich extracts from sea buckthorn fruit oil have been observed to lower weight in hypercholesterolemic hamsters and mitigate blood sugar elevation associated with dyslipidemia. A meta-analysis has highlighted that dietary intake of sea buckthorn berries and their extracts significantly improves triglyceride (TC), low-density lipoprotein (LDL), and high-density lipoprotein cholesterol (HDL-c) levels in subjects with hyperlipidemia, attributed its phytochemicals, particularly flavonoids and β-sitosterol (Guo et al. 2017a, b, c). Notably, dried sea buckthorn berries have been found to reduce LDL levels, especially in overweight and obese women with higher cardiovascular disease (CVD) risk (Larmo et al. 2013). Furthermore, clinical trials with overweight and obese female on sea buckthorn berry diets, reported a significant decrease in waist circumference and a moderate weight reduction (Lehtonen et al. 2011). Moreover, consuming sea buckthorn berries directly has been linked to improved postprandial blood glucose levels in diabetic patients, enhancing their glycemic profile and reducing plasma insulin concentration after 30 and 45 min (Mortensen et al. 2018).

3.3.2 Anti-depressant and Neuroprotective Effect

Major depressive disorder, a significant metal health challenge, is anticipated to become a leading cause of death and disability globally. Recent studies have investigated the antidepressant properties of sea buckthorn. Research have found that rodents were administered an aqueous fruit extract derived from sea buckthorn (40 mg/kg body weight) over 2 weeks. This treatment significantly increased the application of sea buckthorn, suggesting a notable antidepressant-like effects in animal models and may serve as a potential resource for natural psychotherapeutic agents against depression (Batool et al. 2009). Furthermore, sea buckthorn has shown promise in neurodegenerative conditions like Alzheimer's disease, which is a neurodegenerative condition characterized by extracellular amyloid-β (Aβ) deposition and the presence of neurofibrillary tangles due to Tau protein hyperphosphorylation (Zhang et al. 2022). Research has found that sea buckthorn can effectively eliminate intracellular Aβ deposits, with sea buckthorn powder at a concentration of 1.5 g/mL demonstrating the highest efficacy. This finding may be attributed to the higher levels of antioxidants present in the sea buckthorn berry powder. These antioxidants may provide neuroprotective effects by mitigating Aβ-induced toxicity and prevent cell death by exerting a neuroprotective effect. Therefore, sea buckthorn holds promise as a potential therapeutic agent for the treatment of Alzheimer's disease (Dong et al. 2020). Additionally, sea buckthorn flavonoids have been observed to modulate insulin signaling pathways by stimulating insulin receptor substrate (IRS)/AKT activation, reducing protein tyrosine phosphatase 1B (PTP1B) expression, and normalizing insulin ERK/CREB/BDNF signaling pathways, suggesting potential benefits in attenuating high fat diet (HFD)-induced cognitive impairment, and effectively preventing memory loss (Mulati et al. 2020). Studies have also noted improvements of sea buckthorn in epileptiform activity in the cerebral cortex and hippocampus in iron-induced epilepsy rats, reduced anxiety-like behavior, and improved memory impairment and histological damage in animal models (Ladol and Sharma 2021). The neuroprotective effects of sea buckthorn, including its effect on insulin resistance and neuroinflammation through the removal of Aβ deposits, and inhibition of Aβ-induced toxicity. These properties are likely linked to the presence of flavonoids and other antioxidant compounds. However, clinical trials are necessary to confirm these effects in humans.

3.3.3 Antioxidant Mechanisms and Efficacy

Numerous studies have established the antioxidant activity of sea buckthorn, both in vitro and in vivo. The phenolic fraction from sea buckthorn fruits has demonstrated the ability to inhibit hydrogen peroxide (H_2O_2) or H_2O_2/Fe-stimulated plasma lipid peroxidation and protein carbonylation. In fact, protein carbonylation is a relatively stable biomarker of oxidative stress. The phenolic constituents of sea buckthorn fruit reduced the concentration of carbonyl groups in plasma proteins treated with H_2O_2 or H_2O_2/Fe. Specifically, when plasma was treated with a 50 g/mL concentration of sea buckthorn phenolic fractions for 60 min, the inhibition rate of plasma lipid peroxidation reached 60% (Olas et al. 2016). Further, in vitro trials have shown that sea buckthorn extract, used alone or in conjunction with atorvastatin for the treatment of hyperlipidemia, effectively reduces the oxidative damage caused by lipid peroxidation (Mohamed et al. 2020). Additionally, sea buckthorn leaf extract has been observed to decrease intracellular oxidative stress in a dose-dependent manner, enhancing the viability of neuronal PC-12 cells and the integrity of their membranes (Cho et al. 2017).

3.3.4 *Hepatoprotective Activity and Mechanism*

Sea buckthorn extract and oil exhibit significant hepatoprotective activities, attributed to their rich carotenoid content, and as an important source of bioavailable lutein (Tudor et al. 2020). These carotenoids, such as β-carotene, lycopene, lutein, and β-cryptoxanthin, have been shown to mitigate oxidative stress and regulate hepatocyte lipid metabolism. Biochemical and histopathological studies indicate that sea buckthorn flavonoid extract effectively improves biomarkers related to non-alcoholic fatty liver disease, such as triglycerides, total cholesterol, low-density lipoprotein cholesterol, aspartate aminotransferase, and alanine aminotransferase. Its therapeutic effects of sea buckthorn have been compared favorably to curcumin (Guo et al. 2021). Furthermore, sea buckthorn fruit fermentation solution has been found to regulate hepatic lipid metabolism and oxidative stress, partly by modulating the composition of the gut microbiota, thus offering protection against alcoholic liver disease and exerting a hepatoprotective effect (Ran et al. 2021). Further research demonstrates that the bioactive compounds present in sea buckthorn can impede the activation of hepatic stellate cells, diminish inflammation by reducing levels of cytokines, and limit the progression of fibrosis induced by bile duct ligation (BDL), all in a dose-dependent manner in rats. Moreover, sea buckthorn reduces liver injury, and inflammation and restored liver function (Zhang et al. 2018). It has been reported that a flavonoid extract derived from sea buckthorn seed residues decreases the quantity of adipocytes in the livers of obese mice, leading to a substantial reduction in HFD-induced fatty infiltration in the liver tissue. Additionally, the expression of PPARγ in the liver and adipose tissue is reduced, resulting in decreased fat accumulation (Hao et al. 2020). Overall, the flavonoids and carotenoids in sea buckthorn contribute to the hepatoprotective effects, primarily through the regulation of lipid metabolism and oxidative stress and the reduction in inflammatory factor levels, warranting further in vivo exploration of the hepatoprotective activity of sea buckthorn.

3.3.5 *Dermatological Benefits in Treating Skin Diseases*

Sea buckthorn is increasingly recognized for its efficacy in the treatment of skin diseases, particularly those exacerbated by metabolic processes and environmental factors like ultraviolet radiation (UVB) (Wang and Wu 2019). Clinical studies have indicated that the extract of sea buckthorn, particularly rich in polyphenols, effectively reduces melanin levels in patients with melasma. In a controlled study, participants applied formulations containing sea buckthorn extracts on one cheek and a placebo to the other over 12 weeks. The results indicated a significant reduction in the level of melanin among the sea buckthorn-treated patients compared to the placebo group. The difference between the pre- and post-treatment levels of melanin was statistically significant (Khan et al. 2013). Moreover, sea buckthorn has been reported to have a wide range of dermatological effects. A clinical trial involving patients with mild to moderate psoriasis demonstrated improvements in the Psoriasis Area Severity Index (PASI) score and Dermatology Life Quality Index (DLQI) scores after treatment with topical sea buckthorn fruit extract, further substantiating its therapeutic potential (Boca et al. 2019).

3.3.6 Anti-psoriatic and Anti-atopic Dermatitis Property

Sea buckthorn demonstrates notable anti-psoriatic and anti-atopic dermatitis properties in animal models. In a study using the 12-*O*-tetradecanoylphorbol-13-acetate (TPA)-induced psoriasis-like lesion CD-1 mouse model, both oral (100 and 200 mg/kg) and topical (20 μL) administration of sea buckthorn oil significantly reduced ear edema ($34.05 \pm 7.65\%$, and $30.45 \pm 8.90\%$, respectively) and decreased ear biopsy weights. These anti-inflammatory and anti-psoriatic properties are potentially linked to the high levels of fatty acids in sea buckthorn oil, which may inhibit reactive nitrogen and downregulate nuclear factor kappa-B (NF-κB) protein and pro-inflammatory cytokines (Balkrishna et al. 2019).

3.3.7 Cardiovascular Health Benefit

Concerning cardiovascular health, sea buckthorn oil (SBT oil) has been studied for its protective effects against myocardial injury. CVD refers to the heart, brain, and systemic tissue of ischemic or hemorrhagic disease caused by hyperlipidemia, blood viscosity, atherosclerosis, and high blood pressure (Yang et al. 2018). Dietary incorporation of SBT oil has shown promising results in maintaining of hemodynamic functions and lipid peroxide metabolism, as well as scavenging free radicals and improving the antioxidant systems. This indicates that the efficacy of SBT oil as a safe and potent antioxidant in animals, suggesting potential benefits for those who are at the risk of myocardial injury (Malik et al. 2011).

References

Ali S, Khan AS, Malik AU, Shaheen T, Shahid M (2018) Prestorage methionine treatment inhibits postharvest enzymatic browning of cold stored 'Gola' litchi fruit. Postharvest Biol Technol 140:100–106

Arif S, Ahmed S, Shah A, Hassan L, Awan S, Hamid A et al (2010) Determination of optimum harvesting time for vitamin C, oil and mineral elements in berries sea buckthorn (*Hippophaë rhamnoides*). Pak J Bot 42:3561–3568

Bal L, Meda V, Naik S, Satya S (2011) Sea buckthorn berries: a potential source of valuable nutrients for nutraceuticals and cosmoceuticals. Food Res Int 44:1718–1727

Balkrishna A, Sakat S, Joshi K, Joshi K, Sharma V, Ranjan R et al (2019) Cytokines driven anti-inflammatory and anti-psoriasis like efficacies of nutraceutical sea buckthorn (*Hippophaë rhamnoides*) oil. Front Pharmacol 10:1186

Batool F, Shah AH, Ahmed SD, Haleem DJ (2009) Oral supplementation of Sea buckthorn (*Hippophaë rhamnoides L. spp. Turkestanica*) fruit extract modifies haloperidol induced behavioral deficits and increases brain serotonin metabolism. J Food Drug Anal 17(4):257–263

Biel W, Jaroszewska A (2017) The nutritional value of leaves of selected berry species. Sci Agric 74:405–410

Boca A, Ilies R, Saccomanno J, Pop R, Vesa S, Tataru A et al (2019) Sea buckthorn extract in the treatment of psoriasis. Exp Ther Med 17:1020–1023

Cho C, Jang H, Lee M, Kang H, Heo H, Kim D (2017) Sea buckthorn (*Hippophaë rhamnoides L.*) leaf extracts protect neuronal PC-12 cells from oxidative stress. J Microbiol Biotechnol 27:1257–1265

Cui Q, Liu JZ, Wang LT et al (2018) Sustainable deep eutectic solvents preparation and their efficiency in extraction and enrichment of main bioactive flavonoids from sea buckthorn leaves. J Clean Prod 184:826–835

de la Rosa LA, Moreno-Escamilla JO, Rodrigo-Garcıa J, Alvarez-Parrilla E (2019) Chapter 12: Phenolic compounds. In: Yahia EM (ed) Postharvest physiology and biochemistry of fruits and vegetables. Woodhead Publishing, Duxford, pp 253–271

Dong K, Fernando W, Durham R, Stockmann R, Jayatunga D, Jayasena V (2020) A role of sea buckthorn on Alzheimer's disease. Int J Food Sci Technol 55:3073–3081

Dragović-Uzelac V, Bursać Kovačević D, Levaj B, Pedisić S, Mezak M, TomljenovićAna A (2009) Polyphenols and antioxidant capacity in fruits and vegetables common in the Croatian diet. Agric Conspec Sci 74:175–179

Eggersdorfer M, Wyss A (2018) Carotenoids in human nutrition and health. Arch Biochem Biophys 652:18–26

Green RC, Low NH (2013) Physicochemical composition of buffaloberry (*Shepherdia argentea*), chokecherry (*Prunus virginiana*) and sea buckthorn (*Hippophaë rhamnoides*) fruit harvested in Saskatchewan. Can J Plant Sci 93(6):1143–1153

Guo RX, Chang XX, Guo XB, Brennan CS, Li T, Fu X, Liu RH (2017a) Phenolic compounds, antioxidant activity, antiproliferative activity and bioaccessibility of Sea buckthorn (*Hippophaë rhamnoides L.*) berries as affected by in vitro digestion. Food Funct 8(11):4229–4240

Guo RX, Guo XB, Li T et al (2017b) Comparative assessment of phytochemical profiles, antioxidant and antiproliferative activities of Sea buckthorn (*Hippophaë rhamnoides L.*) berries. Food Chem 221:997–1003

Guo XF, Yang B, Cai WW, Li D (2017c) Effect of sea buckthorn (*Hippophaë rhamnoides L.*) on blood lipid profiles: a systematic review and meta-analysis from 11 independent randomized controlled trials. Trends Food Sci Technol 61:1–10

Guo Z, Cheng J, Zheng L, Xu W, Xie Y (2021) Mechanochemical-assisted extraction and hepatoprotective activity research of flavonoids from sea buckthorn (*Hippophaë rhamnoides L.*) pomaces. Molecules 26:7615

Hao Y, Zhou F, Dong J, Wang Y, Lang Z, Li S et al (2020) Study on the role of flavonoids derived extract from seed residues of *Hippophaë rhamnoides* on high-fat diet induced obese mice. J King Saud Univ Sci 32:1597–1603

Hou ZQ, Zhao L, Wang YT, Liao XJ (2019) Purification and characterization of superoxide dismutases from sea buckthorn and chestnut rose. J Food Sci 84(4):746–753

Hu N, Suo Y, Zhang Q, You J, Ji Z, Wang A, Han L, Lv H, Ye Y (2015) Rapid, selective, and sensitive analysis of triterpenic acids in *Hippophaë rhamnoides L.* using HPLC with pre-column fluorescent derivatization and identification with post-column APCI-MS. J Liq Chromatogr Relat Technol 38(4):451–458

Hussain M, Ali S, Awan S, Hussain M, Hussain I (2014) Analysis of minerals and vitamins in sea buckthorn (*Hippophaë rhamnoids*) pulp collected from Ghizer and Skardu districts of Gilgit-Baltistan. Int J Biosci 8:144–152

Ighodaro OM, Akinloye OA (2018) First line defence antioxidants-superoxide dismutase (SOD), catalase (CAT) and glutathione peroxidase (GPX): their fundamental role in the entire antioxidant defence grid. Alex J Med 54(4):287–293

Ji M, Gong X, Li X, Wang C, Li M (2020) Advanced research on the antioxidant activity and mechanism of polyphenols from *Hippophaë* species – a review. Molecules 25:917

Johnson E (2002) The role of carotenoids in human health. Nutr Clin Care 5:56–65

Kenan G, Seyithan T, Hakan AH, Omer A, Nuray O, Fatih A, Mustafa G (2010) Vitamin E and *Hippophea rhamnoides L.* extract reduce nicotine-induced oxidative stress in rat heart. Cell Biochem Funct 28(4):1972–1989

Khan BA, Akhtar N, Braga VA (2013) Anti-aging effects of *Hippophaë rhamnoides* emulsion on human skin. Trop J Pharm Res 11(6):419–430

Kim JS, Kwon YS, Sa YJ, Kim MJ (2011) Isolation and identification of sea buckthorn (*Hippophaë rhamnoides*) phenolics with antioxidant activity and a-glucosidase inhibitory effect. J Agric Food Chem 59(1):138–144

Kuhkheil A, Badi H, Mehrafarin A, Abdossi V (2017) Chemical constituents of sea buckthorn (*Hippophaë rhamnoides L.*) fruit in populations of central Alborz mountains in Iran. Pharm Res 4:1–12

Ladol S, Sharma D (2021) The effects of *Hippophaë rhamnoides* in neuroprotection and behavioral alterations against iron-induced epilepsy. Epilepsy Res 175:106695

Larmo PS, Kangas AJ, Soininen P, Lehtonen HM, Suomela JP, Yang B, Viikari J, Ala-Korpela M, Kallio HP (2013) Effects of sea buckthorn and bilberry on serum metabolites differ according to baseline metabolic profiles in overweight women: a randomized crossover trial. Am J Clin Nutr 98(4):941–951

Lee Y, Jang H, Park K, Kim S, Kim J, Kim J et al (2021) Phytochemical analysis of the fruits of sea buckthorn (*Hippophaë rhamnoides*): identification of organic acid derivatives. Plan Theory 10:860

Lehtonen HM, Suomela JP, Tahvonen R, Yang B, Venojarvi M, Viikari J, Kallio H (2011) Different berries and berry fractions have various but slightly positive effects on the associated variables of metabolic diseases on overweight and obese women. Eur J Clin Nutr 65(3):394–401

Li TS (2002) Product development of sea buckthorn. In: Janick J, Whipkey A (eds) Trends in new crops and new uses. ASHS Press, Alexandria, pp 393–398

Li CY, Zhang JJ, Zhao CJ et al (2018) Separation of the main flavonoids and essential oil from Sea buckthorn leaves by ultrasonic/microwave assisted simultaneous distillation extraction. R Soc Open Sci 5(7):180133

Li N, Hu Y, Ge X (2021) Determination of contents of total flavonoids and total polyphenols in *Hippophaë rhamnoides L.* from different origins and their antioxidant activity. Biochem Eng J 38:64–68

Liu S, Xiao P, Kuang Y, Hao J, Huang T, Liu E (2021) Flavonoids from sea buckthorn: a review on phytochemistry, pharmacokinetics and role in metabolic diseases. J Food Biochem 45:e13724

Lopez-Huertas E, Palma JM (2020) Changes in glutathione, ascorbate, and antioxidant enzymes during olive fruit ripening. J Agric Food Chem 68(44):12221–12228

Ma SW, Gao Y, Zong HF et al (2019a) Study on preparation and antioxidant activity of mung bean peptide beverage. J Univ Sci Technol Liaoning 42(6):440–447

Ma XY, Yang W, Marsol-Vall A, Laaksonen O, Yang BR (2019b) Analysis of flavour compounds and prediction of sensory properties in sea buckthorn (*Hippophaë rhamnoides L.*) berries. Int J Food Sci Technol 55(4):1705–1715

Ma C, Zheng X, Deng S, Erdake A, He H (2019c) Determination and analysis of amino acids in fruits of *Hippophaë rhamnoide*. Prot For Sci Technol 9:39–40

Ma Z, Sun Q, Chang L, Peng J, Zhang M, Ding X (2022) A natural anti-obesity reagent derived from sea buckthorn polysaccharides: structure characterization and anti-obesity evaluation in vivo. Food Chem 375:131884

Malik S, Goyal S, Ojha SK, Bharti S, Nepali S, Kumari S, Singh V, Arya DS (2011) Seabuckthorn attenuates cardiac dysfunction and oxidative stress in isoproterenol-induced cardiotoxicity in rats. Int J Toxicol 30(6):671–680

Mohamed EA, Tulcan C, Alexa E, Morar D, Dumitrescu E, Muselin F, Radulov I, Puvaca N, Cristina RT (2020) Sea buckthorn and grape extract might be helpful and sustainable phytoresources as associated hypolipidemic agents-preliminary study. Sustain For 12(21)

Mortensen MW, Spagner C, Cuparencu C, Astrup A, Raben A, Dragsted LO (2018) Sea buckthorn decreases and delays insulin response and improves glycaemic profile following a sucrose containing berry meal: a randomized, controlled, crossover study of Danish sea buckthorn and strawberries in overweight and obese male subjects. Eur J Nutr 57(8):2827–2837

Mulati A, Ma S, Zhang H, Ren B, Zhao B, Wang L et al (2020) Sea-buckthorn flavonoids alleviate high-fat and high-fructose diet-induced cognitive impairment by inhibiting insulin resistance and neuroinflammation. J Agric Food Chem 68:5835–5846

Nazir F, Bashir M (2017) Chemical and antioxidant properties of Sea buckthorn (*Hippophaë rhamnoides*). Pharma Innov J 6(12):173–176

Odgerel U, Islam M, Kitamura Y, Kokawa M, Odbayar T (2021) Effect of micro wet milling process on particle sizes, antioxidants, organic acids, and specific phenolic compounds of whole sea buckthorn (*Hippophaë rhamnoides L.*) juices. J Food Process Preserv 45:1–14

Olas B, Kontek B, Malinowska P, Żuchowski J, Stochmal A (2016) *Hippophaë rhamnoides* L. fruits reduce the oxidative stress in human blood platelets and plasma. Oxidative Med Cell Longev 2016:1–8

Pop R, Weesepoel Y, Socaciu C, Pintea A, Vincken J, Gruppen H (2014) Carotenoid composition of berries and leaves from six Romanian sea buckthorn (*Hippophaë rhamnoides L.*) varieties. Food Chem 147:1–9

Pop EA, Diaconeasa ZM, Fetea F, Bunea A, Dulf F, Pintea A, Socaciu C (2015) Carotenoids, tocopherols and antioxidant activity of lipophilic extracts from sea buckthorn berries (*Hippophaë rhamnoides*), apricot pulp and apricot kernel (*Prunus armeniaca*). Bull Univ Agric Sci Vet Med Cluj-Napoca Food Sci Technol 72(2):169–176

Qiao TS, Geng JQ, Feng JZ, Xu WY (1988) Nutrient content of sea buckthorn fruit. J Nutr 2:168–172

Ran B, Guo C, Li W, Li W, Wang Q, Qian J et al (2021) Sea buckthorn (*Hippophaë rhamnoides L.*) fermentation liquid protects against alcoholic liver disease linked to regulation of liver metabolome and the abundance of gut microbiota. J Sci Food Agric 101:2846–2854

Raudonis R, Raudone L, Janulis V, Viskelis P (2014) Flavonoids in cultivated berries of sea buckthorn (*Hippophaë rhamnoides L.*). Planta Med 80:24

Ren QC, Li XH, Li QY et al (2019) Total flavonoids from sea buckthorn ameliorates lipopolysaccharide/cigarette smoke-induced airway inflammation. Phytother Res 33(8):2102–2117

Saeidi K, Alirezalu A, Akbari Z (2016) Evaluation of chemical constitute, fatty acids and antioxidant activity of the fruit and seed of sea buckthorn (*Hippophaë rhamnoides L.*) grown wild in Iran. Nat Prod Res 30:366–368

Segliòa D, Krasnova I, Grygier A, Radziejewska-Kubzdela E, Rudzińska M, Górnaś P (2021) Unique bioactive molecule composition of sea buckthorn (*Hippophaë rhamnoides L.*) oils obtained from the peel, pulp, and seeds via physical "solvent-free" approaches. J Am Oil Chem Soc 98:1009–1020

Shobako N, Ohinata K (2020) Anti-hypertensive effects of peptides derived from rice bran protein. Nutrients 12(10):3060

Tan L, Zhao J, Ma J, Ji T, Dong Q, Shen J (2018) Analysis of nutritional compositions and nutritional quality evaluation in different parts of Yushu Hippophaë (*Hippophaë rhamnoides L.* subsp. sinensis). Nat Prod Res 30:807–816

Tkacz M, Wojdyło A, Rudzińska M, Oszmiański J, Golis T (2015) Analysis of lipophilic and hydrophilic bioactive compounds content in sea buckthorn (*Hippophaë rhamnoides L.*) berries. J Agric Food Chem 63:4120–4129

Tkacz K, Wojdyło A, Turkiewicz IP, Bobak Ł, Nowicka P (2019) Anti-oxidant and anti-enzymatic activities of sea buckthorn (*Hippophaë rhamnoides L.*) fruits modulated by chemical components. Antioxidants 8:618

Tudor C, Bohn T, Iddir M, Dulf F, Focsan M, Ruginã D et al (2020) Sea buckthorn oil as a valuable source of bioaccessible xanthophylls. Nutrients 12:76

Vaitkevièienë N, Jarienë E, Danilèenko H, Kulaitienë J, Mažeika R, Hallmann E et al (2019) Comparison of mineral and fatty acid composition of wild and cultivated sea buckthorn berries from Lithuania. J Elem 24:1101–1113

Wang X, Wu J (2019) Modulating effect of fatty acids and sterols on skin aging. J Funct Foods 57:135–140

Wang K, Xu Z, Liao X (2021) Bioactive compounds, health benefits and functional food products of sea buckthorn: a review. Crit Rev Food Sci Nutr 5:1–22

Wu ZJ, Ruan CJ, Li HY, Xie ML, M M, X TT (2016) Study on fruit soluble sugar and organic acid components of 12 sea buckthorn cultivars. J Northwest For Coll 31(4):106–112

Yang B, Kallio H (2001) Fatty acid composition of lipids in sea buckthorn (*Hippophaë rhamnoides L.*) berries of different origins. J Agric Food Chem 49(4):1939–1947

Yang H, Tian T, Wu D, Guo D, Lu J (2018) Prevention and treatment effects of edible berries for three deadly diseases: cardiovascular disease, cancer and diabetes. Crit Rev Food Sci Nutr 59(12):1903–1912

Yuan H, Shi F, Meng L et al (2018) Effect of sea buckthorn protein on the intestinal microbial community in streptozotocin-induced diabetic mice. Int J Biol Macromol 107:1168–1174

Zadernowski R, Nowak-Polakowska H, Lossow B, Nesterowicz J (1997) Sea-buckthorn lipids. J Food Lipids 4(3):165–172

Zadernowski R, Naczk M, Czaplicki S, Rubinskiene M, Szałkiewicz M (2005) Composition of phenolic acids in sea buckthorn (*Hippophaë rhamnoides L.*) berries. J Am Oil Chem Soc 82:175–179

Zhang W, Yan J, Duo J, Ren B, Guo J (1989) Preliminary study of biochemical constitutions of berry of sea buckthorn growing in Shanxi province and their changing trend. In: Paper presented at the proceedings of international symposium on sea buckthorn (*H. rhamnoides L.*), Xian, China, October 19–23

Zhang G, Liu Y, Liu P (2018) Active components from sea buckthorn (*Hippophaë rhamnoides L.*) regulate hepatic stellate cell activation and liver fibrogenesis. J Agric Food Chem 66:12257–12264

Zhang P, Ji H, Hu Q (2022) Research progress in clinical treatment of Alzheimer's disease and potential drugs from natural products. Acta Pharm Sin B 2022:1–21

Zhou WN, Yuan ZZ, Li G et al (2018) Isolation and structure determination of a new flavone glycoside from seed residues of seabuckthorn (*Hippophaë rhamnoides L.*). Nat Prod Res 32(8):892–897

Chapter 4
Active Substances and Antioxidant Activity of Different Solvent Extracts of Sea Buckthorn Leaves

Yaoran Li, Xiaobo Wei, Peiyi Zhu, and Jinhu Tian

Abstract This chapter provides an overview of the main compounds found in sea buckthorn leaves, including vitamins, minerals, sugars, organic acids, amino acids, carotenoids, and polyphenols. Additionally, it delves into the biological activity of sea buckthorn leaf extract, highlighting its antioxidant properties when used as a food additive. Furthermore, the chapter explores potential applications and prospects for sea buckthorn leaf extract in various fields.

Keywords Sea buckthorn leaves · Active substances · Antioxidant activity · Different solvent extracts

Abbreviations

ISA	International Sea Buckthorn Association
DW	dry weight
FW	fresh weight
HRP	*Hippophae rhamnoides* polysaccharide
Nrf-2	nuclear factor erythroid-2-related factor 2
HO-1	heme oxygenase-1
SOD	superoxide dismutase
GSH-Px	glutathione peroxidase
MDA	malondialdehyde
DPPH	1,1-Diphenyl-2-picrylhydrazyl radical 2

Y. Li (✉)
Food College, Ningxia University, Yinchuan, China

X. Wei
Food Science and Technology, Ningxia University, Yinchuan, Ningxia, China

P. Zhu · J. Tian
Department of Food Science and Nutrition, Zhejiang University, Hangzhou, China

J. Tian et al. (eds.), *Sea Buckthorn*, https://doi.org/10.1007/978-981-97-9865-0_4

GAE gallic acid equivalent
PPO polyphenol oxidase
POD peroxidase

4.1 Introduction

Sea buckthorn (*Hippophae rhamnoides* L.) contains nearly 200 nutritional and bioactive compounds, including vitamins and minerals, sugar alcohols, polyunsaturated fatty acids, carotenoids, polyphenols, and superoxide dismutase. Therefore, sea buckthorn is known as a "medicine food homology" fruit categorized by China's National Health Commission (Wang et al. 2022a). It is widely used by the food sector in the creation including breads, yogurts, jams, drinks, and teas (Gatlan and Gutt 2021; Ghendov-Mosanu et al. 2020). The medicinal value of sea buckthorn has been recorded in the Tibetan Medical Classic dating back to as early as the first half of the eighth century. Sea buckthorn has been extensively exploited in the folklore treatment of slow digestion, malformation of stomach, cardiovascular problems, liver injury, skin diseases, and ulcers (Suryakumar and Gupta 2011). In recent years, there have been numerous reports on the pharmacological activities of sea buckthorn, including anticancer, anti-inflammatory, antimicrobial, and antiviral activities, and its ability to act in cardiovascular protection (Masoodi et al. 2020; Olas 2016; Redei et al. 2019). There is no doubt that sea buckthorn has great medicinal and therapeutic potential, which may be attributed to the fact that it contains several vitamins, carotenoids, polyphenols, and fatty acids (Pop et al. 2014; Stobdan et al. 2010).

The ecological and economic importance of sea buckthorn cannot be overstated. The International Sea Buckthorn Association (ISA) was founded in 1999 by China, India, Canada, and other countries to promote the role of sea buckthorn in environmental conservation, economic development, and human health. In recent years, more countries have recognized the medicinal potential of sea buckthorn, and many countries have begun to acknowledge and promote the sea buckthorn industry. As of December 2020, sea buckthorn has been dispersed to 52 nations worldwide, covering a total area of 2.33 million hm^2, with China accounting for approximately 2.1 million hm^2 (Wang et al. 2022b). With the increased growth and production of sea buckthorn, more attention should be directed at its development and exploitation. As a result, the purpose of this chapter is to provide industrial producers and researchers with a complete understanding of the phytochemistry, nutritional and bioactive components, health advantages, and dietary uses of sea buckthorn.

4.2 Morphology of Sea Buckthorn Leaves

Sea buckthorn is a deciduous tree or shrub belonging to the *Elaeaceae* family and the genus *Hippophae* L. It grows to a height of 1–8 m, with some plants reaching up to 18 m. The lanceolate or linear leaves are 3–8 cm long and less than 7 mm wide. The leaves have a dark gray top surface and a silver-gray beneath surface.

4.3 Nutrients and Bioactive Compounds

The leaves of sea buckthorn are high in minerals and bioactive compounds. Their intensity is determined by the growing conditions, agricultural technology, and climate. The chemical composition of sea buckthorn leaves is listed in Table 4.1 according to the published data (Jaroszewska and Biel 2017).

4.3.1 Vitamins and Minerals

The detrimental effects of free radicals are mitigated by antioxidants, including vitamins A, C, and E, which are abundant in fruits and vegetables. The contents of vitamins ($n = 3$) in sea buckthorn leaves are listed in Table 4.2 (Jaroszewska and Biel 2017).

4.3.2 Carbohydrates

The polysaccharide isolated from sea buckthorn fruit has antioxidant ability and anti-fatigue capacity (Ni et al. 2013). Wang et al. (2018) reported the protective effect of *Hippophae rhamnoides* polysaccharide (HRP) against acetaminophen-induced hepatotoxicity and its mechanism and demonstrated that HRP could protect against acetaminophen-induced hepatotoxicity in mice by activating the nuclear factor erythroid-2-related factor 2 (Nrf-2)/heme oxygenase-1 (HO-1)-SOD-2 signaling pathway. Nrf-2/heme oxygenase-1 (HO-1)-SOD-2 signaling pathway

Table 4.1 Chemical composition ($n = 3$) of sea buckthorn leaves (g/kg dry weight) (drawn by the author)

Item	Content
Moisture (g kg^{-1} fresh weight)	680.0 ± 70.3
Dry matter	953.4 ± 0.9
Crude protein	140.9 ± 0.9
Crude ash	40.9 ± 1.0
Crude fat	54.4 ± 4.6
Crude fiber	104.8 ± 10.6
Nitrogen free extract	655.0 ± 14.8

Table 4.2 Content of vitamins ($n = 3$) in sea buckthorn leaves (drawn by the author)

Item	Content (mg/100 g DW)
Tocopherols	40.98 ± 4.9
L-ascorbic acid	2217.3 ± 188
Thiamine	1.40 ± 0.1
Riboflavin	5.86 ± 0.3
Niacin	4.86 ± 0.4

protects against acetaminophen-induced hepatotoxicity in mice. The transcription factor Nrf2 is a major regulator of the adaptive response to oxidative stress and regulates the expression of several cytoprotective genes, such as antioxidants, phase II enzymes, and membrane transport proteins (Hoang et al. 2019). HRP has a higher capacity for $O_2^{-}\cdot$ scavenging compared with vitamin C (VC). GSH-Px activity in the liver, brain tissue, and serum significantly reduced the MDA content in the liver, brain tissue, and serum. Thus, HRP can effectively alleviate and improve aging through antioxidant effects. It has been reported that HRP can scavenge 1,1-Diphenyl-2-picrylhydrazyl radical 2 (DPPH) radicals, –OH radicals and O_2^{-} radicals, and inhibits spontaneous lipid peroxidation in mouse liver homogenates and liver lipid peroxidation induced by CCl_4, H_2O_2, Fe^{2+}, and VC.

4.3.3 Organic Acids and Amino Acids

Sea buckthorn is rich in amino acids, which are indispensable to the human body. Amino acids are the basic units that make up proteins and are closely related to life activities. Seventeen amino acids, including seven essential amino acids (threonine, valine, methionine, isoleucine, leucine, phenylalanine, and lysine), have been detected in sea buckthorn fruits, leaves, branches, and seeds. The amino acid content in sea buckthorn seeds is 18.63%, in leaves 15.41%, in branches 11.62%, and in fruits 6.89% (Wang et al. 2022b). The content of aspartic acid and glutamic acid was highest in sea buckthorn fruits, leaves, and branches, with 1.11% and 1.24% in fruits, 2.42% and 1.60% in leaves, and 3.71% and 0.97% in branches. Sea buckthorn seeds have the highest levels of tyrosine and glutamic acid, with 4.72% and 3.42%, respectively (Wang et al. 2022b).

4.3.4 Carotenoids

Sea buckthorn fruit contains a high concentration of carotenoid, which gives it a distinctive orange hue. Carotenoids are primarily antioxidant, although they also have additional functions. For instance, β-carotene is a precursor of vitamin A, and lutein/zeaxanthin is an eye macular pigment (Eggersdorfer and Wyss 2018). Carotenoids are thought to reduce the risk of certain diseases, including cancer and eye disease (Johnson 2010). The content of carotenoids varies considerably in different species and parts of sea buckthorn. For example, in a study on Russian sea buckthorn, the average content of total carotenoids in eight cultivars was 11 mg/100 g FW (Teleszko et al. 2015). Another study found that the total carotenoid concentration in six Romanian sea buckthorn kind (*H. rhamnoides ssp. carpatica*) berries ranged from 53 to 97 mg/100 g dry weight and the total carotenoid content in the leaves ranged from 3.5 to 4.2 mg/100 g dry weight (Pop et al. 2014). The major pigment in sea buckthorn is β-carotene. β-Carotene is found in 15–55% of the berry and 26–34% of the pericarp, pulp, and seed oil (Seglina et al. 2021). In addition, carotenoids include α-carotene, γ-carotene, cis-γ-carotene, lycopene, *cis*-lycopene, β-cryptoxanthin, and others.

4.3.5 Polyphenols

Polyphenols are the principal antioxidant chemicals found in sea buckthorn. The polyphenol content of the fruit has been observed to vary from 12.36 to 34.6 mg GAE/g (GAE, gallic acid equivalents), which is greater than that of strawberries (1.12 mg GAE/g), mandarins (1.16 mg GAE/g), oranges (1.27 mg GAE/g), blueberries (2.19 mg GAE/g), and sour cherries (2.56 mg GAE/g) (Wang et al. 2022b). According to a recent analysis, approximately 100 polyphenolic components have been extracted and identified from sea buckthorn (Ji et al. 2020). Polyphenols are mostly composed of phenolic acids and flavonoids. Sea buckthorn berries contain 17 phenolic acids. Salicylic acid is the most common phenolic acid found in berries, accounting for 55–74.3% of total phenolic acids (Zadernowski et al. 2005). However, according to another study, gallic acid is the primary phenolic acid in sea buckthorn (Ji et al. 2020).

Flavonoids may play a role in the prevention of chronic illnesses such as heart disease, diabetes, and cancer (Ciesarova et al. 2020). Total phenols and flavonoid glycosides in sea buckthorn extracts show antioxidant and antiproliferative properties, according to Guo et al. (2017). Sea buckthorn contains 95 flavonoids, including 75 flavonols, two dihydroflavonoids, six catechins, nine anthocyanidins, one leucoanthocyanidin, one proanthocyanidin, and one chalcone (Liu et al. 2021). Flavonols are the primary constituents of flavonoids and may be found mostly in the glycosylated forms of quercetin, isorhamnetin, and kaempferol (Liu et al. 2021). Flavonols vary in concentration from 463.14 mg to 893.92 mg/100 g DW, which accounts for approximately 99% of total phenolic components (Tkacz et al. 2019). Geographical considerations, meteorological circumstances, and berry varietals have a substantial impact on the concentration and composition of polyphenolic chemicals. According to the published literature, Fig. 4.1 depicts the chemical structures of the principal phenolic compounds found in sea buckthorn (Wang et al. 2022b).

4.4 Antioxidant Activity

Many studies have proven the antioxidant activity of sea buckthorn in vitro and in vivo, and Table 4.3 shows the health advantages of sea buckthorn leaves. The phenolic contents of sea buckthorn berries reduce hydrogen peroxide (H_2O_2) or

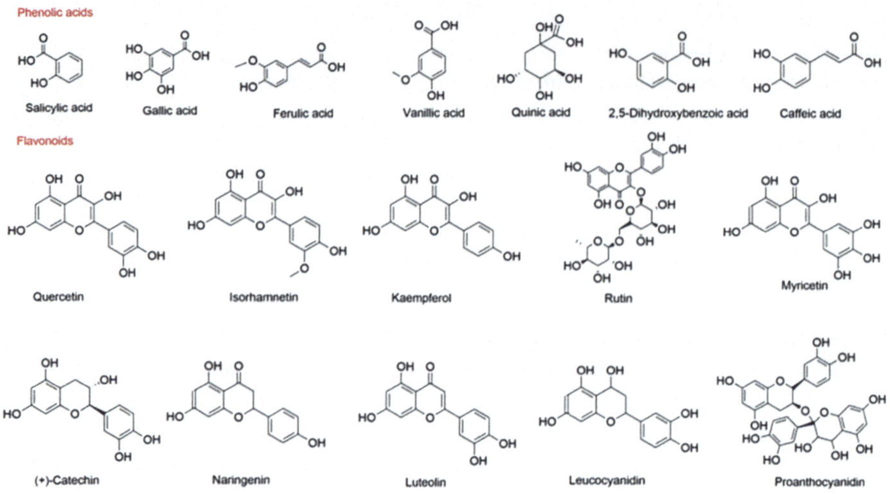

Fig. 4.1 Structure of the main phenolic compounds in sea buckthorn. (Drawn by the author)

H_2O_2/Fe-stimulated plasma lipid peroxidation and protein carbonylation. Protein carbonylation is a durable indicator of oxidative stress. Sea buckthorn berry phenolic components decreased the carbonyl content of plasma proteins treated with H_2O_2 or H_2O_2/Fe. When plasma was treated for 60 min with a 50 g/mL fraction of sea buckthorn phenol, lipid peroxidation was reduced by up to 60%. In vitro studies have demonstrated that using sea buckthorn extracts with or without atorvastatin to treat hyperlipidemia helps to minimize oxidative stress (Mohamed et al. 2020). Furthermore, in a dose-dependent manner, sea buckthorn leaf extract reduced intracellular oxidative stress, enhancing neuronal PC-12 cell survival and membrane integrity (Cho et al. 2017).

4.5 Extraction of Active Substances from Sea Buckthorn Leaves

Plant extracts are substances rich in phenols, flavonoids, alkaloids, terpenoids, and other functional components extracted from roots, stems, leaves, flowers, or fruits of plants. Different substances need to be extracted through appropriate specific methods (Ahmad Mir et al. 2018; Yong et al. 2013). Based on the differences in their functional components, plant extracts can be classified into six major groups, including alkaloids, flavonoids, polysaccharides, glycosides, organic acids, and volatile oils (Nikmaram et al. 2018). In addition, depending on the extraction method, plant extracts can be divided into two categories: water-soluble and oil-soluble. Water-soluble substances refer to plant extracts that are soluble in water, alcohols, or glycerol, while oil-soluble substances mainly refer to plant essential oils.

Table 4.3 Health benefits of sea buckthorn leaves (drawn by the author)

Sea buckthorn	Effective concentration/time	Study type	Experimental model	Main results	Bioactive compounds	References
Antioxidant						
Sea buckthorn leaf extracts	5, 10, 20 μg/mL	In vitro	PC-12 cells	↓ Relative proportion of total apoptotic PC-12 cells	Ellagic acid, gallic acid, isorhamnetin, kaempferol, and quercetin	Serban et al. (2019)
Anticancer						
Leaf aqueous extract	3.12, 6.25, 12.5, 25, 50 μg/mL	In vitro	LNCaP and C4-2 cell	↓ Proliferation and migration of prostate cancer cells	–	Masoodi et al. (2020)
Leaf extract	6.2, 62 μg/mL	In vitro	Rat C6 glioma cells	↓ Intracellular ROS ↑ Pre-apoptosis in rat C6 glioma cells	Phenolics	Kim et al. (2017)
Anti-inflammatory						
Sea buckthorn branches, berries, and leaf extracts	10 μg/mL	In vitro	RAW 264.7 macrophages	NO inhibition rates increased from 73 to 98	Phenolic compound	Dudau et al. (2021)
Sea buckthorn leaves extracts	0.05, 5, 50 μg/mL	In vivo	Mouse peritoneal macrophages	↓ Pro-inflammatory cytokine level (TNF-α, IFN-γ and IL-6)	Tannins, proteins, and carbohydrate groups	Tanwar et al. (2018)
Antimicrobial						
Sea buckthorn leaf extracts	5%	In vitro	Common skin and wound pathogens	↓ Gram-positive bacteria (*Staphylococcus aureus, S. epidermidis, S. intermedius, and S. pyogenes*)	–	Verma et al. (2011)
Sea buckthorn berries and leaves extracts	6 mg/mL	In vitro	Staphylococcus aureus (MRSA)	Inhibits the growth of MRSA	–	Shaikh (2016)
Sea buckthorn leaf extract	100 mg/kg body weight	In vitro	Healthy Swiss albino mice	↑ RVNA titers and CTL population ↑ Memory T cells, plasma cells	Isorhamnetin and other flavonoids	Singh et al. (2017)

The composition and functionality of substances contained in different plants or different parts of the same plant vary greatly. Therefore, the selection of plants determines the anti-browning ability of plant extracts. Among the different plant species, Wessels et al. (2014) investigated the effect of 36 different plant extracts on browning of fresh-cut apples, among which pumpkin seed extract had the strongest anti-browning ability (Wessels et al. 2014); among herbs and teas, hibiscus showed strong antioxidant ability; among medicinal and food plants, geranium root extract had the best inhibitory effect on browning, and only slight color change was observed throughout the measurement process. Meng et al. (2020) found that the total phenolic content of peanut coat extract was significantly higher than that of peanut shell extract and the scavenging rate of DPPH radicals by peanut coat extract was nearly 200 times higher than that of peanut shell extract, and the scavenging rate of ABTS (2,2'-azino-bis(3-ethylbenzothiazoline-6-sulfonic acid) radicals by peanut coat was 30 times higher than that of peanut shell (Meng et al. 2020). Pineapple heart, pulp, and peel were also effective in inhibiting browning of fresh-cut apples, but pineapple heart extract was more effective than peel and pulp extract (Supapvanich et al. 2012). Many studies have suggested that phenolic compounds in plants are the main anti-browning substances, among which fruit shells/skins, seed coats, or plant leaves/skins are rich in phenolic substances (e.g., chlorogenic acid, anthocyanins, and carotenoids). Rice bran (Sukhonthara et al. 2016), mangosteen shells (Yu et al. 2014), and potato peels (Venturi et al. 2019) all have good inhibitory effects on the browning of freshly cut fruits and vegetables. In addition, fruits, flower buds, and seeds are where substances are absorbed and assimilated, which means that they get more nutrients and active ingredients than other plant parts, such as lilac buds (Park et al. 2019), grape seeds (Altunkaya and Gokmen 2012), and pineapple (Supapvanich et al. 2012) also have good inhibitory browning properties.

4.5.1 Antioxidant Properties of Plant Extracts

Antioxidant activity is an important indicator for evaluating the anti-browning ability of plant extracts. The determination of DPPH, ABTS, and FRAP (ferric ion reducing antioxidant power) radical scavenging activities is often used to evaluate the antioxidant capacity of plant extracts. Akbari et al. (2019) determined the DPPH radical scavenging activity of pomegranate, pomegranate peel, kiwi, and grape extracts and found that pomegranate peel aqueous extract and pomegranate juice had stronger DPPH radical scavenging ability, and pomegranate juice contained more anthocyanins, total phenols, and flavonoids than kiwi and grape (Akbari et al. 2019). A positive correlation between phenolic content and DPPH and ABTS radical scavenging activity was also found in antioxidant studies on the extracts of pine needles, seaweed, peanut shells, and peanut coats (Augusto et al. 2016; Yu et al. 2014). In addition, some minerals (e.g., selenium, manganese, zinc, etc.), which are abundant in plants, are also natural antioxidant active substances, but their anti-browning effect on fresh-cut fruits and vegetables has rarely been reported.

4.5.2 *Polyphenolic Extracts*

Phenolic compounds are considered the main antioxidant active substances in plant extracts, and some of them have inhibitory effects on browning-related enzymatic activities. These substances have the potential for chelation with pro-oxidant metal groups, and their efficient free radical scavenging ability can scavenge reactive oxygen species in the body to inhibit polyphenol oxidase (PPO), peroxidase (POD) enzyme activity, and membrane lipid oxidation (Liu et al. 2019). Wessels et al. (2014) investigated the anti-browning properties of several plant extracts on fresh-cut apples and confirmed that their inhibitory ability may be related to the antioxidant activity of their phenolic compounds, which may directly affect PPO activity through competitive or non-competitive inhibitors (Wessels et al. 2014). Fattahifar et al. (2018) found that gallic acid and m-triacontanol were the major phenolics in pistachio extracts and exhibited strong tyrosinase inhibitory effects. Polyphenols include not only major antioxidants but also natural PPO inhibitors that can inhibit browning-related enzyme activities either competitively or non-competitively (Chen et al. 2021).

4.5.3 *Organic Acid Extracts*

While a suitable pH environment is necessary for enzymes to exhibit their optimal activity, organic acids can inhibit the browning of fruits and vegetables by reducing pH to reduce or inactivate browning-related enzyme activity. In unripe grapes, organic acid content does not vary seasonally, but with grape variety and ripeness, and increases its ability to inhibit oxidase activity by lowering pH (Fevrier et al. 2017). In addition, some organic acids are structurally like phenolic substrates and can form complexes with enzymes or chelate with copper ions at the enzyme active site to directly inhibit PPO or POD activity. Among them, oxalic acid can inhibit PPO and POD enzyme activity by chelating, removing copper ions, or lowering the pH of the treated product (Ali et al. 2020).

4.5.4 *Volatile Oil Extracts*

Essential oils are volatile active substances extracted from the roots, seeds, and flowers of plants and are rich in phenylpropanoids, which have a variety of properties such as antioxidant and antibacterial. Plant essential oils can be used as a natural, non-toxic, and non-hazardous phytochemical additive that contains many phenolic compounds that help protect cells from oxidative damage caused by free radicals. Rodriguez-Garcia et al. (2016) found that Prilosec coating with 1–10% oregano essential oil was effective in reducing the activity of PPO and POD and in maintaining the appearance and color of cabbage during storage (Rodriguez-Garcia et al. 2016). Chen et al. (2017) found that the essential oil of clove buds could inhibit the occurrence of browning in fresh-cut lettuce, and its main component, eugenol, could competitively inhibit PPO enzyme activity (Chen et al. 2017).

4.5.5 Other Compounds

Minerals (e.g., selenium, copper, manganese, and zinc) and vitamin E, vitamin C, and vitamin A present in plants are natural antioxidants. Among them, selenium (Se) is considered to be the most relevant mineral for antioxidant function, and there are at least 35 selenium-based antioxidant proteins, including selenoprotein P, five glutathione peroxidases and three thioredoxin reductases (Nikmaram et al. 2018). In contrast, tocopherols (vitamin E) act as membrane-bound antioxidants, capturing lipid peroxyl radicals generated from unsaturated fatty acids under "oxygen stress" conditions (Mcdowell et al. 2007). Ascorbic acid (vitamin C), a common antioxidant, has been widely used to control browning in fresh-cut fruits and vegetables.

4.6 Food Applications

Adding sea buckthorn fruit powder to wheat bread extended its shelf life by 1–3 days. It could improve bread's antioxidant and sensory properties (Ghendov-Mosanu et al. 2020). The free radical scavenging activity of white wine increased from 28.4% to 55.8% when 0.8 g/L sea buckthorn leaf powder was added. The reducing power of white wines increased from 35.3% to 62.1% when sea buckthorn leaves were added (as measured by the amount of reduced iron ions in the antioxidant capacity test). The overall phenolic content of white wines increased from 11% to 23.7%, and the color intensity increased from 39.9% to 50.7%, both of which contributed to the antioxidant capacity of sulfite-free wine (Tzachristas et al. 2020). Sea buckthorn leaves have a high antioxidant content and can be used to boost the antioxidant capacity of wines (Ul Haq et al. 2022).

4.7 Conclusion and Outlooks

Sea buckthorn is a rare and important plant found all over the world, spanning more than 2 million hectares of land. Sea buckthorn has almost 200 bioactive components with considerable health advantages, and in vivo, in vitro, and clinical research conducted over the last five years have proven these benefits. Antioxidant, anti-inflammatory, hypolipidemic, anti-obesity, anticancer, antibacterial, and antiviral properties, as well as dermatologic, neuroprotective, and hepatoprotective benefits, are among the bioactivities. Furthermore, sea buckthorn has a high economic value in the food industry and is a promising source of bioactive components in the diet.

Sea buckthorn has a promising future as a multifunctional commercial and ecological plant. This ancient plant possesses potent medicinal and synergistic capabilities and has made several contributions to humanity. Sea buckthorn has the remarkable potential to aid economic growth and benefit the environment. Further study should focus on the following areas to rationally create and utilize sea buckthorn resources: (1) Improve mechanical harvesting and preservation procedures. Sea buckthorn is a berry shrub with low hand harvesting efficiency. Because of their high water content, the berries are readily squeezed, resulting in distortion and mold. It is required to increase the harvesting stage's efficiency. (2) Isolate and identify more specific bioactive compounds to further study their health-promoting mechanisms; (3) conduct more clinical trials to validate the benefits of sea buckthorn on human health; (4) use sea buckthorn to prevent and treat diseases; (5) develop sea buckthorn-based functional foods and other products; and (6) adhere to the paradigm of merging economy and environment to build the sea buckthorn business in a scientific, reasonable, and sustainable manner.

References

Ahmad Mir S, Dar BN, Albas Wani A, Ahmad Shah M (2018) Effect of plant extracts on the techno-functional properties of biodegradable packaging films. Trends Food Sci Technol 80:141–154

Ali S, Khan AS, Anjum MA, Nawaz A, Naz S, Ejaz S, Hussain S (2020) Effect of postharvest oxalic acid application on enzymatic browning and quality of lotus (Nelumbo nucifera Gaertn.) root slices. Food Chem 312:126051

Altunkaya A, Gokmen V (2012) Effect of grape seed extract on phenolic profile and browning of fresh-cut lettuce (L. SATIVA). J Food Biochem 36:268–274

Augusto A, Simoes T, Pedrosa R, Silva SFJ (2016) Evaluation of seaweed extracts functionality as post-harvest treatment for minimally processed Fuji apples. Innovative Food Sci Emerg Technol 33:589–595

Akbari E, Gholami M, Cyrus G (2019) Shelf-life and quality attributes in fresh-cut pear cv. Shahmive treated with different kinds of antioxidants. J Food Sci Technol 56:3998–4008

Chen XN, Ren LP, Li ML, Qian J, Fan JF, Du B (2017) Effects of clove essential oil and eugenol on quality and browning control of fresh-cut lettuce. Food Chem 214:432–439

Chen C, Jiang AL, Liu CH, Wagstaff C, Zhao QQ, Zhang YH, Hu WZ (2021) Hydrogen sulfide inhibits the browning of fresh-cut apple by regulating the antioxidant, energy and lipid metabolism. Postharvest Biol Technol 175:111487

Cho CH, Jang H, Lee M, Kang H, Heo HJ, Kim DO (2017) Sea Buckthorn (Hippophae rhamnoides L.) leaf extracts protect neuronal PC-12 cells from oxidative stress. J Microbiol Biotechnol 27:1257–1265

Ciesarova Z, Murkovic M, Cejpek K, Kreps F, Tobolkova B, Koplik R, Belajova E, Kukurova K, Dasko L, Panovska Z, Revenco D, Burcova Z (2020) Why is sea buckthorn (Hippophae rhamnoides L.) so exceptional? A review. Food Res Int 133:109170

Dudau M, Codrici E, Tarcomnicu I, Mihai S, Popescu ID, Albulescu L, Constantin N, Cucolea I, Costache T, Rambu D, Enciu AM, Hinescu ME, Tanase C (2021) A fatty acid fraction purified from Sea Buckthorn seed oil has regenerative properties on normal skin cells. Front Pharmacol 12:737571

Eggersdorfer M, Wyss A (2018) Carotenoids in human nutrition and health. Arch Biochem Biophys 652:18–26

Fattahifar E, Barzegar M, Gavlighi HA, Sahari MA (2018) Evaluation of the inhibitory effect of pistachio (Pistacia vera L.) green hull aqueous extract on mushroom tyrosinase activity and its application as a button mushroom postharvest anti-browning agent. Postharvest Biol Technol 145:157–165

Fevrier H, Le Quere JM, Le Bail G, Guyot S (2017) Polyphenol profile, PPO activity and pH variation in relation to colour changes in a series of red-fleshed apple juices. Lwt-Food Sci Technol 85:353–362

Gatlan AM, Gutt G (2021) Sea Buckthorn in plant based diets. An analytical approach of Sea Buckthorn fruits composition: nutritional value, applications, and health benefits. Int J Environ Res Public Health 18:8986

Ghendov-Mosanu A, Cristea E, Patras A, Sturza R, Padureanu S, Deseatnicova O, Turculet N, Boestean O, Niculaua M (2020) Potential application of Hippophae Rhamnoides in wheat bread production. Molecules 25:1272

Guo RX, Guo XB, Li T, Fu X, Liu RH (2017) Comparative assessment of phytochemical profiles, antioxidant and antiproliferative activities of Sea buckthorn (Hippophae rhamnoides L.) berries. Food Chem 221: 997-1003

Hoang TT, Johnson DA, Raines RT, Johnson JA (2019) Angiogenin activates the astrocytic Nrf2/ antioxidant-response element pathway and thereby protects murine neurons from oxidative stress. J Biol Chem 294:15095–15103

Jaroszewska A, Biel W (2017) Chemical composition and antioxidant activity of leaves of mycorrhized sea-buckthorn (Hippophae rhamnoides L.). *Chilean*. J Agric Res 77:155–162

Ji MY, Gong X, Li X, Wang CC, Li MH (2020) Advanced research on the antioxidant activity and mechanism of polyphenols from Hippophae species-A review. Molecules 25:917

Johnson EJ (2010) The role of carotenoids in human health. Nutr Clin Care 5:56–65

Kim SJ, Hwang E, Yi SS, Song KD, Lee HK, Heo TH, Park SK, Jung YJ, Jun HS (2017) Sea Buckthorn leaf extract inhibits glioma cell growth by reducing reactive oxygen species and promoting apoptosis. Appl Biochem Biotechnol 182:1663–1674

Liu X, Wang T, Lu YZ, Yang Q, Li Y, Deng XD, Liu Y, Du XR, Qiao LP, Zheng JX (2019) Effect of high oxygen pretreatment of whole tuber on anti-browning of fresh-cut potato slices during storage. Food Chem 301:125287

Liu SY, Xiao PT, Kuang YJ, Hao JH, Huang TQ, Liu EH (2021) Flavonoids from sea buckthorn: a review on phytochemistry, pharmacokinetics and role in metabolic diseases. J Food Biochem 45:e13724

Masoodi KZ, Wani W, Dar ZA, Mansoor S, Anam-ul-Haq S, Farooq I, Hussain K, Wani SA, Nehvi FA, Ahmed N (2020) Sea buckthorn (Hippophae rhamnoides L.) inhibits cellular proliferation, wound healing and decreases expression of prostate specific antigen in prostate cancer cells in vitro. J Funct Foods 73:104102

Mcdowell LR, Wilkinson N, Madison R, Felix T (2007) Vitamins and minerals functioning as antioxidants with supplementation considerations. s173
Mohamed EA, Tulcan C, Alexa E, Morar D, Dumitrescu E, Muselin F, Radulov I, Puvaca N, Cristina RT (2020) Sea Buckthorn and grape extract might be helpful and sustainable Phyto-resources as associated Hypolipidemic agents-preliminary study. Sustain For 12:9297
Meng W, Shi J, Zhang X, Lian H, Wang Q, Peng Y (2020) Effects of peanut shell and skin extracts on the antioxidant ability, physical and structure properties of starch-chitosan active packaging films. Int J Biol Macromol 152:137–146
Ni WH, Gao TT, Wang HL, Du YZ, Li JY, Li C, Wei LX, Bi HT (2013) Anti-fatigue activity of polysaccharides from the fruits of four Tibetan plateau indigenous medicinal plants. J Ethnopharmacol 150:529–535
Nikmaram N, Budaraju S, Barba FJ, Lorenzo JM, Cox RB, Mallikarjunan K, Roohinejad S (2018) Application of plant extracts to improve the shelf-life, nutritional and health-related properties of ready-to-eat meat products. Meat Sci 145:245–255
Olas B (2016) Sea buckthorn as a source of important bioactive compounds in cardiovascular diseases. Food Chem Toxicol 97:199–204
Park JB, Kang JH, Song KB (2019) Clove bud essential oil emulsion containing benzethonium chloride inactivates Salmonella Typhimurium and Listeria monocytogenes on fresh-cut pak choi during modified atmosphere storage. Food Control 100:17–23
Pop RM, Weesepoel Y, Socaciu C, Pintea A, Vincken JP, Gruppen H (2014) Carotenoid composition of berries and leaves from six Romanian Sea buckthorn (*Hippophae rhamnoides* L.) varieties. Food Chem 147:1–9
Redei D, Kusz N, Rafai T, Bogdanov A, Burian K, Csorba A, Mandi A, Kurtan T, Vasas A, Hohmann J (2019) 14-Noreudesmanes and a phenylpropane heterodimer from sea buckthorn berry inhibit Herpes simplex type 2 virus replication. Tetrahedron 75:1364–1370
Rodriguez-Garcia I, Cruz-Valenzuela MR, Silva-Espinoza BA, Gonzalez-Aguilar GA, Moctezuma E, Gutierrez-Pacheco MM, Tapia-Rodriguez MR, Ortega-Ramirez LA, Ayala-Zavala JF (2016) Oregano (*Lippia graveolens*) essential oil added within pectin edible coatings prevents fungal decay and increases the antioxidant capacity of treated tomatoes. J Sci Food Agric 96:3772–3778
Seglina D, Krasnova I, Grygier A, Radziejewska-Kubzdela E, Rudzinska M, Gornas P (2021) Unique bioactive molecule composition of sea buckthorn (*Hippophae rhamnoides* L.) oils obtained from the peel, pulp, and seeds via physical "solvent-free" approaches. J Am Oil Chem Soc 98:1009–1020
Serban MC, Serban A, Ursoniu S, Dragan S (2019) Systematic review on the potential of sea buckthorn hippophae rhamnoides l. for a possible novel enriched bread for the patients with cardiovascular diseases. Atherosclerosis 287:E285–E285
Shaikh RS (2016) Antibacterial activity of sea buckthorn (*Hippophae rhamnoides* L.) against methicillin resistant Staphylococcus aureus (MRSA). Pak J Pharm Sci 29:1711–1713
Singh D, Jayashankar B, Mishra KP, Tanwar H, Madhusudana SN, Belludi AY, Tulsawani R, Singh SB, Ganju L (2017) Adjuvant activity of ethanol extract of Hippophae rhamnoides leaves with inactivated rabies virus antigen. Pharm Biol 56:25–31
Stobdan T, Chaurasia OP, Korekar G, Mundra S, Ali Z, Yadav A, Singh SB (2010) Attributes of Seabuckthorn (*Hippophae rhamnoides* L.) to meet nutritional requirements in high altitude. Def Sci J 60:226–230
Sukhonthara S, Kaewka K, Theerakulkait C (2016) Inhibitory effect of rice bran extracts and its phenolic compounds on polyphenol oxidase activity and browning in potato and apple puree. Food Chem 190:922–927
Supapvanich S, Prathaan P, Tepsorn R (2012) Browning inhibition in fresh-cut rose apple fruit cv. Taaptimjaan using konjac glucomannan coating incorporated with pineapple fruit extract. Postharvest Biol Technol 73:46–49
Suryakumar G, Gupta A (2011) Medicinal and therapeutic potential of Sea buckthorn (*Hippophae rhamnoides* L.). J Ethnopharmacol 138:268–278

Tanwar H, Shweta Singh D, Singh S, Ganju L (2018) Anti-inflammatory activity of the functional groups present in Hippophae rhamnoides (Seabuckthorn) leaf extract. Inflammopharmacology 26:291–301

Teleszko M, Wojdylo A, Rudzinska M, Oszmianski J, Golis T (2015) Analysis of lipophilic and hydrophilic bioactive compounds content in Sea Buckthorn (*Hippophae rhamnoides* L.) Berries. J Agric Food Chem 63:4120–4129

Tkacz K, Wojdylo A, Turkiewicz IP, Bobak L, Nowicka P (2019) Anti-oxidant and anti-enzymatic activities of Sea Buckthorn (*Hippophae rhamnoides* L.) fruits modulated by chemical components. Antioxidants 8:618

Tzachristas A, Pasvanka K, Liouni M, Calokerinos AC, Tataridis P, Proestos C (2020) Effect of *Hippophae rhamnoides* l. leaves treatment on the antioxidant capacity, total phenol content and sensory profile of moschofilero wines vinified with and without added sulphites. Appl Sci Basel 10:3444

Ul Haq SA, Mir MA, Lone SM, Banoo A, Shafi F, Mir SA, Bhat JIA, Rashid R, Wani SH, Masoodi TH, Khan MN, Nehvi FA, Masoodi KZ (2022) Explicating genetic diversity based on ITS characterization and determination of antioxidant potential in sea buckthorn (*Hippophae* spp.). Mol Biol Rep 49:5229–5240

Venturi F, Bartolini S, Sanmartin C, Orlando M, Taglieri I, Macaluso M, Lucchesini M, Trivellini A, Zinnai A, Mensuali A (2019) Potato peels as a source of novel green extracts suitable as antioxidant additives for fresh-cut fruits. Appl Sci Basel 9:2431

Verma H, Chahota R, Palial A, Sharma M (2011) Antibacterial properties of seabuckthorn (*Hippophae rhamnoides* L.) leaf extracts against common skin and wound bacteria. Indian J Veterin Res (India) 20:38–41

Wang X, Liu JR, Zhang XH, Zhao SM, Zou K, Xie JM, Wang XX, Liu CY, Wang JL, Wang YZ (2018) Seabuckthorn berry polysaccharide extracts protect against acetaminophen induced hepatotoxicity in mice via activating the Nrf-2/HO-1-SOD-2 signaling pathway. Phytomedicine 38:90–97

Wang KW, Xu ZZ, Liao XJ (2022a) Bioactive compounds, health benefits and functional food products of sea buckthorn: a review. Crit Rev Food Sci Nutr 62:6761–6782

Wang Z, Zhao FL, Wei PP, Chai XY, Hou GG, Meng QG (2022b) Phytochemistry, health benefits, and food applications of sea buckthorn (*Hippophae rhamnoides* L.): a comprehensive review. Front Nutr 9:1036295

Wessels B, Damm S, Kunz B, Schulze-Kaysers N (2014) Effect of selected plant extracts on the inhibition of enzymatic browning in fresh-cut apple. J Appl Bot Food Qual 87:16–23

Yong P, Yan W, Yunfei L (2013) Development of tea extracts and chitosan composite films for active packaging materials. Int J Biol Macromol 59:282–289

Yu ZL, Zhang Z, Zeng WC (2014) Investigation of antibrowning activity of pine needle (*Cedrus deodara*) extract with fresh-cut apple slice model and identification of the primary active components. Eur Food Res Technol 239:669–678

Zadernowski R, Naczk M, Czaplicki S, Rubinskiene M, Szlakiewicz M (2005) Composition of phenolic acids in sea buckthorn (*Hippophae rhamnoides* L.) berries. J Am Oil Chem Soc 82:175–179

Chapter 5
Total Flavonoids and Total Phenolics Content in Different Parts of Sea Buckthorn

Huiyan Liu, Haitian Fang, Jinhu Tian, Yujun Zeng, and Shuifang Mao

Abstract Sea buckthorn (*Hippophae rhamnoides* L.) is a type of deciduous shrub or tree that goes by several names, including seaberry, sea buckthorn, and Siberian pineapple. There are around 200 beneficial and bioactive substances found in sea buckthorn. The pharmacological properties of sea buckthorn, including its antiviral, anti-inflammatory, anticancer, and antibacterial properties, as well as its cardiovascular protective effects, have been the subject of numerous studies in recent years. Sea buckthorn is rich in phenolic compounds with strong antioxidant potential and contains high levels of vitamins, carotenoids, polyphenols, and fatty acids.

Keywords Sea buckthorn · Flavonoids · Phenolics

Abbreviation

FW fresh weight
DW dry weight
GAE gallic acid equivalents
TPC total phenolic content
TFC total flavonoid content

H. Liu
Food College, Ningxia University, Yinchuan, China

H. Fang (✉)
Food Science and Technology, Ningxia University, Yinchuan, Ningxia, China
e-mail: fanght@nxu.edu.cn

J. Tian · Y. Zeng · S. Mao
Department of Food Science and Nutrition, Zhejiang University, Hangzhou, China

J. Tian et al. (eds.), *Sea Buckthorn*, https://doi.org/10.1007/978-981-97-9865-0_5

5.1 The Bioactive Compounds of Sea Buckthorn

Sea buckthorn (*Hippophae rhamnoides* L.) is a species of deciduous shrub or tree, also called Siberian pineapple, sea, seaberry, sand thorn, and sallow thorn (Singh 2018). A "natural vitamin treasure house" and a "source of nutrition and health care," sea buckthorn has been reported to have about 200 nutritious and bioactive substances (Wang et al. 2021, 2022). Many studies have examined sea buckthorn's pharmacological properties in the past few years, including its antiviral, anticancer, anti-inflammatory, and antimicrobial characteristics, as in the prevention of cardiovascular diseases (Masoodi et al. 2020; Olas 2016; Rédei et al. 2019; Tanwar et al. 2018; Verma et al. 2011). Sea buckthorn's high concentration of vitamins, carotenoids, polyphenols, and fatty acids may contribute to its undeniable medical and therapeutic potential (Pop et al. 2014; Solà Marsiñach and Cuenca 2019; Stobdan et al. 2010; Yang et al. 2016).

Sea buckthorn berries contain an abundance in phenolic compounds, with flavonols and condensed tannins being the primary polyphenolic constituents in the berries (Rösch et al. 2003; Teleszko et al. 2015), and this provides them with a strong antioxidant potential. Our analysis demonstrated a significant phenolic content (2.79 g GAE/kg) in sea buckthorn berries, which is in agreement with previous discoveries by Kitrytė et al. (2017). In addition, ascorbic acid, polyphenols, and carotenoids function together to provide sea buckthorn berries with their antioxidant potential (Gao et al. 2000).

5.2 Analysis of Total Phenolics and Flavonoids

5.2.1 Polyphenols in Sea Buckthorn

Polyphenols are among the most abundant and broadly distributed secondary metabolites in plants (de la Rosa et al. 2019). In the work of Belščak-Cvitanović et al. (2018), the taxonomy of phenol rings frequently depends on their quantity and structural components. Phenolic acids and flavonoids constitute the majority of the polyphenols found in sea buckthorn (Guo et al. 2017a; Ma et al. 2019). The total phenolic content (TPC) of sea buckthorn ranges from 32.93 to 1417 mg/100 g FW (Green and Low 2013; Ivanišová et al. 2017; Rop et al. 2014). Sea buckthorn's principal antioxidant components are polyphenols. The polyphenol content of the fruit has been observed to range from 12.36 to 34.6 mg GAE/g, which is in excess of that in oranges (1.27 mg GAE/g), blueberries (2.19 mg GAE/g), mandarins (1.16 mg GAE/g), sour cherries (2.56 mg GAE/g), and strawberries (1.12 mg GAE/g) (Dragović-Uzelac et al. 2009; Kuhkheil et al. 2017; Li et al. 2021). A recent investigation found that sea buckthorn has produced approximately 100 identified and isolated polyphenolic components (Ji et al. 2020). Seventeen phenolic acids have been discovered in sea buckthorn berries, with salicylic acid being the most widespread, accounting for 55–74.3% of the total (Zadernowski et al. 2005). Nevertheless, another investigation found that gallic acid is the primary phenolic acid in sea buckthorn fruit and leaves. Besides, twelve phenolic acids have been discovered in sea buckthorn, including *p*-coumaric acid, gallic acid, and caffeic acid as the predominant components (Bittová et al. 2014).

5.2.2 *Flavonoid Distribution in Sea Buckthorn*

The most prevalent phenolic chemicals in plant-based diets are known as flavonoids (Tsimogiannis and Oreopoulou 2019). Flavonols and flavanols are the two forms of flavonoids found in sea buckthorn. According to Ciesarová et al. (2020), these substances may be able to prevent cancer, heart disease, and diabetes, among other chronic illnesses. Guo et al. (2017b) discovered that phenolic and flavonoid glycosides of sea buckthorn extracts have antioxidant and antiproliferative properties. Raudonis et al. (2014) investigated total flavonoid content (TFC) in 11 sea buckthorn species cultivated in Lithuania, ranging from 385 to 616 μg/g FW.

Flavonols are the main constituents of flavonoids and are mainly found in glycosylated variants of quercetin, isorhamnetin, and kaempferol (Liu et al. 2021). Flavonol content is in the range of 463.14 mg–893.92 mg/100 g DW, accounting for approximately 99% of the TFC (Tkacz et al. 2019). Geographical considerations, meteorological circumstances, and berry varieties all have a substantial influence on the concentration and composition of polyphenols.

5.3 Variability in Phenolic and Flavonoid Content

5.3.1 *Seasonal and Geographical Variations*

The quantity of natural chemicals in plants fluctuates throughout time and between ecosystems. For example, the concentration of phenolic acids and flavonoids in honeysuckle flowers has been found to change at six different growth stages (Kong et al. 2017). Pan et al. (2017) found that the number of flavonoids extracted from *Abelmoschus manihot* was greatly regulated by the harvest time. The TPC of the Pyrola incarnata plant varied seasonally (Yao XiaoHui et al. 2016). Nevertheless, little has been discovered about the fluctuation of flavonoids and phenolic contents in sea buckthorn leaves and fruits. It was discovered that the monthly fluctuations in total flavonoid concentration in leaves and fruits were not the same. The variance analysis revealed that the TFC of sea buckthorn leaves and fruits fluctuated substantially across the growing season. Flavonoids are important bioactive components in sea buckthorn fruits, with total flavonoids levels ranging from 46.16 mg/g (July) to 30.28 mg/g (August). The TFC of the leaves and fruits was highest in July and lowest in June (for the leaves) and August (for the fruits). During the growth season, leaves had a higher total flavonoid concentration than fruits. In August, leaves had a total flavonoid concentration thirteen times higher than fruits. The TFC in leaves ranged from 308.11 mg/g (June) to 405.79 mg/g (July). The TFC in leaves and fruits changed considerably between sampling dates and followed a marked monthly patterning. The TFC in leaves was consistently higher than in fruits. The TFC in leaves was highest in August (19.37 mg/g), followed by July (17.91 mg/g), and lowest in June (13.30 mg/g). The TPC in fruits was highest in July (8.24 mg/g), followed by November, and the lowest in August (4.30 mg/g). The findings of the research were as follows: (1) the contents of flavonoids and phenolic compounds in sea buckthorn leaves and fruits change during the growing season; (2) the antioxidant activities of fruit and leaf extracts change during the growing season; and (3) certain meteorological factors are associated with the accumulation of active compounds in sea buckthorn leaves and fruits. The TFC and antioxidant capability of sea buckthorn leaves were greater than those of fruits. The TFC and TPC of sea buckthorn leaf and fruit extracts fluctuated dramatically over the growing season, which supported our initial hypothesis. The third hypothesis states that the effects of temperature and length of sunshine on the accumulation of phenolic compounds and flavonoids were minimal, and the general trend appears to support the fourth hypothesis (Yang et al. 2023).

5.3.2 Genotypic Effects and Extraction Methods

Another research investigation discovered that isorhamnetin derivatives are widely considered to be the major polyphenol constituents (more than 65% of total flavonols) in whole berries of *Elongaceae* family species from various geographical locations (Fatima et al. 2015; Ma XueYing et al. 2016; Tkacz et al. 2019). Nevertheless, the research of plant anatomical components revealed that I derivatives vary more than Q and K derivatives, yet Q derivatives are more concentrated in most fractions (skins, berries, branches, and leaves). Therefore, the highest levels of I-3-O-glucoside, I-3-O-rutinoside, Q-3-O-glucoside, and I-3-O-glucoside-7-O-rhamnoside were detected in skins, but in flesh, it was predominantly Q-3-O-glucoside and, to a lesser extent, the pericarp-specific flavonols, which depended on the berry variety. I-3-O-glucoside-7-O-rhamnoside was also found to be the main flavonol in seeds and endocarp, which had not previously been reported. The primary flavonoid glycosides of sea buckthorn seeds from the Indian Himalayan Region were determined by Arimboor and Arumughan (2012) to be Q-3-O-rutinoside and I-3-O-rutinoside; however, these monoglucosides were not detected in the seeds or branches of this study. The branches and leaves were considerably high in Q-3-O-glucoside-7-O-rhamnoside ($p < 0.05$), and the majority of the leaves were likewise plentiful in Q-3-O-glucoside and Q-3-O-rutinoside (except for 'Prozrocznaja'). Rutin was found in addition to leaves, skins, meat, and endocarp (save for 'Tatiana'), in contrast to cultivars cultivated in Canada, including 'Golden Rain', which was also examined in this study, where rutin was found solely in leaves

(Fatima et al. 2015). It is a prominent flavonoid in Romanian sea buckthorn berries (Criste et al. 2020) as well as Czech sea buckthorn berries and leaves (Bittová et al. 2014). The sea buckthorn sections are sorted by total flavonol levels: branches < endocarp < seeds < flesh < skins < leaves (mean total levels between 41.34 mg and 921.74 mg/100 g DW). The most prevalent flavonols were discovered in leaves and skins, with leaves containing the majority of Q derivatives and skins containing I derivatives. Branches had an average of 17 to 22 times less flavonol than leaves and skins. In similarity to Romanian sea buckthorn leaves (Pop et al. 2013), Q-3-O-rhamnoside was not detected in leaves and branches, although it was the most highly concentrated flavonol in leaves and shoots investigated by Criste et al. (2020) and Bittová et al. (2014). With the exception of the endocarp ($r = 0.55$), the concentration of Q and I derivatives within each individual fraction of sea buckthorn exhibited a significant association, with a Pearson correlation coefficient (r) ranging from 0.82 to 0.95. For skins, flesh, branches, and leaves, the ratio of the sum of Q derivatives to I derivatives was more than 1.0 (1.3, 1.6, 1.6, and 4.2, respectively). The cultivar 'Prozrocznaja' had approximately 15-fold more Q and I derivatives in skins and approximately fourfold in flesh than those portions of 'Tatiana', which were the least flavonoid-rich. Conversely, seeds of 'Tatiana' and 'Podarok Sadu' had the greatest Q and I derivatives (82.17 mg of Q compounds and 150.24 mg of I derivatives per 100 g DW). The content of flavonols in the branches and leaves of "Maryja" is significantly higher than in other varieties (9.47 mg in branches of I derivatives and 1035.21 mg in leaves of Q derivatives, per 100 g DW), which was 20-fold higher than in other varieties (Tkacz et al. 2019). Total phenolic components averaged 0.51% to 9.32% of the dry weight of the pulp and branches. The order of TPC for each part of sea buckthorn is as follows: flesh < endocarp < skins < seeds < leaves < branches (507.16 mg–9318.65 mg/100 g DW). Previous research simply indicated that TPC in leaves were several times higher than those of entire sea buckthorn berries, other types of berries, and pome fruits, setting aside their anatomical sections (Bittová et al. 2014; Criste et al. 2020; Teleszko and Wojdyło 2015). However, a single investigation found that branches contain 4.6-fold more phenolic chemicals than leaves, 6-fold more than seeds, and 7-fold more than skins. The fractions had three to four times more of these chemicals than the flesh and endocarp. The total sum of mono-, di-, oligo-, and polymeric flavan-3-ols in the endocarp and seeds constituted for more than 80% of total phenolic compounds, with shoots accounting for more than 99%. The meat contained the majority of flavonols, accounting for around 60% of TPC. Most skins and leaves had a 1:1 ratio of flavan-3-ols to flavonols, although the quantities of total phenolic compounds varied by cultivar. Another study found that the total antioxidant capacity and TPC of sea buckthorn leaves increased during leaf growth, peaking in samples taken at the end of July. Nevertheless, if the maximum amount of a single major phenolic constituent is required for a specific bioactive effect, the amount of each phenolic compound must be addressed, as not every compound peaks in fully mature leaves. The overall output of bioactive chemicals is determined by the cultivar and leaf yields, both of which are important factors in commercial production. A detailed examination of the phenolic chemicals contained in sea buckthorn leaves may give significant insights and deserve future exploration (Morgenstern et al. 2014).

Many studies have been conducted on the TPC, TFC, and total antioxidant capacity of different sections of sea buckthorn. In comparison to the leaf, stem, and pulp, the seed has significantly more antioxidants and phenolics (Korekar et al. 2011a, b; Michel et al. 2012; Sharma et al. 2008). Quantification of the chemicals in seeds that promote health, however, has only been explored in a small number of samples or in top selections. Specific plant genotypes and the interplay of growing conditions determine the TPC and total antioxidant capacity (Scalzo et al. 2005). While studies have shown the significance of genotype in influencing TPC and total antioxidant capacity in certain fruits (Bajpai et al. 2014; Connor et al. 2002; Girish Korekar et al. 2011a, b; Leccese et al. 2012), this impact has not been studied in detail in SBT. Previous research in wild genotypes and natural populations revealed genotypic influence and heterogeneity in the antioxidant activity of the fleshy section of SBT berries (Ercisli et al. 2007; Korekar et al. 2014). According to the study, the TPC in the methanol extraction cycle (first cycle) was substantially greater than the acetone extraction cycle (second cycle). The first extraction cycle had a mean TPC of 108.3 ± 33.2 mg GAE/g DW, whereas the second cycle with acetone yielded 2.6 ± 1.7 mg GAE/g DW. The average TPC was 110.8 ± 33.3 mg GAE/g DW. The TPC of sea buckthorn varied from 32 to 208 mg GAE/g across individuals in the study. As a result, the TPC varied from 1 to 6.5-fold. The TPC data is largely distributed in the highest quartile of the population. Dry hazelnut, walnut, and pistotal antioxidant capacity with seed coatings contain 425, 589, and 461 mg GAE/100g, respectively (Dolkar et al. 2017; Arcan and Yemenicioğlu 2009). The flavonoid concentration in SBT seed varied from 8 to 38 mg QE/g dry weight. In the first cycle of methanol extraction, the mean TFC was 19.9 ± 6.2 mg QE/g. In the second cycle of acetone extraction, the value was 1.5 ± 0.3 mg QE/g. The average total flavonoid concentration was 21.4 ± 6.2 mg QE/g DW. Nassourou et al. (2016) found that cowpea (*Vigna unguiculata* L.) seed coat consists of 3.6–4.5 mg rutin equivalent of total flavonoids for each gram of DW. The study of nine varieties of sea buckthorn berries and leaves with different ripening times by free radical scavenging and photo-chemiluminescence assays yielded very valuable information on the effect of TPC and TFC on antioxidant activity (Sytařová et al. 2020). Berries of cultivar 'Bojan' (0.70 g GAE/kg) and leaves of 'Hergo' (1.88 g GAE/kg) showed the lowest TPC (total phenolic compound content) among the specimens under investigation. Conversely, the highest levels were found in Dorana berries and leaves, with respective values of 3.62 g GAE/kg and 3.72 g GAE/kg. The majority of the examined cultivars had TPC values that exceeded the 1.75 g GAE/kg FW standard for Canadian Indian Summer cultivars' berries (Araya-Farias et al. 2011). Nevertheless, sea buckthorn leaves from Poland were discovered to have a higher TPC of 8.58 g/kg DW (Jaroszewska and Biel 2017). The variability in phenolic levels is the result of a number of factors, including location type and harvest year. The TPC of Iranian sea buckthorn berries varied by harvest year, ranging from 21.58 g/kg in 2014 to 32.16 g/kg in 2015 (Kuhkheil et al. 2017). The sea buckthorn berry cultivars in another Czech locality, Žabčice, had varying TPC values when harvested in different years, indicating a clear influence of the local climate at that particular harvest time. The study found that the total phenolic concentration in Hergo was 1.15 g GAE/kg, but berries of the same cultivar harvested in 2011–2012 had a documented TPC of 9.65 g GAE/kg (Rop et al. 2014). Interestingly, sea buckthorn leaves from Ladakh had a TPC of 83.08 g GAE/kg DW, which was double that of berries (Stobdan et al. 2013). From June to August, the TPC of sea buckthorn leaves in South Korea increased from 86.2 to 116.7 g GAE/kg DW due to harvest time (Cho et al. 2017). In terms of total flavonoid concentration (TFC), the lowest values were found in Tytii berries (0.55 mg RE/kg) and leaves from the 'Dorana' and 'Slovan' cultivars (−14.40 mg RE/kg). Dorana berries had the greatest TFC of 4.11 mg RE/kg, whereas TFC in leaves ranged from 14.40 mg RE/kg ('Dorana' and 'Slovan') to 49.58 mg RE/kg (Bojan). Similar to TPC, berries of cultivar Hergo from Žabčice had a greater total flavonoid concentration of 4.98 g RE/kg (Rop et al. 2014) contrasting with the content of 1.35 g RE/kg in the same cultivar. Likewise, confirmed total flavonoid levels in leaves were significantly lower than reported amounts in leaves from South Korea, which ranged from 16.0 to 21.8 g catechin (C)/kg depending on harvest season. A remarkably high total flavonoid concentration of 2546.2 mg quercetin/kg DW was found in Poland sea buckthorn leaves (Jaroszewska and Biel 2017).

References

Araya-Farias M, Makhlouf J, Ratti C (2011) Drying of seabuckthorn (*Hippophae rhamnoides L.*) berry: impact of dehydration methods on kinetics and quality. Dry Technol 29(3):351–359

Arcan I, Yemenicioğlu A (2009) Antioxidant activity and phenolic content of fresh and dry nuts with or without the seed coat. J Food Compos Anal 22(3):184–188

Arimboor R, Arumughan C (2012) HPLC-DAD-MS/MS profiling of antioxidant flavonoid glycosides in sea buckthorn (*Hippophae rhamnoides L.*) seeds. Int J Food Sci Nutr 63(6):730–738

Bajpai PK, Warghat AR, Dhar P, Kant A, Srivastava RB, Stobdan T (2014) Variability and relationship of fruit color and sampling location with antioxidant capacities and bioactive content in Morus alba L. fruit from trans-Himalaya, India. LWT-Food Sci Technol 59(2):981–988

Belščak-Cvitanović A, Durgo K, Huđek A, Bačun-Družina V, Komes D (2018) Overview of polyphenols and their properties. In: Polyphenols: properties, recovery, and applications, pp 3–44

Bittová M, Krejzová E, Roblová V, Kubáň P, Kubáň V (2014) Monitoring of HPLC profiles of selected polyphenolic compounds in sea buckthorn (*Hippophaë rhamnoides L.*) plant parts during annual growth cycle and estimation of their antioxidant potential. Open Chem 12(11):1152–1161

Cho CH, Jang H, Lee M, Kang H, Heo HJ, Kim DO (2017) Sea buckthorn (*Hippophae rhamnoides L.*) leaf extracts protect neuronal PC-12 cells from oxidative stress. J Microbiol Biotechnol 27(7):1257–1265

Ciesarová Z, Murkovic M, Cejpek K, Kreps F, Tobolková B, Koplík R, Belajová E, Kukurová K, Daško Ľ, Panovská Z (2020) Why is sea buckthorn (*Hippophae rhamnoides L.*) so exceptional? A review. Food Res Int 133:109170

Connor AM, Luby JJ, Tong CB, Finn CE, Hancock JF (2002) Genotypic and environmental variation in antioxidant activity, TPC, and anthocyanin content among blueberry cultivars. J Am Soc Hortic Sci 127(1):89–97

Criste A, Urcan AC, Bunea A, Pripon Furtuna FR, Olah NK, Madden RH, Corcionivoschi N (2020) Phytochemical composition and biological activity of berries and leaves from four Romanian Sea Buckthorn (*Hippophae rhamnoides L*) varieties. Molecules 25(5):1170

de la Rosa LA, Moreno-Escamilla JO, Rodrigo-García J, Alvarez-Parrilla E (2019) Phenolic compounds. In: Postharvest physiology and biochemistry of fruits and vegetables, pp 253–271

Dolkar P, Dolkar D, Angmo S, Kumar B, Stobdan T (2017) Variability in phenolics, flavonoids and antioxidants in Seabuckthorn (*Hippophae rhamnoides L.*) seed from nine trans-Himalayan natural population. J Berry Res 7(2):109–116

Dragović-Uzelac V, Bursać Kovačević D, Levaj B, Pedisić S, Mezak M, Tomljenović A (2009) Polyphenols and antioxidant capacity in fruits and vegetables common in the croatian diet. Agric Conspec Sci 74(3):175–179

Ercisli S, Orhan E, Ozdemir O, Sengul M (2007) The genotypic effects on the chemical composition and antioxidant activity of sea buckthorn (*Hippophae rhamnoides L.*) berries grown in Turkey. Sci Hortic 115(1):27–33

Fatima T, Kesari V, Watt I, Wishart D, Todd JF, Schroeder WR, Paliyath G, Krishna P (2015) Metabolite profiling and expression analysis of flavonoid, vitamin C and tocopherol biosynthesis genes in the antioxidant-rich sea buckthorn (*Hippophae rhamnoides L.*). Phytochemistry 118:181–191

Gao X, Ohlander M, Jeppsson N, Björk L, Trajkovski V (2000) Changes in antioxidant effects and their relationship to phytonutrients in fruits of sea buckthorn (*Hippophae rhamnoides L.*) during maturation. J Agric Food Chem 48(5):1485–1490

Green RC, Low NH (2013) Physicochemical composition of buffaloberry (*Shepherdia argentea*), chokecherry (*Prunus virginiana*) and sea buckthorn (*Hippophae rhamnoides*) fruit harvested in Saskatchewan, Canada. Can J Plant Sci 93(6):1143–1153

Guo R, Chang X, Guo X, Brennan CS, Li T, Fu X, Liu RH (2017a) Phenolic compounds, antioxidant activity, antiproliferative activity and bioaccessibility of Sea buckthorn (*Hippophaë rhamnoides L.*) berries as affected by in vitro digestion. Food Funct 8(11):4229–4240

Guo R, Guo X, Li T, Fu X, Liu RH (2017b) Comparative assessment of phytochemical profiles, antioxidant and antiproliferative activities of Sea buckthorn (*Hippophaë rhamnoides L.*) berries. Food Chem 221:997–1003

Ivanišová E, Mittuchová K, Mareček J, Frančáková H (2017) Small berries–attractive source of bioactive compounds for consumers. In: Agrobiodiversity for improving nutrition, health and life quality

Jaroszewska A, Biel W (2017) Chemical composition and antioxidant activity of leaves of mycorrhized sea-buckthorn (*Hippophae rhamnoides L.*). Chile J Agric Res 77(2):155–162

Ji M, Gong X, Li X, Wang C, Li M (2020) Advanced research on the antioxidant activity and mechanism of polyphenols from *Hippophae* species—a review. Molecules 25(4):917

Kitrytė V, Povilaitis D, Kraujalienė V, Šulniūtė V, Pukalskas A, Venskutonis PR (2017) Fractionation of sea buckthorn pomace and seeds into valuable components by using high pressure and enzyme-assisted extraction methods. LWT-Food Sci Technol 85:534–538

Kong D, Li Y, Bai M, Deng Y, Liang G, Wu H (2017) A comparative study of the dynamic accumulation of polyphenol components and the changes in their antioxidant activities in diploid and tetraploid *Lonicera japonica*. Plant Physiol Biochem 112:87–96

Korekar G, Stobdan T, Arora R, Yadav A, Singh SB (2011a) Antioxidant capacity and phenolics content of apricot (*Prunus armeniaca L.*) kernel as a function of genotype. Plant Foods Hum Nutr 66:376–383

Korekar G, Stobdan T, Singh H, Chaurasia O, Singh S (2011b) Phenolic content and antioxidant capacity of various solvent extracts from seabuckthorn (*Hippophae rhamnoides L.*) fruit pulp, seeds, leaves and stem bark. Acta Aliment 40(4):449–458

Korekar G, Dolkar P, Singh H, Srivastava RB, Stobdan T (2014) Variability and the genotypic effect on antioxidant activity, total phenolics, carotenoids and ascorbic acid content in seventeen natural population of Seabuckthorn (*Hippophae rhamnoides L.*) from trans-Himalaya. LWT-Food Sci Technol 55(1):157–162

Kuhkheil A, Naghdi Badi H, Mehrafarin A, Abdossi V (2017) Chemical constituents of sea buckthorn (*Hippophae rhamnoides L.*) fruit in populations of central alborz mountains in Iran. Res J Pharmacogn 4(3):1–12

Leccese A, Bartolini S, Viti R (2012) From genotype to apricot fruit quality: the antioxidant properties contribution. Plant Foods Hum Nutr 67:317–325

Li N, Hu Y, Ge X (2021) Determination of contents of total flavonoids and total polyphenols in *Hippophae rhamnoides L.* from different origins and their antioxidant activity. Chem Bioeng 38:64–68

Liu S, Xiao P, Kuang Y, Hao J, Huang T, Liu E (2021) Flavonoids from sea buckthorn: a review on phytochemistry, pharmacokinetics and role in metabolic diseases. J Food Biochem 45(5):e13724

Ma XueYing MX, Laaksonen O, Zheng Jie ZJ, Yang Wei YW, Trépanier M, Kallio H, Yang BaoRu YB (2016) Flavonol glycosides in berries of two major subspecies of sea buckthorn (*Hippophaë rhamnoides L.*) and influence of growth sites. Food Chem 200:189

Ma X, Moilanen J, Laaksonen O, Yang W, Tenhu E, Yang B (2019) Phenolic compounds and antioxidant activities of tea-type infusions processed from sea buckthorn (*Hippophaë rhamnoides*) leaves. Food Chem 272:1–11

Masoodi KZ, Wani W, Dar ZA, Mansoor S, Anam-ul-Haq S, Farooq I, Hussain K, Wani SA, Nehvi FA, Ahmed N (2020) Sea buckthorn (*Hippophae rhamnoides L.*) inhibits cellular proliferation, wound healing and decreases expression of prostate specific antigen in prostate cancer cells in vitro. J Funct Foods 73:104102

Michel T, Destandau E, Le Floch G, Lucchesi ME, Elfakir C (2012) Antimicrobial, antioxidant and phytochemical investigations of sea buckthorn (*Hippophaë rhamnoides L.*) leaf, stem, root and seed. Food Chem 131(3):754–760

Morgenstern A, Ekholm A, Scheewe P, Rumpunen K (2014) Changes in content of major phenolic compounds during leaf development of sea buckthorn (*Hippophaë rhamnoides L.*). Agric Food Sci 23(3):207–219

Nassourou MA, Njintang YN, Noubissié TJ-B, Nguimbou RM, Bell JM (2016) Genetics of seed flavonoid content and antioxidant activity in cowpea (*Vigna unguiculata L. Walp.*). Crop J 4(5):391–397

Olas B (2016) Sea buckthorn as a source of important bioactive compounds in cardiovascular diseases. Food Chem Toxicol 97:199–204

Pan X, Du L, Tao J, Jiang S, Qian D, Duan J (2017) Dynamic changes of flavonoids in Abelmoschus manihot different organs at different growth periods by UPLC–MS/MS. J Chromatogr B 1059:21–26

Pop RM, Socaciu C, Pintea A, Buzoianu AD, Sanders MG, Gruppen H, Vincken JP (2013) UHPLC/PDA–ESI/MS analysis of the main berry and leaf flavonol glycosides from different Carpathian Hippophaë rhamnoides L. varieties. Phytochem Anal 24(5):484–492

Pop RM, Weesepoel Y, Socaciu C, Pintea A, Vincken J-P, Gruppen H (2014) Carotenoid composition of berries and leaves from six Romanian Sea buckthorn (*Hippophae rhamnoides L.*) varieties. Food Chem 147:1–9

Raudonis R, Raudone L, Janulis V, Viskelis P (2014) Flavonoids in cultivated berries of sea buckthorn (*Hippophaë rhamnoides L.*). Planta Med 80(16):LP24

Rédei D, Kúsz N, Rafai T, Bogdanov A, Burián K, Csorba A, Mándi A, Kurtán T, Vasas A, Hohmann J (2019) 14-Noreudesmanes and a phenylpropane heterodimer from sea buckthorn berry inhibit herpes simplex type 2 virus replication. Tetrahedron 75(10):1364–1370

Rop O, Ercişli S, Mlcek J, Jurikova T, Hoza I (2014) Antioxidant and radical scavenging activities in fruits of 6 sea buckthorn (*Hippophae rhamnoides L.*) cultivars. Turk J Agric For 38(2):224–232

Rösch D, Bergmann M, Knorr D, Kroh LW (2003) Structure-antioxidant efficiency relationships of phenolic compounds and their contribution to the antioxidant activity of sea buckthorn juice. J Agric Food Chem 51(15):4233–4239

Scalzo J, Politi A, Pellegrini N, Mezzetti B, Battino M (2005) Plant genotype affects total antioxidant capacity and phenolic contents in fruit. Nutrition 21(2):207–213

Sharma UK, Sharma K, Sharma N, Sharma A, Singh HP, Sinha AK (2008) Microwave-assisted efficient extraction of different parts of Hippophae rhamnoides for the comparative evaluation of antioxidant activity and quantification of its phenolic constituents by reverse-phase high-performance liquid chromatography (RP-HPLC). J Agric Food Chem 56(2):374–379

Singh B (2018) Indian sea buckthorn. In: New age herbals: resource, quality and pharmacognosy, pp 29–54

Solà Marsiñach M, Cuenca AP (2019) The impact of sea buckthorn oil fatty acids on human health. Lipids Health Dis 18(1):1–11

Stobdan T, Chaurasia OP, Korekar G, Mundra S, Ali Z, Yadav A, Singh SB (2010) Attributes of Seabuckthorn (*Hippophae rhamnoides L.*) to meet nutritional requirements in high altitudes. Defen Sci J 60:226

Stobdan T, Korekar G, Srivastava B (2013) Nutritional attributes and health application of seabuckthorn (*Hippophae rhamnoides L.*)-a review. Curr Nutr Food Sci 9(2):151–165

Sytařová I, Orsavová J, Snopek L, Mlček J, Byczyński Ł, Mišurcová L (2020) Impact of phenolic compounds and vitamins C and E on antioxidant activity of sea buckthorn (*Hippophaë rhamnoides L.*) berries and leaves of diverse ripening times. Food Chem 310:125784

Tanwar H, Shweta SD, Singh SB, Ganju L (2018) Anti-inflammatory activity of the functional groups present in Hippophae rhamnoides (Seabuckthorn) leaf extract. Inflammopharmacology 26:291–301

Teleszko M, Wojdyło A (2015) Comparison of phenolic compounds and antioxidant potential between selected edible fruits and their leaves. J Funct Foods 14:736–746

Teleszko M, Wojdyło A, Rudzińska M, Oszmiański J, Golis T (2015) Analysis of lipophilic and hydrophilic bioactive compounds content in sea buckthorn (*Hippophae rhamnoides L.*) berries. J Agric Food Chem 63(16):4120–4129

Tkacz K, Wojdyło A, Turkiewicz IP, Bobak Ł, Nowicka P (2019) Anti-oxidant and anti-enzymatic activities of sea buckthorn (*Hippophaë rhamnoides L.*) fruits modulated by chemical components. Antioxidants 8(12):618

Tsimogiannis D, Oreopoulou V (2019) Classification of phenolic compounds in plants. In: Polyphenols in plants, pp 263–284

Verma H, Chahota R, Palial A, Sharma M (2011) Antibacterial properties of seabuckthorn (*Hippophae rhamnoides L.*) leaf extracts against common skin and wound bacteria. Indian J Veterin Res 20(1):38–41

Wang XX, Liu T, Liu Y, Liu Y, Li J et al (2021) Research progress on the relative theory of medicinal plants of *Hippophae L.* World Chinese Med 16:2217–2227

Wang K, Xu Z, Liao X (2022) Bioactive compounds, health benefits and functional food products of sea buckthorn: a review. Crit Rev Food Sci Nutr 62(24):6761–6782

Yang F, Suo Y, Chen D, Tong L (2016) Protection against vascular endothelial dysfunction by polyphenols in sea buckthorn berries in rats with hyperlipidemia. Biosci Trends 10(3):188–196

Yang H, Yang S, Chen X, Zhang J, Zhang Y (2023) Dynamic changes in flavonoid, phenolic, and polysaccharide contents in leaves and fruits of sea buckthorn during the growing season in southeastern Tibet plateau. Sci Hortic 307:111497

Yao XiaoHui YX, Zhang ZhenBin ZZ, Song Peng SP, Hao JingYi HJ, Zhang DongYang ZD, Zhang YiFei ZY (2016) Different harvest seasons modify bioactive compounds and antioxidant activities of Pyrola incarnata. Ind Crop Prod 94:405

Zadernowski R, Naczk M, Czaplicki S, Rubinskiene M, Szałkiewicz M (2005) Composition of phenolic acids in sea buckthorn (*Hippophae rhamnoides L.*) berries. J Am Oil Chem Soc 82(3):175–179

Chapter 6
Chemical Composition and Pharmacological Effects of Flavonoids and Phenolic from Sea Buckthorn

Xizhe Fu and Jinhu Tian

Abstract Sea buckthorn is a valuable plant resource, rich in flavonoids and phenolics, which possess various physiological activities and medicinal values. This chapter primarily discusses the chemical composition and pharmacological effects of these flavonoids and phenolics found in sea buckthorn. Flavonoid compounds encompass flavones, flavonols, isoflavones, anthocyanins, and their derivatives. The phenolic compounds in sea buckthorn fruits predominantly include flavonoids, phenolic acids, proanthocyanidins, and other phenolic compounds. These compounds play significant roles in antioxidant activity, combating obesity and diabetes, antitumor activity, and osteoporosis. Given these characteristics, there are broad prospects for the application of flavonoids and phenolics in fields such as food, medicine, and other health care products.

Keywords Flavonoids · Polyphenols · Chemical composition · Biological activity

Abbreviations

AD	Alzheimer's disease
BMP	bone morphogenetic protein
DPPH	2,2-diphenyl-1-picrylhydrazyl
EGCG	epigallocatechin gallate
EBTB	butylated *t*-butyl ether
MTT	microdilution assay
RP-HPLC-DAD	reverse-phase high-performance liquid chromatography
TFH	total flavonoids

X. Fu
School of Food Science and Technology, Shihezi University, Shihezi, China

J. Tian (✉)
Department of Food Science and Nutrition, Zhejiang University, Hangzhou, China

J. Tian et al. (eds.), *Sea Buckthorn*, https://doi.org/10.1007/978-981-97-9865-0_6

6.1 General Introduction

Sea buckthorn (*Hippophaë rhamnoides* L.) is a perennial deciduous shrub or small tree, which belongs to the *Elaeagnaceae* family and can sustain extreme cold (−43 °C) and hot (40 °C) temperatures, drought, alkalinity, and salinity due to the extensive and substantial root system and its nitrogen-fixing characteristic (Patra, Borugadda, and Dalai 2022). Sea buckthorn is wildly distributed in Asia, Europe, and North America and mainly grows in China, India (Himalayan region), Poland, Russia, and Canada (Dabrowski et al. 2022; Patra et al. 2022; Sharma and Kalkal 2018). The area of sea buckthorn in China also ranks first in the world, accounting for more than 90% of the total area of sea buckthorn in the world. In China, sea buckthorn mainly grows in the transition zone from forest to grassland (meadow) in north and southwest of China and has effectively improved the ecological environment and played a role in preventing soil erosion and mitigating the damage caused by wind and sand (Yang et al. 2023).

The fruits of sea buckthorn are rich in vitamins, sugars, fatty acids, amino acids, carotenoids, organic acids, phenolic acids, flavonoids, tannins, triterpenes, steroids, proanthocyanidins, and alkaloids (Chen et al. 2023; Guo 2019; Huang et al. 2023) and are currently widely used in the processing of various high-quality beverages and nutritional health products, including fruit juice drinks, solid drinks, jams, and gelatin soft candy. The yellow pigment and carotenoid in the fruit can be used as coloring additives in cookies and other baked pastries (Ma et al. 2017; Guo 2019). The leaves of sea buckthorn contain various vitamins, fatty acids, flavonoids, phenolic acids, polysaccharide, amino acid, and alkaloid, while seeds of sea buckthorn possess steroid compounds, fatty acids, and phenolic compounds (Xie 2022; Yang 2022).

Plant polyphenols are secondary kind of metabolites in plants and their content in plants is only lower than cellulose, hemicellulose, and lignin; however, various biological properties of polyphenols have been reported including antioxidant, anti-inflammatory, antiproliferative, anti-allergic, antiviral, anticarcinogenic, neuroprotective and antimicrobial properties (Gomez-Mejia et al. 2019; Huang et al. 2023). Sea buckthorn is rich in polyphenols, among which phenolic acids include gallic acid, ferulic acid, caffeic acid, *p*-coumaric acid and flavonoids contain isorhamnetin, quercetin, and their derivatives (Yang et al. 2023). With the deepening of research, flavonoids and phenols of sea buckthorn have been found to play a crucial role in antioxidant, anti-inflammatory, protection of cardiovascular and cerebrovascular, inhibition of cancer and other important biological activities (Cui 2020).

6.2 Chemical Composition of Phenolic Substances in Sea Buckthorn

6.2.1 Chemical Composition of Flavonoids in Sea Buckthorn

The phenolic substances in sea buckthorn fruit mainly include flavonoids, phenolic acids, proanthocyanidins, and other phenolic compounds. Flavonoids are one of the most important phenolic components in sea buckthorn. Shi et al. investigated the

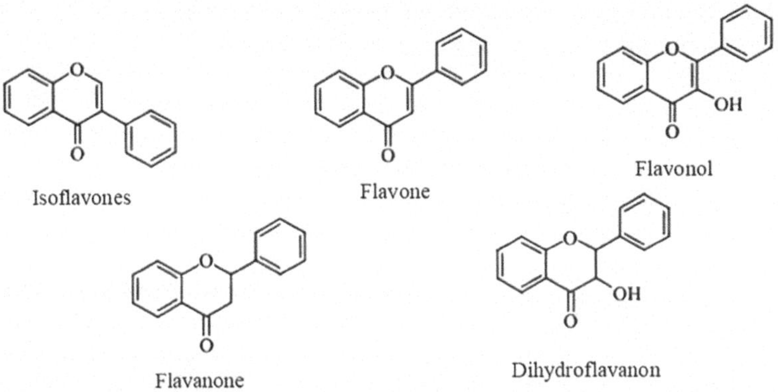

Fig. 6.1 The skeletal diagram of flavonoids. (Drawn by the author)

main bioactive compounds of fruits, leaves, branches, roots, and seeds of sea buckthorn and found that flavonoids accounted for a large proportion (Hao et al. 2019).

Flavonoids are a class of low-molecular-weight benzopyranone compounds, which are characterized by a C6–C3–C6 skeleton, consisting of two phenolic rings (A and B) linked by a central pyran ring (C) that contains an oxygen atom, and widely appeared in herbs and are abundant in fruits and vegetables (Aiziguli 2022) (Fig. 6.1).

Based on the differences in chemical structure, flavonoids contain flavones, flavonols, flavanols, flavanone, isoflavones, anthocyanins, chalcones, and their derivatives, especially forming polymers with other sugars (flavonoid glycoside) or methylated derivatives. Flavonols are distinguished from flavanones by their 3-position hydroxyl and double bonds formed by 2-position carbon and 3-position carbon. Flavonoids are usually hydroxylated at the 3, 5, 7, 2, 3′, 4′, and 5′ positions (Yang 2022). The most common flavonoids in sea buckthorn are isorhamnetin, kaempferol, quercetin, myricetin, and their glycoside compounds. Glycosyl mainly includes glucose, rhamnose, arabinose, galactose, and sophorose, which are mostly in the form of 3-O-glycoside, while a small amount is linked by 5-O-glycoside and 7-O-glycoside pattern (Guo 2019). Isorhamnetin derivatives accounted for more than 65% of the total flavonol, and quercetin derivatives accounted for more than 25% of the total flavonol (Liu et al. 2021). Flavanols in sea buckthorn mainly include catechins and epicatechins (Guo 2019).

The type, proportion, and composition of flavonoids contained in each part of sea buckthorn are affected by many factors, such as origin, picking time and variety. Based on the previous report, the fruits and leaves of sea buckthorn were the main parts containing flavonoids and the flavonoid types in fruits were more than those in leaves; however, the flavonoid content in leaves was higher than that in fruits (Xie et al. 2012). Besides, the season with hot temperatures usually results in higher flavonoids and phenolic content (Yang et al. 2023) (Table 6.1).

6.2.2 Chemical Composition of Other Phenolics in Sea Buckthorn

Sea buckthorn polyphenols are both abundant and diverse, yet their forms of existence pose challenges for absorption by the body. The primary phenolic substances in sea buckthorn fruit encompass flavonoids, phenolic acids, proanthocyanidins, among others (Guo 2019). Specifically, the phenolic acids found in sea buckthorn comprise gallic acid, protocatechuic acid, ferulic acid, salicylic acid, hydroxybenzoic acid, *p*-coumaric acid, vanillic acid, caffeic acid, and caffeoylquinic acid. Notably, conjugated phenolic acids exist as complexes with esters and glycosides (Li et al. 2022). Gunenc et al. identified nine distinct phenolic compounds—namely gallic acid, protocatechuic acid, *p*-hydroxybenzoic acid, vanillic acid, salicylic acid, *p*-coumaric acid, cinnamic acid, caffeic acid, and ferulic acid—from the fruits and leaves of Central Asian sea buckthorn using RP-HPLC-DAD analysis technology. Sea buckthorn leaves contain a substantial number of polyphenols, with content reaching up to 12%, whereas its fruits exhibit fewer types of polyphenols. To date, three structural types of gallotannins, galloyl tannins, and gallotannin have been

Table 6.1 Flavonoids compounds in sea buckthorn identified in previous literature (Guo 2019) (drawn by the author)

Category	Number	Compounds
Flavonol	1	Isorhamnetin
	2	Kaemferol
	3	Myricetin
	4	Morin
	5	Quercetin
Flavanone	6	Pinocembrin
	7	Naringenin
	8	Hesperetin
	9	Dihydromyricetin
	10	Eriodictyol
	11	Taxifolin
Flavonol	12	Catechin
	13	Epicatechin
	14	Epigallocatechin
	15	Gallocatechin

(continued)

Table 6.1 (continued)

Category	Number	Compounds
Flavonoid glycoside	16	Isorhamnetin-3-rutinoside
	17	Isorhamnetin-3-glucoside
	18	Isorhamnetin-3-galactoside
	19	Isorhamnetin-7-rhamnoside
	20	Isorhamnetin-3-sophoroside-7-rhamnoside
	21	Isorhamnetin-3-glucoside-7-rhamnoside
	22	Isorhamnetin-3-rhamnosylglucoside
	23	Isorhamnetin-3-coumaroylglu-cosylglucoside-7-rhamnoside
	24	Isorhamnetin-3-hydroxyferuloyl-glucosyl-glucoside-7-rhamnoside
	25	Isorhamnetin-3-glucoside-7-diglucoside
	26	Isorhamnetin-3-pentoside-7-rhamnoside
	27	Isorhamnetin-3-rutinoside-7-rhamnoside
	28	Isorhamnetin-3,7-diglucoside
	29	Isorhamnetin-3-acyloglucoside-glucoside-7-rhamnoside
	30	Quercetin-3-rutinoside
	31	Quercetin-3-glucoside
	32	Quercetin-3-galactoside
	33	Quercetin-3-glucopyranoside
	34	Quercetin-3-sophoroside-7-rhamnoside
	35	Quercetin diglucosylrhamnoside
	36	Quercetin-3-coumaroylglu-cosylglucoside-7-rhamnoside
	37	Quercetin dirhamno-sylglucoside
	38	Quercetin dimethyl ether 3-hydroxyferuloyl-glucosylglucoside-7-rhamnoside
	39	Quercetin diglucosylrhamnoside
	40	Quercetin-3-glucoside-7-rhamnoside
	41	Quercetin-3-glucopyranosyl-7-rhamnopyranoside
	42	Quercetin-glucosiderhamnoside-7-rhamnoside
	43	Quercetin dimethyl ether 3-caf-feoylglucosylglucoside-7-rhamnoside
	44	Kaempferol-3-O-(6′-O-coumaryl) glycoside
	45	Kaempferol-3-hydroxyvanilloylglucosylglucoside-7-rhamnoside
	46	Kaempferol-3-glucoside
	47	Kaempferol-3-rutinoside-7-rhamnoside
	48	Kaempferol-3-glucosyl-glucoside-7-rhamnoside
	49	Kaempferol-3-rutinoside
	50	Myricetin-3-rutinoside
	51	Myricetin methyl ether rutinoside
	52	Myricetin dimethyl ether rutinoside
	53	1-Feruloyl-β-D-glucopyranoside
	54	Narcissoside
	55	Hyperoside
	56	Quercitrin
	57	Isoquercitrin
	58	Rutin
Anthocyanins	59	Procyanidin B2
	60	Procyanidin B1
Chalcones	61	Naringenin chalcone
	62	Phloretin

predominantly isolated and identified from sea buckthorn leaves. These include sea buckthornin A, sea buckthornin B, galloylglucose, lignans, and more than a dozen hydrolyzable tannins (Gunenc et al. 2016).

6.3 Methods for Extracting Phenols and Flavonoids in Sea Buckthorn

Flavonoids mainly exist in the form of polar glycosides, which can be extracted with aqueous ethanol solution. The factors affecting flavonoid extraction include extraction time, extraction temperature, pressure, and solvent selection. Traditional extraction technologies such as Soxhlet extraction require long extraction time and high solvent cost. To overcome these problems, more extraction technologies such as microwave-assisted extraction, ultrasound-assisted extraction, and enzyme-assisted extraction have been gradually applied (Cui et al. 2020; Dabrowski et al. 2022; Gomez-Mejia et al. 2019).

6.3.1 Microwave-Assisted Extraction Technology

The principle of microwave-assisted extraction technology is using microwaves with a wavelength of 1 mm to 1 m and at a frequency of 300 MHz to 300 GHz to heat up the solvent (Yang et al. 2022), which ruptures the cell membrane quickly and improves the mass transfer rate and thus increase the amount of material extraction (Dabrowski et al. 2022). Factors influencing microwave extraction include solvent type, microwave time, and extraction temperature (Cui et al. 2020; Guo 2019). Feng et al. (2016) conducted an investigation into the optimal process for microwave-assisted extraction of sea buckthorn leaf polysaccharides. The extraction conditions were as follows: an extraction time of 60 s, a microwave power of 250 W, an extraction temperature of 80 °C, and an extraction duration of 30 min. And the impact of microwave-assisted extraction on the yield of sea buckthorn polysaccharides was also examined; meanwhile, the findings indicated that the yield of sea buckthorn polysaccharides increased proportionally with the rise in microwave power ($P < 0.05$). The optimal conditions for the extraction of sea buckthorn polysaccharides were identified as a microwave power of 250 W, an extraction duration of 60 s, and an extraction temperature of 80 °C. Under these conditions, the yield of the extracted polysaccharides peaked. The yield of sea buckthorn polysaccharides was significantly superior to that of the control group under identical conditions ($P < 0.05$). Furthermore, the content of polyphenols in sea buckthorn leaves was also markedly higher than that in the control group under the same conditions ($P < 0.05$). The extraction was conducted at a rate of 600 W, utilizing a solid–liquid ratio of 1:60.4 g/mL. This process was executed twice, resulting in an overall extraction yield of 14.95%. With the increase of temperature, the solubility of the compound in the solvent increases, applying pressure on the cell wall, contributing to the release of the desired compound; in addition, the viscosity of the solvent used decreases accordingly, thus improving the solubility and mobility, but it should be noted that high temperature and high power will lead to the degradation of the desired compound (Huang et al. 2023). Simultaneously, when using flammable solvents in microwave-assisted extraction techniques, it is necessary to prevent rapid increases in solvent temperature that can lead to fire (Dabrowski et al. 2022). Asofiei et al. (2016) conducted an investigation into the microwave-assisted extraction of polyphenols from sea buckthorn leaves. The results indicated that both the adjuvant extract and whole leaf extract could effectively extract phenolic compounds. The extraction time was significantly reduced, with the microwave-assisted method requiring only 7.5 min compared to 30 min for the conventional method. All the procedures were conducted at a temperature of 90 °C. The optimal extraction was achieved using a liquid-to-material ratio of 20:1 and employing 50% ethanol as the extraction solvent, coupled with microwave extraction of polyphenols (Xiang et al. 2020). The findings indicated that the total polyphenol content in the extract escalated with an increase in microwave power ($P < 0.05$). The peak total polyphenols content was recorded at a microwave power of 900 W for a duration of 3 min. Notably, there were significant differences in the total polyphenol contents among the four groups ($P < 0.05$), and these differences were also pronounced between the two groups ($P < 0.05$). Furthermore, the total polyphenol content in the extract consistently increased with the escalating microwave power ($P < 0.05$). The phenolic extract exhibited the highest antioxidant capacity (Asofiei et al. 2016).

Compared with traditional technology, the extraction amount of flavonoids increases four times. In addition, the extraction time and the use of the required extraction agent were also reduced. Among the extractions under the same conditions, the microwave-assisted extraction took less time than the solvent extraction (Liu et al. 2021).

6.3.2 Ultrasound-Assisted Extraction Technique

Ultrasound-assisted extraction technology uses ultrasound with a frequency range of 20 kHz to 2 MHz to produce cavitation, make the bubbles in the extract explosive burst, produce local high pressure and high temperature, and destroy the plant cell wall in order to enhance the interaction between cell components and solvent, improve the quality transfer, and then increase the extraction efficiency (Cui et al. 2020; Dabrowski et al. 2022; Gomez-Mejia et al. 2019). During the test, the ultrasonic extraction method can be applied to extract plant materials by using ultrasonic reactor and ultrasonic cleaner, or by direct means such as ultrasonic probe. The benefits of ultrasound-assisted extraction technology included its capacity to significantly enhance extraction efficiency, lower the necessary reaction temperature, decreasing the required solvent dosage, and reducing the solubility of target compounds. Additionally, this equipment operated with simplicity and offered a low overall cost (Huang et al. 2023; Patra et al. 2022). Factors influencing the efficiency of ultrasound-assisted extraction include time, temperature, and ultrasound frequency. The effect of ultrasound time on extraction yield is a quadratic function rather than a linear relationship, and the extraction rate increases with increasing extraction time, but it tends to decrease after a certain time due to prolonged heating, leading to flavonoid degradation (Sharma et al. 2018). Sanwal et al. (2022) found that ultrasound-assisted extraction effectively enhanced the ability to extract anthocyanins from fruit and increased the stability of polymeric anthocyanins in the extract. Sanwal et al. (2022) investigated the impact of silymarin on hepatotoxicity induced by butylated *t*-butyl ether (EBTB) in rats. The extraction process was facilitated through ultrasonic assistance within an alcohol solution. The optimum extraction conditions were: 40% ethanol solution, ultrasound frequency of 20 kHz, room temperature, and 15 min reaction time. The yield of the extract was 83.97% (w/w). The phenolic content of the extract was 16.03% (w/w). The antioxidant activity of the extract was determined by DPPH assay. The results showed that the extract exhibited a significant antioxidant activity against DPPH radicals with IC50 value of 1.99 mM. The extraction time was 15 min at room temperature, and the flavonoid extraction rate increased to 5.034%.

6.3.3 Enzyme-Assisted Extraction Technique

Protein or carbohydrate-bound polyphenols inside and on the cell wall are not easily extracted. The enzyme-assisted extraction destroys or soft the cell wall by adding the enzyme to the extraction medium to promote material release, thus increasing the yield rate (Tang et al. 2016). Enzyme-assisted extraction usually adopts cellulase, pectinase, and protease. Pectase is often used in juice production, but due to the expensive price of enzymes, they will be combined with other auxiliary extraction technologies such as ultrasonic, microwave, and supercritical fluid, which can effectively reduce the cost. Enzyme-assisted extraction is an environmentally friendly process and widely used in water instead of organic solvents and chemicals, which has the characteristics of low energy requirement and preventing the degradation of extracted substances. Particle size, enzyme-to-sample ratio, pH, time, temperature, and enzyme concentration are key control factors affecting enzyme-assisted extraction and need to be optimized according to the substrate material. Tkacz et al. studied the enzymatic-assisted extraction of flavonoids from sea buckthorn peel using Box–Behnken experimental design and response surface methodology. The best conditions were: 73.3 IU/g, temperature 53.7, pH = 4.14. Under this condition, the amount of total flavonoid extracted was the highest, up to 8.02 mg/g (Tkacz et al. 2021).

The extraction rates of sea buckthorn flavonoids vary depending on the method and conditions employed. Notably, microwave and ultrasound techniques, when used as auxiliary methods, yield different extraction rates. While the auxiliary extraction process could effectively disrupt the cell wall tissue of sea buckthorn leaves and enhance the extraction rate, it needed high-quality equipment. Consequently, this method is not conducive to large-scale production (Zhu 2021). On the other hand, enzymatic extraction required optimal temperature conditions to ensure the protection of flavonoid components during enzymolysis, but it yielded a relatively low extraction rate. Consequently, this study employed a straightforward solvent extraction method to extract total flavonoids from sea buckthorn leaves (Ran et al. 2021).

6.4 Biological Activity and Pharmacological Effects of Flavonoids and Phenolics in Sea Buckthorn

6.4.1 Role in Plants

Flavonoids are polyphenolic secondary metabolites, which are the most common pigments in plants other than chlorophyll and carotenoids, and often present in glycosylated derivatives that can give flowers, color, and aroma to leaves and fruits. Flavonoids play a pivotal role in plant ecology, serving as visual cues for pollinators and aiding plants in combating infections caused by detrimental stressors such as bacteria and viruses. Additionally, their unique astringency serves as a defensive mechanism. To resist harmful damage to plants, not only can flavonoids act as catalysts during the photoreaction phase of photosynthesis, and it is also a regulator of the phosphorylation reaction ion channel at the same time. Flavonoids possess the ability to scavenge reactive oxygen species, which are produced by the photosynthetic electron transport system. This process contributes to their role in stress protection. Additionally, flavonoids exhibit ultraviolet absorption characteristics, thereby offering protection to plants against damage caused by UV radiation (Xie et al. 2023).

6.4.2 Antioxidant Activity

Flavonoids exhibit potent lipid antioxidant properties against both reactive oxygen and morphological nitrogen-induced oxidation. As an oxidant and scavenger of reactive oxygen, this physiological activity is associated with its chemical structure, specifically the radical base. The mechanism of action involves scavenging free radicals through several processes: complexation with iron ions to inhibit radical generation; reaction with superoxide anions to prevent the initiation of free radical reactions; and combination with lipid peroxidation bases to halt lipid peroxidation processes (Zhou et al. 2014). Su et al. (2024) commissioned a study on sea buckthorn's effects on mice using the subacute aging model. The results indicated that flavonoids in sea buckthorn possess significant antioxidant capacity, potentially contributing to anti-aging effects. This mechanism may be attributed to their ability to inhibit brain activity, reduce free radical reactions, enhance human immune function, and elevate cellular nucleic acid metabolism levels. Antioxidants are a class of substances that can significantly delay or prevent the oxidation of the substrate. Flavonoids through chelating metal ions, decomposition of lipid peroxide, destroy the oxidation chain reaction, reduce local oxygen concentration and inhibition of NADH oxidase, mitochondrial amber oxidase, glutathione S-transferase, and microsomal monooxygenase activity of different mechanisms, such as inhibiting the generation of reactive oxygen species, removing free radicals, and displaying antioxidant activity (Kumar et al. 2013). The antioxidant capacity of flavonoids depends on the type of functional group and its arrangement on the skeleton of the class compound, and the number of hydroxyl groups produced is an important factor affecting their antioxidant capacity. Since flavonoids mainly exist in the form of flavonoid glycosides, the difference in the position and number of flavonoid molecules and attached sugar molecules will also affect their antioxidant properties, resulting in stronger antioxidant capacity than glycosides (Shen et al. 2021). The phenolic components in sea buckthorn fruits are the major contributors to their overall antioxidant activity.

6.4.3 Cardiovascular Disease

Atherosclerotic plaques, which are often the result of endothelial injury, frequently precipitate the onset of cardiovascular disease. The primary contributors to this endothelial injury are inflammation and oxidative stress. Cytokines and chemokines, as regulatory proteins, can lead to excessive production when they disrupt the gradient balance. This disruption also stimulates the production of reactive oxygen species, thereby instigating chronic inflammation. Over time, this chronic inflammation can further promote the onset of cardiovascular disease (Xu et al. 2011). Flavonoids have been shown to effectively mitigate endothelial injury induced by inflammation and oxidative stress. Concurrently, anthocyanins possess the ability to scavenge hydroxyl radicals and superoxide ions. Furthermore, flavonoids modulate the pathogenesis of inflammation by diminishing the concentration of reactive oxygen species, thereby inhibiting the progression of chronic inflammation. Notably, animals subjected to high cholesterol and a high-fat diet exhibit a reduction in blood pressure, leading to a significant improvement in endothelial dysfunction. Total flavonoids serve as crucial raw materials in the clinical treatment of cardiovascular diseases. Their preventive impact on heart disease and associated complications has garnered recognition both domestically and internationally. Tang et al. (2016) demonstrated that sea buckthorn extract exhibits significant anti-myocardial ischemia and hypoxia properties, reduces cholesterol levels in the body, enhances blood rheology, prevents cholesterol deposition in the arterial wall, and mitigates the onset of atherosclerosis. Sea buckthorn flavonoids possess a robust capacity to scavenge reactive oxygen species within organisms, thereby exerting anti-myocardial ischemia, anti-arrhythmia effects, reducing myocardial infarction area, alleviating angina pectoris, and improving cardiac function (Xu et al. 2011).

A high intake of flavonoid-rich fruits and vegetables significantly reduced the risk of cardiovascular disease development, and with increased flavonoid intake, mortality from cardiovascular disease and stroke decreased. In a study conducted by Feng et al. involving 42- to 60-year-old men with elevated intakes of five flavonoids (flavone-3-alcohol, flavonols, anthocyanins, flavane, and flavonoids) and a history of stroke risk and cardiovascular mortality, it was observed that high concentrations of flavone-3-ol and flavone significantly reduced the incidence of ischemic stroke over the course of the follow-up period. This suggested that substantial consumption of flavonoids can substantially decrease the mortality rate associated with cardiovascular disease (Feng et al. 2016).

6.4.4 Obesity and Diabetes Mellitus

Overweight and obesity can lead to metabolic disorders, reduce immunity, and promote the production of chronic inflammation such as diabetes, hypertension, and cardiovascular diseases (Dong et al. 2017). Flavonoids play a positive role in promoting glucose metabolism and inhibiting adipose tissue accumulation, and it has an extremely high ability to inhibit lipase activity (Tkacz et al. 2020). In cell culture and animal model test, catechin can inhibit lipogenesis, proliferation, differentiation and absorption, reduce fat mass and body weight, reduce the plasma triglycerides, free fatty acids, cholesterol, glucose, insulin and leptin levels, can also increase the effect of β-oxidation and thermogenesis, and has the potential of anti-obesity (Pap et al. 2021).

Diabetes, the most prevalent metabolic disease globally, is characterized by hyperglycemia resulting from either a deficiency in insulin production or its nonfunctionality. This elevated blood sugar levels induce disorders in protein and lipid metabolism. Furthermore, they exert irreversible impacts on the vascular system. Over time, these effects can culminate in damage and dysfunction of various organs such as nerves, kidneys, eyes, blood vessels, and the heart. Flavonoids can improve β-cell function, promote insulin secretion, increase insulin sensitivity, improve glucose metabolism, maintain blood glucose levels, reduce cholesterol synthesis and reduce triglyceride levels, and have positive effects in the prevention and treatment of diabetes and its complications. Waglewska et al. (2022) found that the addition of epigallocatechin gallate (EGCG) to rat cells improved the motility of β- cells and insulin secretion and reduced insulin resistance caused by high levels of free fatty acids in plasma. Tkacz et al. (2021) investigated the impact of total flavonoids derived from sea buckthorn on patients with diabetes, both those with and without heart failure. The findings indicated that these flavonoids not only significantly influence blood glucose levels but also have a positive effect on glycated hemoglobin. The study also demonstrates a reduction in the concentration of serum, late-stage glycation end products within the myocardium, and an enhanced inhibition of late-stage glycation within the same.

6.4.5 Antitumor Activity

Flavonoids in sea buckthorn act on colon cancer, liver cancer, breast cancer, and lung cancer and have extensive inhibitory effects on the growth of various cancer cells. Cancer is a heterogeneous disease triggered by cellular homeostasis and function generating irreversible damage, and loss of apoptotic function, gene mutations, oxidative stress, and hypoxia can all cause cancer (Tkacz et al. 2021). Grey et al. (2010) discovered that sea buckthorn fruit exhibited significant properties. The extraction process was executed, and the resultant extract demonstrated varying degrees of inhibitory effects on the proliferation of both human colon cancer cells and liver cancer cells.

At present, the incidence of cancer continues to increase around the world. The ideal anti-cancer drug should have the ability to inhibit tumor growth or kill cancer cells and have low side effects on the human body. Flavones, as natural products, are ideal choices in the chemoprevention and treatment of cancer. Flavonoids primarily affect cancer cells through several mechanisms: they initiate apoptosis, inhibit proliferation signaling pathways, and impede angiogenesis. These effects include the inhibition of protein tyrosine kinases, phosphatidylinositol-3-kinase, and topoisomerase. Additionally, flavonoids can downregulate mutant p53 proteins, induce cell cycle arrest, and modulate antiangiogenic mechanisms. Ciesarová et al. (2020) found that tin and myricetin in flavonols and apigenin and luteolin in flavonoids were potent inhibitors of protein tyrosine kinases. Dong et al. (2020) investigated the impact of sea buckthorn in its unconcentrated form on β-amyloid treated cells. Sea buckthorn has been demonstrated to significantly reduce intracellular β-amyloid content and enhance cell viability, thereby validating its potential therapeutic applications. There is an escalating interest in the field of Alzheimer's disease (AD) research, particularly in identifying potential therapeutic agents. One such agent that has garnered significant attention is the intercalated drug name. This compound exhibits promising properties, including its ability to target specific molecular pathways involved in AD progression and its demonstrated efficacy in animal models of the disease. However, further investigation is required to fully understand its mechanism of action and potential side effects. Additionally, clinical trials are needed to assess its safety and efficacy in human patients. Despite these challenges, the discovery of effective treatments for AD remains a critical objective in the field of neuroscience and medicine. With total flavonoids (TFH) used in hepatoma cell line BEL-7402 cells for 48 h, the cancer cells gradually became round, with nuclear sequestration and bud protrusions on the cell surface, showing typical apoptotic morphological changes. Apoptosis was measured by flow cytometry, and obvious subdiploid apoptosome peaks appeared in the drug-treated group before GO/G1 phase, indicating that TFH could induce apoptosis in HCC cells. TFH in the concentration range of 10–1000 ug/mL was able to induce characteristic changes of growth-inhibitory effects and apoptosis in Bcap-37 in human estrogen-dependent breast cancer cells. The tumor suppression rate in the high-dose group was as high as 75.1%. Using gene microarray to examine the expression of genes regulating cell proliferation, cell apoptosis, and entry into the cell cycle, it was found that TFH downregulated the expression of DNA replication-related kinases (DNA polymerases α, PCNA, topology isomerase, etc.) and the cyclin-dependent kinases CDK 1, C-myc, and C-fos in Bcap-37 cells and can upregulate the expression of pro-apoptotic genes Caspase-9, AFDD, and BAI. It is speculated that TFH suppresses breast cancer proliferation by regulating the expression of cell cycle-related genes to arrest the cell interval, inhibits DNA replication and DNA replication, transcription, enhancing the apoptotic sensitivity of breast cancer cells, and then indirectly induces cells into the apoptotic cascade, and fat synthase expression was effectively inhibited. The occurrence of many breast cancers is manifested by the overexpression of intracellular adipose synthase, and the results indicate that proanthocyanidin promotes apoptosis in breast cancer cells through the lipid metabolism regulation pathway of fat synthase inhibition (Lee et al. 2022).

Markkinen et al. (2022) compared the effects of sea buckthorn extract obtained by ethyl acetate extraction and ethanol extraction on the proliferation of HepG2 and colon Caco-2 cells; both extracts inhibited the proliferation of two cancer cell lines and higher antiproliferative activity of the ethyl acetate moiety than that of the products of the ethanol method. This result was related to the differences in the variety of compounds therein. The components in the ethyl acetate extract were mainly proanthocyanidins and large amounts of corosacid, and it suggested that flavonoid ruglycosides are less inhibitory to cancer cell activity than proanthocyanidins and corosacid. Yogendra Kumar et al. (2013) conducted a study on the broad-spectrum antibacterial properties of phenolic components in sea buckthorn leaves against *Escherichia coli* and *Salmonella enterica*. The antimicrobial activity of these phenolics was assessed using the agar diffusion method, revealing significant inhibitory activity against both *E. coli* and *S. enterica*, with an IC50 value ranging from 0.123 to 0.496 μM. Additionally, the broth microdilution assay (MTT) was employed to evaluate the antimicrobial activity of these phenolics, which also demonstrated significant inhibitory activity against *E. coli* and *S. enterica*, with an IC50 value of 0.123–0.496 μM. *Staphylococcus aureus*, among others, exhibited significant inhibitory effects.

Similar to this conclusion, previous studies also found that the main anti-proliferative substances in flavonoids were rhamsin, lyretin, and kaempferol, followed by three glucosides and some phenolic acids. Flavone rutcoside and glucoside had no significant effect on the proliferation of liver cancer cells HepG2, breast cancer cells MCF-7, and MDA-MB-231 and colon cancer cells Caco-2. The combination of epigallocatechin gallate (EGCG) and a variety of flavonoids in cancer treatment can produce effective inhibition of cancer cells (Chang et al. 2020).

6.4.6 *Osteoporosis*

Osteoporosis and fragility fractures generally plague the elderly, with about half of women and one-fifth of men suffering from fractures due to osteoporosis. The bone is in a continuous modeling and remodeling state to maintain the balance between osteoclasts and osteoblasts, and as individuals age, there was an imbalance in the anabolic and reabsorptive activities of osteoblasts and osteoclasts, leading to a reduction in bone mass. Osteoclast differentiation factor and bone morphogenetic protein (BMP) signaling pathways are used to maintain the activity of osteoclasts and osteoblasts, which also promotes the formation and differentiation of osteoclasts. Flavonoids prevent the bone loss by reducing the generation of osteoclast differentiation factor, upregulating signaling pathways that promote osteoblast function, and inhibiting the activity of MMPases that digest skeletal collagen and affect the activity of osteoblasts (Ting et al. 2011).

Estrogen plays a crucial role in maintaining bone health by inhibiting cytokines and promoting the apoptosis of osteoclasts, and the deficiency of estrogen leads to bone loss and shortens the life span of osteoblasts. Sharma et al. showed that after continuous administration of flavin, women can significantly inhibit bone absorption and increase bone mineral density, and with the increase of flavone intake, bone strength and bone mineral density also increased, which is effective for the treatment of osteoporosis (Sharma et al. 2022). Xie et al. (2023) showed that anthocyanin had a positive promoting effect on BMD in the hip bone and spine, and flavonol had a positive promoting effect on hip BMD. Moreover, catechins and flavanins were inversely associated with the resorptive effects of bone markers, and catechin-rich tea can also prevent hip fractures (Song et al. 2023).

6.5 Application Prospects and Challenges of Flavonoids and Phenolics in Sea Buckthorn

Sea buckthorn is a valuable plant resource that is rich in flavonoids and phenolic substances. These compounds have multiple physiological activities and medicinal values, so they have attracted extensive attention and research. Flavonoids and phenolic substances in sea buckthorn have strong antioxidant capacity, which can effectively remove free radicals and reduce cellular oxidative damage, thus having a significant protective effect against chronic diseases such as cardiovascular disease and cancer. In addition, flavonoids and phenolic substances in sea buckthorn also have anti-inflammatory, anti-bacterial, anti-tumor, and anti-aging biological activities and have a positive impact on improving immune function and promoting the health of the digestive system. These properties offer broad prospects for the application of flavonoids and phenolics in food, medicine, health products, and other fields.

In the field of food, flavonoids and phenolic substances in sea buckthorn can be used as natural antioxidants and food additives for preservation, preservative, pigments, and other aspects to improve the quality and nutritional value of food. At the same time, sea buckthorn extract can also be used in the preparation of fruit juice, jam, candy, dairy products, and other foods to provide consumers with more healthy product choices. In the field of drugs, the antioxidant, anti-inflammatory, and anti-tumor activity of flavonoids and phenolic substances in sea buckthorn make it an important resource for the development of new drugs. These compounds can be applied to the prevention and treatment of cardiovascular diseases, cancer, liver disease, diabetes, and other diseases, bringing new drug options to clinical medicine.

However, there are some challenges to flavonoids and the application of phenols in sea buckthorn. First, flavonoids and phenolic substances in sea buckthorn have the problems of low extraction efficiency and low purity, and more optimized extraction and separation techniques need to be sought. Second, the restricted growth environment and the yields are relatively scarce, resulting in an inadequate supply of flavonoids and phenolics. Therefore, it is necessary to study and popularize the planting technology of sea buckthorn to improve the yield and quality to meet the market demand. In addition, more in-depth studies on pharmacology and toxicology are needed for the application of flavonoids and phenolics in sea buckthorn to ensure their safety and effectiveness.

In conclusion, flavonoids and phenolic substances in sea buckthorn have wide application prospects and play important roles in food and medicine. However, a range of challenges still need to be overcome, including extraction process improvement, planting technology innovation, and further research and promotion. By overcoming these challenges, we can better utilize the potential of flavonoids and phenols in sea buckthorn to make positive contributions to people's health and quality of life.

References

Aiziguli M (2022) Effect and mechanism of seabuckthorn flavonoids on glucolipid metabolism disorder and cognitive impairment induced by high calorie diet in mice, vol Doctor's degree. Northwest A&F University, Yangling

Asofiei I, Calinescu I, Trifan A, David IG, Gavrila AI (2016) Microwave-assisted batch extraction of polyphenols from Sea Buckthorn Leaves. Chem Eng Commun 203(12):1547–1553. https://doi.org/10.1080/00986445.2015.1134518

Chang M, Guo Y, Jiang Z, Shi L, Zhang T, Wang Y, Gong M, Wang T, Lin R, Liu R, Wang Y, Jin Q, Wang X (2020) Sea buckthorn pulp oil nanoemulsions fabricated by ultra-high pressure homogenization process: a promising carrier for nutraceutical. J Food Eng 287:110129. https://doi.org/10.1016/j.jfoodeng.2020.110129

Chen Y, Cai Y, Wang K, Wang Y, (2023) Bioactive compounds in sea buckthorn and their efficacy in preventing treating metabolic syndrome. Foods 12(10):1985. https://doi.org/10.3390/foods12101985

Ciesarová Z, Murkovic M, Cejpek K, Kreps F, Tobolková B, Koplík R, Belajová E, Kukurová K, Daško Ľ., Panovská Z., Revenco D, Burčová, Z (2020) Why is sea buckthorn (Hippophae rhamnoides L.) so exceptional? A review. Food Research International 133:109170. https://doi.org/10.1016/j.foodres.2020.109170

Cui M (2020) The anti-colon cancer active components and molecular mechanism of bound polyphenols from seabuckthorn berries, vol Master degree. Shanxi University, Taiyuan

Cui, T., W. Jie, Y. Yujun, H. Xue, and L. Zhaolin (2020) Characterization and distribution rule of quebrachitol in Hippophae rhamnoides. Journal of Beijing Forestry University 42 (1):121-126. https://doi.org/10.1007/s11694-023-02193-z

Dabrowski G, Czaplicki S, Szustak M, Cichonska E, Gendaszewska-Darmach E, Konopka I (2022) Composition of flesh lipids and oleosome yield optimization of selected sea buckthorn (Hippophae rhamnoides L.) cultivars grown in Poland. Food Chem 369:130921. https://doi.org/10.1016/j.foodchem.2021.130921

Dong R, Su J, Nian H, Shen H, Zhai X, Xin H, Qin L, Han T (2017) Chemical fingerprint and quantitative analysis of flavonoids for quality control of Sea buckthorn leaves by HPLC and UHPLC-ESI-QTOF-MS. J Funct Foods 37:513–522. https://doi.org/10.1016/j.jff.2017.08.019

Dong K, Fernando WMADB, Durham R, Stockmann R, Jayatunga W, Jayasena V (2020) A role of sea buckthorn on Alzheimers disease. Int J Food Sci Technol 55(9):3073–3081. https://doi.org/10.1111/ijfs.14571

Feng L, Yin J, Nie S, Wan Y, Xie M (2016) Fractionation, physicochemical property and immunological activity of polysaccharides from Cassia obtusifolia. Int J Biol Macromol 91:946–953. https://doi.org/10.1016/j.ijbiomac.2016.05.030

Gomez-Mejia E, Rosales-Conrado N, Leon-Gonzalez ME, Madrid Y (2019) Citrus peels waste as a source of value-added compounds: extraction and quantification of bioactive polyphenols. Food Chem 295:289–299. https://doi.org/10.1016/j.foodchem.2019.05.136

Grey C, Widén C, Adlercreutz P, Rumpunen K, Duan R-D (2010) Antiproliferative effects of sea buckthorn (Hippophae rhamnoides L.) extracts on human colon and liver cancer cell lines. Food Chem 120(4):1004–1010. https://doi.org/10.1016/j.foodchem.2009.11.039

Gunenc A, Khoury C, Legault C, Mirrashed H, Rijke J, Hosseinian F (2016) Seabuckthorn as a novel prebiotic source improves probiotic viability in yogurt. LWT Food Sci Technol 66:490–495. https://doi.org/10.1016/j.lwt.2015.10.061

Guo R (2019) Bioactivities and bioavailability of Hippophaë rhamnoids L. phenolics and their anticancer mechanism towards human breast cancer cells both in vitro and in vivo, vol Doctor's degree. South China University of Technology, Guangzhou

Hao W, He Z, Zhu H, Liu J, Kwek E, Zhao Y, Ma KY, He W-S, Chen Z-Y (2019) Sea buckthorn seed oil reduces blood cholesterol and modulates gut microbiota. Food Funct 10(9):5669–5681. https://doi.org/10.1039/C9FO01232J

Huang H, Li Y, Gui F, Yang P, Zhang J, Li W et al (2023) Optimizing the purification process of polyphenols of sea buckthorn seed and its potential freshness effect. LWT 173:114380. https://doi.org/10.1016/j.lwt.2022.114380

Kumar, M. Y., R. Tirpude, D. Maheshwari, A. Bansal, K. Misra (2013) Antioxidant and antimicrobial properties of phenolic rich fraction of Seabuckthorn (Hippophae rhamnoides L.) leaves in vitro. Food chemistry 141(4):3443–3450. https://doi.org/10.1016/j.foodchem.2013.06.057

Lee HHL, Lee CJ, Choi SY, Kim Y, Hur J (2022) Inhibitory effect of sea buckthorn extracts on advanced glycation endproduct formation. Food Chem 373:131364. https://doi.org/10.1016/j.foodchem.2021.131364

Li (2022) The screening of high-producing β-glucosidase lactic acid bacteria in Ningxia's pulp water microorganism, as well as its biotransformation of sea buckthorn fruit polyphenols, is the focus of this study. Yinchuan: Ningxa University

Liu S, Xiao P, Kuang Y, Hao J, Huang T, Liu E (2021) Flavonoids from sea buckthorn: a review on phytochemistry, pharmacokinetics and role in metabolic diseases. J Food Biochem 45(5):e13724. https://doi.org/10.1111/jfbc.13724

Ma X, Laaksonen O, Heinonen J, Sainio T, Kallio H, Yang B (2017) Sensory profile of ethyl β-d-glucopyranoside and its contribution to quality of sea buckthorn (Hippophaë rhamnoides L.). Food Chem 233:263–272. https://doi.org/10.1016/j.foodchem.2017.04.073

Markkinen N, Pariyani R, Jokioja J, Kortesniemi M, Laaksonen O, Yang B (2022) NMR-based metabolomics approach on optimization of malolactic fermentation of sea buckthorn juice with Lactiplantibacillus plantarum. Food Chem 366:130630. https://doi.org/10.1016/j.foodchem.2021.130630

Pap N, Reshamwala D, Korpinen R, Kilpeläinen P, Fidelis M, Furtado MM, SantAna AS, Wen M, Zhang L, Hellström J, Marnilla P, Mattila P, Sarjala T, Yang B, Lima ADS, Azevedo L, Marjomäki V, Granato D (2021) Toxicological and bioactivity evaluation of blackcurrant press cake, sea buckthorn leaves and bark from Scots pine and Norway spruce extracts under a green integrated approach. Food Chem Toxicol 153:112284. https://doi.org/10.1016/j.fct.2021.112284

Patra BR, Borugadda VB, Dalai AK (2022) Microwave-assisted extraction of sea buckthorn pomace and seed extracts as a proactive antioxidant to stabilize edible oils. Bioresour Technol Rep 17:100970. https://doi.org/10.1016/j.biteb.2022.100970

Ran B, Guo CE, Li W, Li W, Wang Q, Qian J, Li H (2021) Sea buckthorn (Hippophae rhamnoides L.) fermentation liquid protects against alcoholic liver disease linked to regulation of liver metabolome and the abundance of gut microbiota. J Sci Food Agric 101(7):2846–2854. https://doi.org/10.1002/jsfa.10915

Sanwal N (2022) Effect of ultrasound-assisted extraction on efficiency, antioxidant activity, and physicochemical properties of sea buckthorn (Hippophae salicipholia) seed oil. LWT 153(0023–6438):112386. https://doi.org/10.1016/j.lwt.2021.112386

Sharma PC, Kalkal, M. (2018) Nutraceutical and Medicinal Importance of seabuckthorn (Hippophae sp.). Therapeutic, probiotic, and unconventional foods. 227–253. https://doi.org/10.1016/B978-0-12-814625-5.00021-2

Sharma M, Hussain S, Shalima T, Aav R, Bhat R (2022) Valorization of seabuckthorn pomace to obtain bioactive carotenoids: An innovative approach of using green extraction techniques (ultrasonic and microwaveassisted extractions) synergized with green solvents (edible oils). Ind Crops Prod 175:114257. https://doi.org/10.1016/j.indcrop.2021.114257

Shen C, Wang T, Guo F, Sun K, Wang B, Wang J, Zhang Z, Zhang X, Zhao Y, Chen Y (2021) Structural characterization and intestinal protection activity of polysaccharides from Sea buckthorn (Hippophae rhamnoides L.) berries. Carbohydr Polym 274:118648. https://doi.org/10.1016/j.carbpol.2021.118648

Song Y, Zhang G, Chen N, Zhang J, He C (2023) Metabolomic and transcriptomic analyses provide insights into the flavonoid biosynthesis in sea buckthorn (Hippophae rhamnoides L.). LWT 187:115276. https://doi.org/10.1016/j.lwt.2023.115276

Su T, Zhao J, Zhu Y, Oyom W, Li S, Xie P, Bi Y, Wei J, George G (2024) Comparison of nutrient composition, phytochemicals and antioxidant activities of two large fruit cultivars of sea buckthorn in Xinjiang of China. Sci Hortic 324:112602. https://doi.org/10.1016/j.scienta.2023.112602

Tang L, Lv H, Li S, Bi H, Gao X, Zhou J (2016) Protective effects of polyphenol extracts from sea Buckthorn (*Hippophaë rhamnoides* L.) on rat hearts. Open J Mol Integ Physiol 06(01):10–18. https://doi.org/10.4236/ojmip.2016.61002

Ting H-C, Hsu Y-W, Tsai C-F, Lu F-J, Chou M-C, Chen W-K (2011) The in vitro and in vivo antioxidant properties of seabuckthorn (Hippophae rhamnoides L.) seed oil. Food Chem 125(2):652–659. https://doi.org/10.1016/j.foodchem.2010.09.057

Tkacz K, Chmielewska J, Turkiewicz IP, Nowicka P, Wojdyło A (2020) Dynamics of changes in organic acids, sugars and phenolic compounds and antioxidant activity of sea buckthorn and sea buckthorn-apple juices during malolactic fermentation. Food Chem 332:127382. https://doi.org/10.1016/j.foodchem.2020.127382

Tkacz K, Wojdyło A, Turkiewicz IP, Nowicka P (2021) Anti-diabetic, anti-cholinesterase, and antioxidant potential, chemical composition and sensory evaluation of novel sea buckthorn-based smoothies. Food Chem 338:128105. https://doi.org/10.1016/j.foodchem.2020.128105

Waglewska E, Misiaszek T, Bazylińska U (2022) Nanoencapsulation of poorly soluble seabuckthorn pulp oil in bile salt-origin vesicles: physicochemical characterization and colloidal stability. Colloids Surf A Physicochem Eng Asp 647:129113. https://doi.org/10.1016/j.colsurfa.2022.129113

Xiang H, Waterhouse D-S, Liu P, Waterhouse GIN, Li J, Cui C (2020) Pancreatic lipase-inhibiting protein hydrolysate and peptides from seabuckthorn seed meal: preparation optimization and inhibitory mechanism. LWT 134:109870. https://doi.org/10.1016/j.lwt.2020.109870

Xie X (2022) Flavonoids-enriched extract from seed of Hippophae rhamnoides L. and its effective components alleviates insulin resistance in C2C12 myotubes, vol Master degree. East China Normal University, Shanghai

Xie J, Lin G, Du Y, Ju H, Suo Y, Zhang Z (2012) Comparism research of total Flavonoids in different parts of Sea Buckthorn in Qinghai. Nat Prod Res Dev 24:45–48, 77

Xie X, Song Y, Bi X, Liu X, Xing Y, Che Z (2023) Characterization of sea buckthorn polysaccharides and the analysis of its regulatory effect on the gut microbiota imbalance induced by cefixime in mice. J Funct Foods 104:105511. https://doi.org/10.1016/j.jff.2023.105511

Xu Y-J, Kaur M, Dhillon RS, Tappia PS, Dhalla NS (2011) Health benefits of sea buckthorn for the prevention of cardiovascular diseases. J Funct Foods 3(1):2–12. https://doi.org/10.1016/j.jff.2011.01.001

Yang J (2022) Study on extraction and decolorization of Flavonoids from Sea Buckthorn leaves, vol Master degree. Shihezi University, Shihezi

Yang W, Kallio H, Yang B (2022) Health promoting properties and sensory characteristics of phytochemicals in berries and leaves of sea buckthorn (Hippophaë rhamnoides). Crit Rev Food Sci Nutr 62(14):3798–3816. https://doi.org/10.1080/10408398.2020.1869921

Yang H, Yang S, Chen X, Zhang J, Zhang Y (2023) Dynamic changes in flavonoid, phenolic, and polysaccharide contents in leaves and fruits of sea buckthorn during the growing season in southeastern Tibet plateau. Sci Hortic 307:111497. https://doi.org/10.1016/j.scienta.2022.111497

Yogendra Kumar MS, Tirpude RJ, Maheshwari DT, Bansal A, Misra K (2013) Antioxidant and antimicrobial properties of phenolic rich fraction of Seabuckthorn (Hippophae rhamnoides L.) leaves in vitro. Food Chem 141(4):3443–3450. https://doi.org/10.1016/j.foodchem.2013.06.057

Zhou G-l, Jiang S-q, Jiang Z-y, Lin Y-c, Zheng C-t. (2014) Effects of Dietary Isoflavone Supplementation on Meat Quality and Oxidative Stability During Storage in Lingnan Yellow Broilers. J Integr Agric 13(2):387–393. https://doi.org/10.1016/S2095-3119(13)60386-X

Zhu Y (2021) Composition analysis and antioxidant activity evaluation of a high purity oligomeric procyanidin prepared from sea buckthorn by a green method. Curr Res Food Sci 4:840

Chapter 7
In Vitro Anti-inflammatory and Antioxidant Activities of Flavonoids Isolated from Sea Buckthorn

Haizhi Huang

Abstract This chapter provides a general introduction to the health benefits of sea buckthorn associated with anti-inflammatory and antioxidant activities. The in vitro anti-inflammatory activity of flavonoids extracts from sea buckthorn, the anti-inflammatory activities of extracts from different tissues (including seeds, berries, and leaves), and the effect of the main flavonoids derived from sea buckthorn including flavone aglycones (i.e., isorhamnetin, quercetin, kaempferol, and myricetin) and flavonol glycosides (i.e., isorhamnetin-3-O-rutinoside and isorhamnetin-3-O-glucoside) are introduced. The antioxidant activities of flavonoids extracted from sea buckthorn and different tissues are discussed. Finally, the effects of the main flavonoids obtained from sea buckthorn on scavenging radicals and alleviating oxidative stress are also introduced.

Keywords Sea buckthorn · Flavonoids · Anti-inflammatory · Antioxidant

Abbreviations

AA	Arachidonic acid
AGEs	Advanced glycation end products
CAA	Intracellular antioxidant activity
CAT	Catalase
COX-2	Cyclooxygenase-2
DPPH	2,2α-diphenyl-1-picrylhydrazil
ERK	Extracellular regulated protein kinases
FRAP	Ferric reducing antioxidant power
GSSG	Oxidized glutathione

H. Huang (✉)
Key Laboratory of Microbiological Metrology, Measurement & Bio-product Quality Security, State Administration for Market Regulation, China Jiliang University, Hangzhou, China
e-mail: huanghz@cjlu.edu.cn

J. Tian et al. (eds.), *Sea Buckthorn*, https://doi.org/10.1007/978-981-97-9865-0_7

HDL	High-density lipoprotein
HO-1	Heme oxygenase-1
HUVECs	Human umbilical vein endothelial cells
I3G	Isorhamnetin-3-O-glucoside
I3R	Isorhamnetin-3-O-rutinoside
IBD	Inflammatory bowel disease
IL-1β	Interleukin-1β
iNOS	Nitric oxide synthase
JNK	c-Jun N-terminal kinase
LDH	Lactate dehydrogenase
LDL	Low-density lipoprotein
LPS	Lipopolysaccharide
MAPK	Mitogen-activated protein kinase
MCP-1	Monocyte chemoattratctant protein-1
MMP	Mitochondrial membrane potential
ORAC	Oxygen radical absorbance capacity
PAR-2	Protease-activated receptor-2
PBMCs	Peripheral blood mononuclear cells
PSC	Peroxyl radical scavenging capacity
PXR	Pro-gestational hormone X receptor
ROS	Reactive oxygen species
SOD	Superoxide dismutase
TBARS	Thiobarbituric acid reactive substance
TEAC	Trolox equivalent antioxidant capacity
TEARS	Thiobarbituric acid reactive substances
TNF-α	Tumor necrosis factor-α
VLDL	Very low-density lipoprotein

7.1 Introduction

Sea buckthorn is rich in nutrients and active compounds, especially flavonoids, which create several health benefits, such as anti-inflammatory, antioxidant, hepatoprotective, anti-cancer, anti-cardiovascular, neuroprotective, hypoglycemic, and hypolipidemic effects (Ma et al. 2023). In fact, previous studies have demonstrated the anti-inflammatory and antioxidant capacities as the two principal pharmacological effects of sea buckthorn, because most biological activities of sea buckthorn were involved in the mediation of inflammatory cytokines and antioxidant enzymes (Ren et al. 2020).

7.2 In Vitro Anti-inflammatory Activity of Flavonoids Extracted from Sea Buckthorn

As the number of patients worldwide increases, various treatments involved in reducing inflammation are being developed. The research and development of dietary plant compounds that can suppress inflammation and regulate the immune system has become a hot spot due to the side effects of long-term use of anti-inflammatory drugs (Wang et al. 2021b). In recent years, natural products found in plants have been used for immunotherapy, gradually replacing various traditional therapies such as drug therapy (Zhang et al. 2013). Plant extracts, as immunomodulators, have many advantages such as simple preparation, strong efficacy, and fewer side effects.

According to prior research, flavonoids and their metabolites can affect the expression of multiple genes and their signaling pathways along with some specific receptors of body tissue cells (Esser et al. 2018). Many studies have shown that flavonoids derived from natural plants engage in a variety of physiological activities, such as anti-inflammatory activity (Wang et al. 2021a). The flavonoids obtained from sea buckthorn (*Hippophae rhamnoides L.*) have gained increasing attention because of their anti-inflammatory, antiviral, hepatoprotective, weight-reducing, and hypoglycemic activities.

In recent years, the immunotherapy with medicinal plants and their bioactive components as immunomodulators has gradually replaced traditional therapies such as drug therapy. As immunomodulators, plant extracts have advantages such as easy access, simple preparation, few side effects, and strong efficacy. Previous studies have shown that the inflammatory reaction of the body is closely related to intestinal microorganisms.

Lipopolysaccharide (LPS), which can stimulate the body to secrete tumor necrosis factor-α (TNF-α), interleukin (IL)-1β, IL-4, IL-6, IL-13, and protease-activated receptor-2 (PAR-2), thus inducing immune disorders, is often used by researchers to stimulate inflammatory responses (Lv et al. 2016). Sea buckthorn flavonoids can reduce the levels of IL-6, TNF-α, and intracellular NO, as induced by LPS in the peritoneal macrophages of mice (Raw 264.7), showing that sea buckthorn flavonoids can inhibit inflammation by regulating the secretion of cytokines and inflammatory factors (Fu et al. 2021). In another study, Jiang et al. (2017) found that sea buckthorn flavonoids might exert inhibitory effects on inflammation by regulating the expression of inflammatory mediators through the MAPK (mitogen-activated protein kinase) and NF-κB pathways.

7.2.1 Anti-inflammatory Activity of Flavonoids Extracted from the Seed

The sea buckthorn seeds encapsulated in bright orange berries contain several flavonoids such as isorhamnetin, kaempferol, and quercetin (Slawinska and Olas 2023). The anti-inflammatory activity of sea buckthorn flavonoids was studied mainly through animal experiments. Researchers have discovered that orally administered flavonoid-enriched seed extract (100–300 mg/kg) partially alleviated glucose intolerance in mice fed a high-fat diet and reduced the infiltration of macrophages into the adipose tissue (Yang et al. 2017). Wang et al. (2016c) have investigated the effects of proanthocyanidins obtained from sea buckthorn seeds against retinal degeneration induced by visible light. They found that the proanthocyanidins derived from sea buckthorn seeds mitigated the increase in inflammatory cytokines (IL-1β, TNF-α, and IL-6) induced by light exposure.

Atopic dermatitis is a chronic inflammatory skin disease associated with the increased production of Th2 cytokines in skin lesions and aberrant IgE production (Brandt and Sivaprasad 2011). Sea buckthorn resides (including seeds, berry flesh, and peel) extracts were found to suppress the Th2 chemokines TARC and MDC through the dose-dependent inhibition of JAK2/STAT1, p38-MAPK, and NF-κB signaling pathways in HaCaT cells whose IFN-γ/TNF-α was activated, suggesting that sea buckthorn resides extracts had a positive effect on atopic dermatitis-like skin lesions. This might be achieved through the inhibition of Th2 chemokines (Hou et al. 2017a). In the case of peptic ulcer disease in both humans and horses, sea buckthorn seed oil has also been shown to possess anti-inflammatory properties (Jastrzab and Skrzydlewska 2019).

Natural killer cells are potent immune modulators because of their ability to produce abundant cytokines that can modulate immune responses, and the NK92-MI cell line shows functional characteristics of active natural killer cells. The total flavonoids of sea buckthorn significantly enhance NK92-MI cell cytotoxicity against K562 cells via upregulated expressions of IL-1α, IL-2, IL-7, IL-15, MCP-1, IFN-γ, TNF-α, and TNF-β and downregulated expressions of IL-16, MIP-1β, CX3CL-1, and MIF (Hou et al. 2017b).

7.2.2 Anti-inflammatory Activity of Flavonoids Extracted from Berries

In sea buckthorn berries, over 98% of flavonoids were flavonols. Isorhamnetin derivatives ranged from 66% to 72% of total flavonols were dominant, followed by quercetin derivatives (Tkacz et al. 2020). Mishra et al. (2008) evaluated the immunomodulatory activity of sea buckthorn (*Hippophae rhamnoides L.*) fruit pulp flavone in human peripheral blood mononuclear cells (PBMCs). The sea buckthorn flavone was found to stimulate the production of IL-6 and TNF-α in PBMCs, suggesting that flavones of sea buckthorn fruit can modulate the expression of several signaling molecules associated with immune function and inflammation, including several cytokines. Another study found that the alcoholic extracts of sea buckthorn fruits were able to arrest the chromium-induced inhibition of lymphocyte proliferation, which showed the immunomodulatory property of sea buckthorn (Geetha et al., 2002).

7.2.3 *Anti-inflammatory Activity of Flavonoids Extracted from Leaves*

The leaves of sea buckthorn are rich in various biologically active compounds including flavonoids (He et al. 2023). From previous research, it is known that the leaf extract of sea buckthorn has a significant anti-inflammatory effect (Ganju et al. 2005). Another study was performed in the same laboratory to demonstrate the possible mechanism in relation to its anti-inflammatory effects (Padwad et al. 2006). NO has been reportedly synthesized in large quantities by inflammatory cells and is associated with the pathogenesis of inflammation. Sea buckthorn leaf extract significantly inhibited the enhanced production of NO induced by LPS in a dose-dependent manner in RAW 264.7 cells. The inhibition of NO production by sea buckthorn leaf extract is suspected to result from its scavenging activity of NO radicals and inhibitory effects on the activation of nitric oxide synthase (iNOS). Some researchers have also focused on the effects of sea buckthorn leaf extract on the inhibition of LPS-stimulated NO production. In this study, treating a mouse alveolar macrophage cell line with supercritical CO_2 extract of sea buckthorn significantly inhibited LPS-induced NO production. It also inhibited the release of LPS-induced expression of CD40, iNOS, cyclooxygenase-2 (COX-2), and the pro-inflammatory cytokines IL-6 and TNF-α by inhibiting NF-κB activation and phosphorylation of p38 MAPK, which suggests that the sea buckthorn leaf extract might be applicable to treat chronic inflammatory diseases because of its anti-inflammatory activity (Jayashankar et al. 2012). Simultaneously, the inhibitory effects of supercritical CO_2 extract of sea buckthorn on inflammation and endotoxemia induced by LPS in vivo were identified by the study conducted in the same laboratory using a mouse model (Jayashankar et al. 2014).

7.3 In Vitro Anti-inflammatory of Main Flavonoids in Sea Buckthorn

To date, phytochemical studies have identified more than 90 types of flavonoids in sea buckthorn, including flavone aglycones and flavone glycosides. As the main chemical components of sea buckthorn, flavonoids have been discovered in the

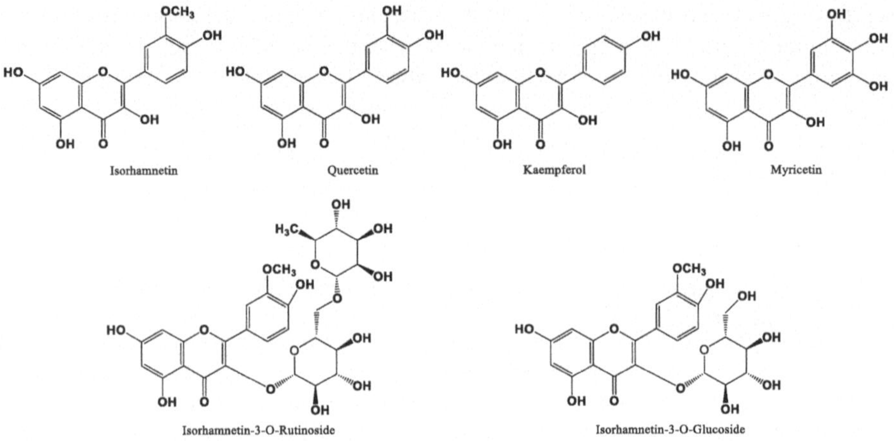

Fig. 7.1 The structure of main flavonoids including flavonol aglycones and flavonol glycosides in sea buckthorn. (Drawn by the author)

leaves, fruits, seeds, branches, and roots of sea buckthorn (He et al. 2023). Isorhamnetin, quercetin, keampferol, myricetin, and their flavonol glycosides, such as isorhamnetin-3-O-rutinoside (I3R) and isorhamnetin-3-O-glucoside (I3G), are the main flavonoids in sea buckthorn, which enable various bioactivities including anti-inflammatory activity (Fig. 7.1).

7.3.1 In Vitro Anti-inflammatory Activity of Isorhamnetin

Isorhamnetin, the most common flavonoid compound extracted from sea buckthorn, has been widely studied for its ability to modulate inflammatory responses. Isorhamnetin exerts anti-inflammatory effects on various diseases, such as osteoarthritis, periodontitis, anti-acute lung injury, anti-tuberculosis, and kidney protection, which makes it capable of regulating the production of inflammatory mediators, cytokines, and reactive oxygens species (ROS) (Gong et al. 2020). NF-κB pathway plays an important role in the anti-inflammatory effect of isorhamnetin. Isorhamnetin could suppress LPS-induced inflammation in human gum fibroblasts by inhibiting the activation of NF-κB, activating the Nrf2 signaling pathway, and upregulating the expression of Nrf2 and HO-1 (Qi et al. 2018). Dou et al. (2014) found that isorhamnetin improved the downregulation of NF-κB signaling and upregulation of xenobiotic metabolism as mediated by the pro-gestational hormone X receptor (PXR), which demonstrates its potential in treating inflammatory bowel disease (IBD). Isorhamnetin may also have therapeutic effects in the treatment of neuroinflammatory diseases. It could inhibit LPS-induced inflammation in BV2 microglia. The mechanisms are related to its effects on inactivating the NF-κB pathway, antagonizing TLR4, and eliminating ROS accumulation (Kim et al. 2019). Isorhamnetin significantly inhibited LPS-induced TNF-α, IL-1β, and IL-6 secretion in both RAW264.7 mouse macrophages and an acute lung injury mouse model (Chi et al. 2016). Shi et al. (2018) investigated the effects of isorhamnetin on dendritic cells, which are recognized as the major target of immunosuppressive therapies. In this study, isorhamnetin significantly downregulated various pro-inflammatory factors like TNF-α, IL-6, IL-1β, and IL-12 in LPS-treated dendritic cells, suggesting that its immunosuppressive capacity may make it suitable for the prevention or treatment chronic inflammation by inhibiting the activation and trafficking of dendritic cells.

7.3.2 In Vitro Anti-inflammatory Activity of Quercetin

As another type of flavonoid that is abundant in sea buckthorn, quercetin has attracted a lot of attention because of its anti-inflammatory activity against many diseases such as rheumatoid arthritis, acute kidney injury, and osteoarthritis (Wang et al. 2022a). A recent study demonstrated the effect of quercetin on excessive inflammation caused by beta-coronavirus, SARS-CoV-2. Quercetin inhibited the inflammation induced by LPS in lung epithelial cells A549 by suppressing the nuclear translocation of NF-κB and reducing NO production and the levels of TNF-α, IL-1, IL-6, and lipocalin-2 (Saeedi-Boroujeni and Mahmoudian-Sani 2021). Quercetin inhibited inflammation in lung epithelial cells A549 induced by LPS by suppressing nuclear translocation of NF-κB, reducing NO production and levels of TNF-α, IL-1, IL-6, and lipocalin-2 (Sul and Ra 2021). In another research, human umbilical vein endothelial cells were used to determine whether quercetin suppressed the inflammation induced by TNF-α and that the mechanisms were involved in NF-κB and AP-1 signaling pathways (Chen et al. 2020). In conclusion, NF-κB signaling pathway plays a key role in the anti-inflammatory activity of quercetin, which is like that of isorhamnetin. Many molecules involved in the inflammatory response are regulated by NF-κB, especially pro-inflammatory cytokines.

7.3.3 In Vitro Anti-inflammatory Activity of Kaempferol

Numerous previous studies have shown that kaempferol exerts a significant anti-inflammatory effect through different mechanisms such as the flavonoids as mentioned above. The NF-κB signaling pathway is related to the occurrence and development of inflammatory diseases (Yang et al. 2021). Kaempferol could reduce the levels of IL-8 in TNF-α-induced HEK 293 cells, a mechanism that involves the NF-κB signaling pathway (Lee et al. 2009). Tang et al. (2015) discovered that kaempferol could suppress NF-κB in ATP plus LPS-stimulated cardiac fibroblasts and inhibit the expression of various inflammatory cytokines including IL-1β, IL-6, IL-18, and TNF-α. The MAPK signaling pathway, which is composed of JNK,

ERK1/2, and p38 pathways, is related to the inflammatory response. Kaempferol inhibits microglial activation in LPS plus ATP-stimulated BV-2 cells through its inhibitory effect on the phosphorylation of p38, MAPK, and JNK, which subsequently suppresses the nuclear translocation of NF-κB to express various pro-inflammatory factors such as IL-1β and IL-18 (Liu et al. 2021). The anti-inflammatory effect of kaempferol is also related to other signaling pathways. Lim et al. (2018) found that kaempferol could inhibit the activation of the NLRP3 inflammasome, which has been confirmed to be related to the occurrence of various inflammatory diseases and effective in alleviating the relevant diseases. Kaempferol could also exert an anti-inflammatory effect by modulating aldosterone signaling (Liu et al. 2014).

7.3.4 In Vitro Anti-inflammatory Activity of Myricetin

Many studies have shown that myricetin produces different anti-inflammatory effects through multiple signal modulations in in vitro models. In bone marrow macrophages, myricitrin significantly suppressed osteoclast genesis induced by the activation of nuclear factor-kappaB ligand (RANKL) and its receptor RANK. Myricetin also attenuated RNAKL/MAPKs (phosphorylation of p38, ERK, and JNK) and AKT signal cascade, indicating the potential therapeutic effect of myricetin on inflammatory osteolysis (Wang et al. 2018). Chen et al. (2019) discovered that myricetin reduced cytokine production and exerted anti-inflammatory effects by regulating the TLR and mTOR signaling pathways in human sebocytes in vitro. NF-κB signaling pathway is also related to the anti-inflammatory effects of myricetin. Myricetin reduced the upregulation of cytokines and chemokines induced by TNF-α and UV light via NF-κB, Akt, and mTOR signaling pathways in keratinocytes, which suggests the potential therapeutic effect of myricetin against skin inflammation (Lee and Lee 2016; Xie and Zheng 2017). Another study conducted by Cho et al. suggested that myricetin upregulated the heme oxygenase-1 (HO-1) induction mediated by Nrf-2, suppressed NF-κB and STAT1 signaling pathways, and inhibited the production of inflammatory markers in LPS-stimulated RAW 264.7 cells (Cho et al. 2016).

7.3.5 In Vitro Anti-inflammatory Activity of Flavone Glycosides in Sea Buckthorn

I3R and I3G are the main flavone glycosides that can be extracted from sea buckthorn. According to the results of some previous in vitro studies, isorhamnetin-3-O-rutinoside has the potential to exert an anti-inflammatory effect. In LPS-stimulated

RAW 264.7 cells, I3R suppressed the secretion of pro-inflammatory cytokines (IL-6 and IL-10) as measured by enzyme-linked immunosorbent assay (ELISA) (Naksuriya et al. 2022). Fu et al. (2022) found that I3R inhibited the level of inflammatory cytokines (TNF-α, IL-1beta, and INL-6) in advanced glycation end products (AGEs)-induced RAW264.7 cells, and demonstrated its effectiveness in improving endothelial dysfunction in AGEs-induced EA. hy926 cells, which suggests its effects against inflammation and vascular endothelial dysfunction. I3G also showed anti-inflammatory properties through in vitro models. Kim et al. found that I3G suppressed the LPS-induced nitric oxide production and the expression of cytokines such as inducible iNOS, NF-κB, and IL-1beta in Raw 264.7 cells, showing the potential of using I3G as an anti-inflammatory agent due to its effect on pro-inflammatory responses (Kim et al. 2009). Another study showed that I3G demonstrated a notable concentration-dependent inhibitory effect on the production of inflammation mediators derived from arachidonic acid (AA) metabolism. They may be involved in COX-1 and 12-LOX pathways (Lesjak et al. 2018). Yang et al. (2013) isolated three new flavonol glycosides, along with 27 known constituents from sea buckthorn leaves, and evaluated their inhibitory activities against NO production in RAW264.7 cells, suggesting that isolated compounds may be the source of anti-inflammatory agents in sea buckthorn leaf extracts.

7.4 In Vitro Antioxidant Activity of Flavonoids Extracted from Sea Buckthorn

In the past few years, sea buckthorn plants have attracted extensive attention because of their ability to grow in the harshest environments. Besides morphological adaptation, this adaptability may also be associated with various antioxidants present in plants. Flavonoids play a key role in the antioxidant effect of sea buckthorn and are the main antioxidant components in sea buckthorn. Their antioxidant capacity is affected by the content of flavonoids in different parts of the body (Ma et al. 2023). The 2,2α-diphenyl-1-picrylhydrazil (DPPH) radical scavenging and the ferric reducing antioxidant power (FRAP) assays were performed to study the antioxidant capacity of different parts of sea buckthorn. The sea buckthorn seed extract showed the most significant antioxidant activity, followed by the root, leaf, and stem extract (Michel et al. 2012). Another study was conducted to investigate the antioxidant activities of the leaves, stems, and fruits of sea buckthorn. The antioxidant capacity of different tissues of sea buckthorn was found to be in the following order: leaf > stem > fruit, and that IC50 of scavenging DPPH free radicals was 217.8, 350.5, and 739.2 μg/mL, respectively (Wang et al. 2016b). Four different subspecies (subsp. *sinensis*, *yunnanensis*, *mongolica*, and *turkestanica*) of sea buckthorn were evaluated for their antioxidant capacity using oxygen radical absorbance capacity

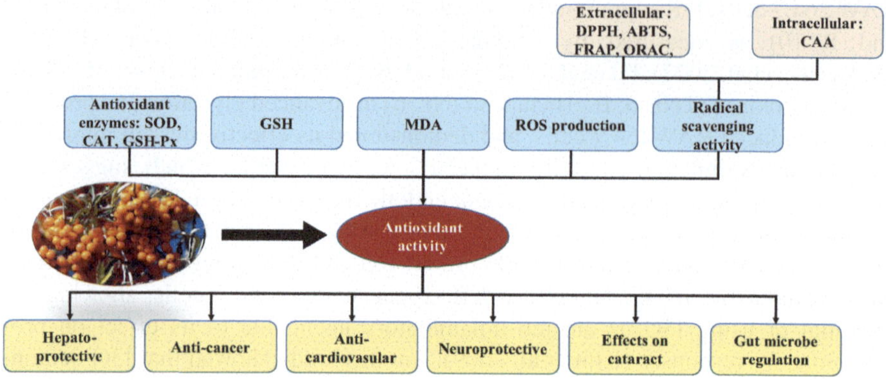

Fig. 7.2 Sea buckthorn showed antioxidant activity via radical scavenging and regulation of antioxidant enzymes, MDA, and ROS levels. (Drawn by the author)

(ORAC), peroxyl radical scavenging capacity (PSC), and intracellular antioxidant activity (CAA) assays. The subsp. *sinensis* showed the strongest antioxidant activity, which is related to the content of flavonoids and phenolic acid. Their ORAC, PSC, and CAA values were 369 μmol Trolox equiv. per g dry weight, 211 vitamin C equiv. per g dry weight, and 186 quercetin equiv. per 100 g dry weight, respectively (Guo et al. 2017b).

Sea buckthorn extract and its isolated flavonoids were found to protect against multiple injuries by suppressing oxidative stress, which is associated with the regulation of various antioxidant enzymes and the reduction of ROS levels (Fig. 7.2). Geetha et al. (2002) determined the antioxidant activity of sea buckthorn by using lymphocytes as the model system. Sea buckthorn extract suppressed the chromium-induced inhibition of lymphocyte proliferation, free radical production, and decrease in glutathione levels. Another study conducted in the same laboratory revealed that the sea buckthorn flavones extract with a concentration of 100 μg/mL not only significantly inhibited *tert*-butyl hydroperoxide-induced oxidative damage, such as cytotoxicity, free radical production, and DNA breaks, but also restored the antioxidant capacity to the same level as control lymphocytes (Geetha et al. 2009). In Raw 264.7 cells, sea buckthorn extract also suppressed *tert*-butyl hydroperoxide-induced cytotoxicity, reactive oxygen species (ROS) production and decrease in mitochondrial membrane potential, which is responsible for the reduced intracellular antioxidant levels (Kumar et al. 2011). In addition, both aqueous and hydroalcoholic extracts of sea buckthorn were observed to protect against hydrogen peroxide and hypoxanthine-xanthine oxidase-induced oxidant damage in BHK-21 cell line (Upadhyay et al. 2010).

7.4.1 In Vitro Antioxidant Activity of Flavonoids Extracted from the Seeds

Sea buckthorn seeds were found to be rich in flavonoids, ranging from 0 to 38 mg quercetin equivalent/g of DW, which shows the free-radical-scavenging activity and ferric reducing antioxidant potential (FRAP) (Dolkar et al. 2017). Arimboor and Arumughan (2012) evaluated the DPPH, 2,2'-azino-bis(3-ethylbenzothiazoline-6-sulfonic acid) (ABTS) radical scavenging, superoxide and hydroxyl radical scavenging, ferric reduction, ferrous chelation, and xanthine oxidase inhibitory capacities of ethanol extract from sea buckthorn seeds. Abundant in flavonoids, the ethanol extract demonstrated its antioxidant properties. After oil recovery, the residue of sea buckthorn seeds was extracted using subcritical water in which the yield of total phenolics and flavonoids reached 36.62 mg gallic acid equivalents/g and 19.98 mg rutin equivalent/g, respectively. In addition, the extract exhibited ABTS radical scavenging activity (32.42 mmol Trolox equivalent/100 g) (Gong et al. 2015).

According to the published in vitro studies, sea buckthorn seed extract could suppress oxidative stress. The sea buckthorn seeds are rich in flavonoids, and 76 types of flavonoids have been identified in the residue of *Hippophae rhamnoides* ssp. sinensis seeds. Its extract could not only suppress the production of reactive oxygen species and malondialdehyde, but also improve the superoxide dismutase activity and glutathione level to reverse alcohol-induced oxidative stress (Wei et al. 2023). Slawinska et al. (2023) studied the effect of two extracts from regular and roasted sea buckthorn seeds on the oxidative stress induced by H_2O_2/Fe^{2+} in plasma. Both extracts that contained various flavonoids inhibited lipid peroxidation and protein carbonylation, while the extract of roasted seeds decreased the oxidation of thiol groups in the plasma treated with H_2O_2/Fe^{2+}.

7.4.2 In Vitro Antioxidant Activity of Flavonoids Extracted from Berries

Ultrasound-assisted and accelerated solvent extraction using 80% acetone (v/v) as extraction solvent was used to extract sea buckthorn berries, and the results showed that the advanced methods were more efficient in extracting various flavonoids such as flavonols, flavan-3-ols, and flavones, resulting in more antioxidant capacity determined by ORAC assay (Culina et al. 2021). According to Dienaite et al. (2020), the sea buckthorn pomace defatted by supercritical CO_2 was subjected to consecutive fractionation with pressurized ethanol and water, the extracts in which flavonoids such as isorhamnetin, quercetin, kaempferol glycosides were the most abundant components, possessed high antioxidant activities by ORAC, ABTS, CAA assays. The antioxidant activity of ethanol extract from the press residues of sea buckthorn berries was investigated. It showed a high radical quenching capacity through TRAP, ORAC, and DPPH assays, suggesting its potential as a source of antioxidants for food-related use (Puganen et al. 2018). Olas et al. (2018) studied the antioxidant properties of phenolic fraction obtained from the sea buckthorn fruits rich in flavonoids. The extract inhibited the plasma lipid peroxidation induced by H_2O_2 and its activity depends on concentration. The antioxidant property of sea buckthorn berries was evaluated through in vitro digestion in comparison with chemical extraction. The CAA of sea buckthorn berries was enhanced by digestion while the extracellular antioxidant capacity of the digesta was lower than that of the extract. The mechanism is that more flavonoids aglycones were released and the cellular uptake might be improved after digestion (Guo et al. 2017a).

Sea buckthorn berries extract has also been reported to inhibit oxidative stress by mediating various antioxidant enzymes. Goel et al. (2005) found out that sea buckthorn berries extract significantly inhibited the radiation-induced increase in superoxide anions, oxidized glutathione (GSSG), thiobarbituric acid reactive substances (TEARS), NADH-ubiquinone oxidoreductase, NADH-cytochrome c oxidoreductase activity, and mitochondrial membrane potential (MMP). Additionally, it suppressed the oxidation of proteins in all periods.

7.4.3 In Vitro Antioxidant Activity of Flavonoids Extracted from Leaves

Phenolic compounds were extracted from sea buckthorn berries and leaves by using acidified aqueous ethanol, while the antioxidant activities of the extracts were evaluated through ORAC, DPPH, and TRAP assays. The leaf extract showed stronger antioxidant activity than the berries extract (5.96–6.90 fold in ORAC, 16.04–29.52 fold in TRAP), which is closely correlated with the content of flavonoids (Tian et al. 2018). A study was carried out to characterize the bioactive compounds in sea buckthorn leaves and sea buckthorn leaf tea. It was found that they were mainly composed of flavonoid glycosides with flavonol (such as isorhamnetin, quercetin, and kaempferol) as aglycone, showing strong antioxidant activities based on DPPH, ABTS, and ORAC results (Wang et al. 2022b). The antioxidant effect of aqueous and hydroalcoholic extracts from sea buckthorn leaves was investigated, and the sea buckthorn leaf extracts exhibited potent antioxidant activity as determined by ABTS, DPPH, and FRAP assays. Both extracts showed cytoprotective activity against hydrogen peroxide and hypoxanthine-xanthine oxidase-induced damage in the BHK-21 cell line (Upadhyay et al. 2010).

Similar to the extracts of sea buckthorn seeds and berries, the sea buckthorn leaf extract has also been reported to suppress oxidative stress. A study was conducted by Skalski et al. (2019a) to evaluate the antioxidant activity of sea buckthorn leaf extract by using various in vitro models. The sea buckthorn leaf extract was observed to exhibit an antioxidant effect against two strong biological oxidants: H_2O_2 and H_2O_2/Fe (the donor of hydroxyl radicals), thus inducing lipid peroxidation and protein carbonylation in human plasma. The antioxidant compounds (including flavonoids with a content of 47.06–66.03 mg/g rutin) were extracted from sea buckthorn leaves by using subcritical water, with antioxidant properties shown against the *tertiary*-butyl hydroperoxide (*tert*-BOOH)-induced oxidative stress in RAW 264.7. The pretreatment of cells with sea buckthorn leaf extract reversed the *tert*-BOOH-induced increase in cytotoxicity, ROS production, and the decrease in mitochondrial membrane potential (Kumar et al. 2011). Sharma et al. (2015) evaluated the antioxidant potential of aqueous and alcoholic extracts from the leaves of *Hippophae salicifolia* (HS), *Hippophae rhamnoides mongolica* (HRM), and *Hippophae rhamnoides turkestanica* (HRT). The supplementation of rats with aqueous extract of HS showed the maximum resistance to cold-hypoxia-restraint (C-H-R) stress-induced hypothermia, which is due to its effects on the level of various biochemical parameters such as MDA, ROS, lactate dehydrogenase (LDH), superoxide dismutase (SOD), GSH/GSSG and catalase (CAT).

7.5 In Vitro Antioxidant of Main Flavonoids of Sea Buckthorn

As mentioned above, the bioactive compounds of sea buckthorn, which are mainly composed of flavonoid aglycone, glycosides, isorhamnetin, quercetin, and kaempferol, are the primary flavonoid aglycones. These flavonols and their glycosides reportedly show not only antioxidant activities according to various extracellular assays such as DPPH, ABTS, TRAP, and ORAC, but also the capacity to inhibit oxidative stress in cellular models through the mediation of antioxidant enzymes and the reduction of ROS.

7.5.1 *In Vitro Antioxidant Activity of Isorhamnetin*

As the most abundant flavonoid aglycone, isorhamnetin has been widely studied by using different chemical methods (such as DPPH, ABTS, FRAP, and ORAC) and cell models for its antioxidant activities and inhibitory effects on oxidative stress (Novaes et al. 2019). Liu et al. (2009) extracted isorhamnetin from sea buckthorn and then purified it. The antioxidant capacities of the pure isolated isorhamnetin were investigated through the DPPH, iron (III)- to iron (II)-reducing, and iron-chelating assays. Compared with ascorbic acid and BHT (as the positive control), high-purity isorhamnetin (92.1%) exhibited strong antioxidant activity. Compared with the other 10 kinds of flavonoids such as tamarixetin, quercetin, myricetin, kaempferol, and EGCG, isorhamnetin showed the highest CAA value (212.02 ± 18.12 lmol QE/100 lmol) and the second highest ORAC value (87.56 ± 1.84 lmol QE/100 lmol) (Blasa et al. 2011).

Isorhamnetin also exhibits inhibitory effects on oxidative stress in cells. Isorhamnetin effectively protected hepatocytes against the oxidative stress induced by NF-E2-related factor 2 (Nrf2) activation and its associated pathway. The pre-treatment of HepG2 cells with isorhamnetin blocked *tert*-butylhydroquinone-simulated ROS production and reversed the GSH depletion induced by *tert*-butylhydroquinone, thus suppressing cell death (Yang et al. 2014). Sun et al. (2013) found out that isorhamnetin pretreatment could counteract various doxorubicin-induced oxidative stress such as the increase of ROS and MDA production, reverse the decrease in SOD, CAT, and GSH-Px activities in H9c2 cardiomyocytes, and inhibit the activation of mitochondrion apoptotic pathway and MAPK pathway. In another research, isorhamnetin was observed to prevent mouse-derived C2C12 myoblasts from H_2O_2-induced cytotoxicity by activating the Nrf2/HO-1 and ERK pathways and suppressing intracellular ROS (Choi 2016). In addition, isorhamnetin reportedly protects against the oxidative modification of high-density lipoprotein (HDL), very low-density lipoprotein (VLDL), and low-density lipoprotein (LDL) as induced by Cu^{2+} (Gong et al. 2020).

7.5.2 In Vitro Antioxidant Activity of Quercetin

Quercetin shows strong antioxidant properties, scavenges free radicals, and binds the transition metal ions related to its phenolic hydroxyl group. It also shows a higher rate of scavenging DPPH and ABTS free radicals than isorhamnodin at the same concentration (Xiao et al. 2012). ORAC, CAA, and PSC assays were performed to investigate some flavonoids including quercetin, while quercetin, whose PSC and CAA values were the highest, exhibited strong antioxidant activity (Wang et al. 2016a).

In recent years, the antioxidant activity of quercetin has been extensively studied, including its effects on GSH, antioxidant enzymes, and ROS as caused by environmental and toxicological factors. Quercetin can increase the antioxidant capacity of the body by regulating GSH levels. After the production of oxygen free radicals, SOD can rapidly capture O^{2-} and convert it into H_2O_2. Previous studies revealed that quercetin induced GSH synthesis and then degraded H_2O_2 by CAT, in which reactive GSH functioned as a hydrogen donor (Kobori et al. 2015; Xu et al. 2019). Celik et al. (2017) evaluated the protective effect of quercetin on the oxidative stress induced by homocysteinemia and found that MDA levels in the group treated with quercetin were significantly reduced, while GSH levels and CAT activity were significantly increased. In another study, quercetin not only improved the resistance of nematodes to acute stress but also extended the life span of nematodes. In addition, quercetin reduced the DNA strand breakage and apoptotic cell death caused by oxidative stress (Kampkoetter et al. 2008).

7.5.3 *In Vitro Antioxidant Activity of Kaempferol*

As the third most abundant flavonoid aglycone in sea buckthorn, kaempferol shows various pharmacological effects, most of which are related to its antioxidant capacity. DPPH, Trolox equivalent antioxidant capacity (TEAC), ORAC, ABTS, and FRAP assays were performed to evaluate the antioxidant activities of some flavonoids including kaempferol, with high antioxidant activity shown in all the assays (7.43 μg/mL, 1.10 mol TE/mol standard, 1.51 mol TE/mol standard, and 15.60 mol TE/mol standard for DPPH IC_{50}, DPPH, TEAC, and ORAC, respectively; 0.8506 μg/mL and 0.0263 mmol Fe^{2+}/μg/mL for ABTS IC_{50} and FRAP, respectively) (Arruda et al. 2018; Tian et al. 2021). Kaempferol exhibited strong cellular antioxidant ability in CAA assay and a CAA value of 59.80 μM of quercetin/100 μM (EC_{50} = 7.74 μM) (Liao et al. 2016).

The protective effects of kaempferol against intracellular ROS were also investigated. The pretreatment conducted with kaempferol significantly not only attenuated the ROS-induced accumulation of MDA, which is a kind of toxic lipid peroxidation product, but also reduced hemolysis in human erythrocytes. The mechanisms were associated with scavenging excessive ROS and preserving the intrinsic antioxidant activities exhibited by enzymes (SOD, CAT, and GPx) at normal levels (Liao et al. 2016). Wu et al. (2020) established a doxorubicin-toxic injury human umbilical vein endothelial cells (HUVECs) model and investigated the protective effect of kaempferol against doxorubicin toxicity. Doxorubicin causes excessive ROS generation, which induces endothelio toxicity and cardiotoxicity. The treatment with kaempferol reduced doxorubicin-induced damage in HUVECs, promoted the expression of mitochondrial Bcl-2 and phosphorylate Bad, increased cell viability and NO content, maintained NAD (+)/NADH and GSH/GSSG balance, and decreased LDH and ROS generation, which demonstrates that kaempferol protected the vascular endothelium against Dox-induced damage by inhibiting oxidative stress and improving mitochondrial function.

7.5.4 *In Vitro Antioxidant Activity of Myricetin*

Many studies have indicated that myricetin possesses antioxidant capacity. The radical scavenging capacities of myricetin and other plant-derived polyhydroxy compounds were analyzed through DPPH and ORAC assays. Myricetin showed high antioxidant capacity with 4.37 μM of DPPH EC_{50} and 6.76 mmol Trolox equivalent/mmol of ORAC value (Bertin et al. 2016). ABTS and FRAP assays were also performed by Ye et al. (2019) to measure the antioxidant activities of myricetin and other six kinds of flavonols. It was found that myricetin showed high radical scavenging capacity with 0.48 mol ascorbic acid equivalent/mol of ABTS value and 2.39 mol ascorbic acid equivalent/mol of FRAP value, suggesting that the structural characteristics (such as *p*-hydroxyl group, *m*-hydroxyl group, methoxy group, glycoside structure) might affect the antioxidant capacity.

Through cell models, myricetin has been reported in previous studies to reduce oxidative stress. According to the in vitro study on MCF-7 breast cancer cells, myricetin reduced the fluorescence activity of DCF, which indicates the decreased oxidation of DCF, and showed its ROS scavenging activity (Barzegar 2016). In the cultured HUVEC cells, myricetin reduced the LDL oxidation induced by H_2O_2 according to the TBARS assay and showed an inhibitory effect against ROS production (Bertin et al. 2016). The mechanisms may be related to the effect of myricetin on the antioxidant enzymes. Oxidative stress has been recognized as an important mediator in the pathogenesis of cataracts. Yang et al. (2020) found out that the treatment carried out with myricetin significantly increased the level of SOD, CAT, and GSH in H_2O_2-stressed human epithelial eye lens cells, thus reducing ROS production.

7.5.5 In Vitro Antioxidant of Flavone Glycosides in Sea Buckthorn

The antioxidant activities of six compounds isolated from sea buckthorn leaf extracts, namely, kaempferol-3-O-β-D-(6″-O-coumarin) glycoside, 1-feruloyl-β-D-glucopyranoside, isorhamnetin-3-O-glucoside, quercetin-3-O-β-D--glucopyranoside, quertetin-3-O-β-D-glucopyranosyl-7-O-α-L-rhamnopyranoside, and isorhamnetin-3-O-rutinoside, were evaluated. Among them, quercetin-3-O-β-D-glucopyranoside showed stronger reducing power and DPPH radical scavenging activity (EC50 = 1.86 μg/mL) (Kim et al. 2011). Chen et al. (2013) isolated isorhamnetin-β-D-glucopyranosyl-7-α-L-rhamnopyranoside (1), quercetin-β-D-glucopyranosyl-7-O-α-L-rhamnopyranoside (2), and kaempferol-β-D-glucopyranosyl-7-O-α-L-rhamnopyranoside) (3) from sea buckthorn berries for the first time. Compounds 1–3 showed antioxidant activities, with the DPPH IC_{50} values of 8.91, 4.26, and 30.90 μM; 2.89, 4.04, and 2.44 Trolox equivalent μM/μM for ABTS assay.

I3G is one of the main flavonol glycosides found in sea buckthorn, whose inhibitory effect on oxidative stress was measured in the previous study. I3G showed its effects against oxidative stress in selenite-simulated rat lenses in vitro by increasing SOD, CAT, and Ca^{2+}-ATPase activities, improving GSH levels, and reducing thiobarbituric acid reactive substance (TBARS) and calcium levels. It is implied that I3G has the potential to retard selenite cataracts through its antioxidant property (Devi et al. 2010). Two new isorhamnetin derivatives, namely isorhamnetin-3-O-β-glucoside-7-O-α-rhamnoside and isorhamnetin-3-O-β-glucoside-7-O-α-(3‴-isovaleryl)-rhamnoside, were isolated from sea buckthorn berries by Skalski et al. (2019b), showing various biological properties against the oxidative stress induced by H_2O_2/Fe (the donor of hydroxyl radicals) in lipid and protein, such as lipid peroxidation, protein carbonylation, and protein thiol groups.

References

Arimboor R, Arumughan C (2012) HPLC-DAD-MS/MS profiling of antioxidant flavonoid glycosides in sea buckthorn (Hippophae rhamnoides L.) seeds. Int J Food Sci Nutr 63(6):730–738

Arruda HS, Pereira GA, Pastore GM (2018) Brazilian Cerrado fruit araticum (*Annona crassiflora* Mart.) as a potential source of natural antioxidant compounds. Int Food Res J 25(5):2005–2012

Barzegar A (2016) Antioxidant activity of polyphenolic myricetin in vitro cell- free and cell-based systems. Mol Biol Res Commun 5(2):87–95

Bertin R, Chen Z, Marin R, Donati M, Feltrinelli A, Montopoli M, Zambon S, Manzato E, Froldi G (2016) Activity of myricetin and other plant-derived polyhydroxyl compounds in human LDL and human vascular endothelial cells against oxidative stress. Biomed Pharmacother 82:472–478

Blasa M, Angelino D, Gennari L, Ninfali P (2011) The cellular antioxidant activity in red blood cells (CAA-RBC): a new approach to bioavailability and synergy of phytochemicals and botanical extracts. Food Chem 125(2):685–691

Brandt EB, Sivaprasad U (2011) Th2 cytokines and atopic dermatitis. J Clin Cell Immunol 2(3)

Celik N, Vurmaz A, Kahraman A (2017) Protective effect of quercetin on homocysteine-induced oxidative stress. Nutrition 33:291–296

Chen C, Xu X-M, Chen Y, Yu M-Y, Wen F-Y, Zhang H (2013) Identification, quantification and antioxidant activity of acylated flavonol glycosides from sea buckthorn (*Hippophae rhamnoides ssp sinensis*). Food Chem 141(3):1573–1579

Chen KC, Yang CH, Li TT, Zouboulis CC, Huang YC (2019) Suppression of Propionibacterium acnes-stimulated proinflammatory cytokines by Chinese bayberry extracts and its active constituent myricetin in human sebocytes in vitro. Phytother Res 33(4):1104–1113

Chen T, Zhang X, Zhu G, Liu H, Chen J, Wang Y, He X (2020) Quercetin inhibits TNF-alpha induced HUVECs apoptosis and inflammation via downregulating NF-kB and AP-1 signaling pathway *in vitro*. Medicine (Baltimore) 99(38):e22241

Chi G, Zhong W, Liu Y, Lu G, Lu H, Wang D, Sun F (2016) Isorhamnetin protects mice from lipopolysaccharide-induced acute lung injury via the inhibition of inflammatory responses. Inflamm Res 65(1):33–41

Cho BO, Yin HH, Park SH, Byun EB, Ha HY, Jang SI (2016) Anti-inflammatory activity of myricetin from Diospyros lotus through suppression of NF-kB and STAT1 activation and Nrf2-mediated HO-1 induction in lipopolysaccharide-stimulated RAW264.7 macrophages. Biosci Biotechnol Biochem 80(8):1520–1530

Choi YH (2016) The cytoprotective effect of isorhamnetin against oxidative stress is mediated by the upregulation of the Nrf2-dependent HO-1 expression in C2C12 myoblasts through scavenging reactive oxygen species and ERK inactivation. Gen Physiol Biophys 35(2):145–154

Culina P, Cvitkovic D, Pfeifer D, Zoric Z, Repajic M, Elez Garofulic I, Balbino S, Pedisic S (2021) Phenolic profile and antioxidant capacity of selected medicinal and aromatic plants: diversity upon plant species and extraction technique. Processes 9(12):2207

Devi VG, Rooban BN, Sasikala V, Sahasranamam V, Abraham A (2010) Isorhamnetin-3-glucoside alleviates oxidative stress and opacification in selenite cataract in vitro. Toxicol In Vitro 24(6):1662–1669

Dienaite L, Pukalskas A, Pukalskiene M, Pereira CV, Matias AA, Venskutonis PR (2020) Phytochemical composition, antioxidant and antiproliferative activities of defatted sea buckthorn (*Hippophae rhamnoides* L.) berry pomace fractions consecutively recovered by pressurized ethanol and water. Antioxidants 9(4):274

Dolkar P, Dolkar D, Angmo S, Kumar B, Stobdan T (2017) Variability in phenolics, flavonoids and antioxidants in Seabuckthorn (*Hippophae rhamnoides* L.) seed from nine trans-Himalayan natural population. J Berry Res 7(2):109–116

Dou W, Zhang J, Li H, Kortagere S, Sun K, Ding L, Ren G, Wang Z, Mani S (2014) Plant flavonol isorhamnetin attenuates chemically induced inflammatory bowel disease via a PXR-dependent pathway. J Nutr Biochem 25(9):923–933

Esser D, Geleijnse JM, Matualatupauw JC, Dower JI, Kromhout D, Hollman PCH, Afman LA (2018) Pure flavonoid epicatechin and whole genome gene expression profiles in circulating immune cells in adults with elevated blood pressure: a randomised double-blind, placebo-controlled, crossover trial. PLoS One 13(4):e194229

Fu Y, Wang Y, Li Y, Zhang X, Yuan P, Xia K, Tan Z (2021) Study on anti-infammatory and anti-oxidant activities of favonoids from *Hippophae rhamnoides* L. China Food Addit 32:67–74

Fu Y, Jia Y, Sun Y, Liu X, Yi J, Cai S (2022) Dietary flavonoids alleviate inflammation and vascular endothelial barrier dysfunction induced by advanced glycation end products *in vitro*. Nutrients 14(5):1026

Ganju L, Padwad Y, Singh R, Karan D, Chanda S, Chopra MK, Bhatnagar P, Kashyap R, Sawhney RC (2005) Anti-inflammatory activity of Seabuckthorn (*Hippophae rhamnoides*) leaves. Int Immunopharmacol 5(12):1675–1684

Geetha S, Sai Ram M, Singh V, Ilavazhagan G, Sawhney RC (2002) Anti-oxidant and immunomodulatory properties of seabuckthorn (Hippophae rhamnoides)--an in vitro study. J Ethnopharmacol 79(3):373–378

Geetha S, Ram MS, Sharma SK, Ilavazhagan G, Banerjee PK, Sawhney RC (2009) Cytoprotective and antioxidant activity of seabuckthorn (*Hippophae rhamnoides* L.) flavones against tert-butyl hydroperoxide-induced cytotoxicity in lymphocytes. J Med Food 12(1):151–158

Goel HC, Gupta D, Gupta S, Garg AP, Bala M (2005) Protection of mitochondrial system by *Hippophae rhamnoides* L. against radiation-induced oxidative damage in mice. J Pharm Pharmacol 57(1):135–143

Gong Y, Zhang X, He L, Yan Q, Yuan F, Gao Y (2015) Optimization of subcritical water extraction parameters of antioxidant polyphenols from sea buckthorn (*Hippophae rhamnoides* L.) seed residue. J Food Sci Technol Mysore 52(3):1534–1542

Gong G, Guan YY, Zhang ZL, Rahman K, Wang SJ, Zhou S, Luan X, Zhang H (2020) Isorhamnetin: a review of pharmacological effects. Biomed Pharmacother 128:110301

Guo R, Chang X, Guo X, Brennan CS, Li T, Fu X, Liu RH (2017a) Phenolic compounds, antioxidant activity, antiproliferative activity and bioaccessibility of Sea buckthorn (*Hippophae rhamnoides* L.) berries as affected by in vitro digestion. Food Funct 8(11):4229–4240

Guo R, Guo X, Li T, Fu X, Liu RH (2017b) Comparative assessment of phytochemical profiles, antioxidant and antiproliferative activities of Sea buckthorn (*Hippophae rhamnoides* L.) berries. Food Chem 221:997–1003

He N, Wang Q, Huang H, Chen J, Wu G, Zhu M, Shao F, Yan Z, Sang Z, Cao L, Wei R, Ma Q (2023) A comprehensive review on extraction, structure, detection, bioactivity, and metabolism of flavonoids from sea buckthorn (*Hippophae rhamnoides* L.). J Food Biochem:4839124

Hou DD, Di ZH, Qi RQ, Wang HX, Zheng S, Hong YX, Guo H, Chen HD, Gao XH (2017a) Sea buckthorn (*Hippophae rhamnoides* L.) oil improves atopic dermatitis-like skin lesions via inhibition of NF-kappaB and STAT1 activation. Skin Pharmacol Physiol 30(5):268–276

Hou D, Wang D, Ma X, Chen W, Guo S, Guan H (2017b) Effects of total flavonoids of sea buckthorn (*Hippophae rhamnoides* L.) on cytotoxicity of NK92-MI cells. Int J Immunopathol Pharmacol 30(4):353–361

Jastrzab A, Skrzydlewska E (2019) Composition and biomedical relevance of sea buckthorn. Acta Pol Pharm 76(6):959–969

Jayashankar B, Mishra KP, Kumar MS, Udayasankar K, Misra K, Ganju L, Singh SB (2012) A supercritical CO(2) extract from seabuckthorn leaves inhibits pro-inflammatory mediators via inhibition of mitogen activated protein kinase p38 and transcription factor nuclear factor-kappaB. Int Immunopharmacol 13(4):461–467

Jayashankar B, Mishra KP, Ganju L, Singh SB (2014) Supercritical extract of Seabuckthorn Leaves (SCE200ET) inhibited endotoxemia by reducing inflammatory cytokines and nitric oxide synthase 2 expression. Int Immunopharmacol 20(1):89–94

Jiang F, Guan H, Liu D, Wu X, Fan M, Han J (2017) Flavonoids from sea buckthorn inhibit the lipopolysaccharide-induced inflammatory response in RAW264.7 macrophages through the MAPK and NF-kappaB pathways. Food Funct 8(3):1313–1322

Kampkoetter A, Timpel C, Zurawski RF, Ruhl S, Chovolou Y, Proksch P, Waetjen W (2008) Increase of stress resistance and lifespan of Caenorhabditis elegans by quercetin. Comp Biochem Physiol B Biochem Mol Biol 149(2):314–323

Kim YA, Kong C-S, Um YR, Lim S-Y, Yea SS, Seo Y (2009) Evaluation of Salicornia herbacea as a potential antioxidant and anti-inflammatory agent. J Med Food 12(3):661–668

Kim JS, Kwon Y-S, Sa YJ, Kim M-J (2011) Isolation and identification of sea buckthorn (*Hippophae rhamnoides*) phenolics with antioxidant activity and alpha-glucosidase inhibitory effect. J Agric Food Chem 59(1):138–144

Kim SY, Jin CY, Kim CH, Yoo YH, Choi SH, Kim GY, Yoon HM, Park HT, Choi YH (2019) Isorhamnetin alleviates lipopolysaccharide-induced inflammatory responses in BV2 microglia by inactivating NF-kappaB, blocking the TLR4 pathway and reducing ROS generation. Int J Mol Med 43(2):682–692

Kobori M, Takahashi Y, Akimoto Y, Sakurai M, Matsunaga I, Nishimuro H, Ippoushi K, Oike H, Ohnishi-Kameyama M (2015) Chronic high intake of quercetin reduces oxidative stress and induces expression of the antioxidant enzymes in the liver and visceral adipose tissues in mice. J Funct Foods 15:551–560

Kumar MS, Dutta R, Prasad D, Misra K (2011) Subcritical water extraction of antioxidant compounds from Seabuckthorn (*Hippophae rhamnoides*) leaves for the comparative evaluation of antioxidant activity. Food Chem 127(3):1309–1316

Lee DH, Lee CS (2016) Flavonoid myricetin inhibits TNF-alpha-stimulated production of inflammatory mediators by suppressing the Akt, mTOR and NF-kappa B pathways in human keratinocytes. Eur J Pharmacol 784:164–172

Lee S, Kim YJ, Kwon S, Lee Y, Choi SY, Park J, Kwon HJ (2009) Inhibitory effects of flavonoids on TNF-alpha-induced IL-8 gene expression in HEK 293 cells. BMB Rep 42(5):265–270

Lesjak M, Beara I, Simin N, Pintac D, Majkic T, Bekvalac K, Orcic D, Mimica-Dukic N (2018) Antioxidant and anti-inflammatory activities of quercetin and its derivatives. J Funct Foods 40:68–75

Liao W, Chen L, Ma X, Jiao R, Li X, Wang Y (2016) Protective effects of kaempferol against reactive oxygen species-induced hemolysis and its antiproliferative activity on human cancer cells. Eur J Med Chem 114:24–32

Lim H, Min DS, Park H, Kim HP (2018) Flavonoids interfere with NLRP3 inflammasome activation. Toxicol Appl Pharmacol 355:93–102

Liu P, Deng T, Hou X, Wang J (2009) Antioxidant properties of isolated isorhamnetin from the sea buckthorn marc. Plant Foods Hum Nutr 64(2):141–145

Liu ZK, Xiao HB, Fang J (2014) Anti-inflammatory properties of kaempferol via its inhibition of aldosterone signaling and aldosterone-induced gene expression. Can J Physiol Pharmacol 92(2):117–123

Liu Z, Yao X, Sun B, Jiang W, Liao C, Dai X, Chen Y, Chen J, Ding R (2021) Pretreatment with kaempferol attenuates microglia-mediate neuroinflammation by inhibiting MAPKs-NF-kappaB signaling pathway and pyroptosis after secondary spinal cord injury. Free Radic Bio Med 168:142–154

Lv H, Ren W, Zheng Y, Wang L, Lu G, Yi P, Ci X (2016) Tenuigenin exhibits anti-inflammatory activity via inhibiting MAPK and NF-kappaB and inducing Nrf2/HO-1 signaling in macrophages. Food Funct 7(1):355–363

Ma QG, He NX, Huang HL, Fu X-M, Zhang Z-L, Shu J-C, Wang QY, Chen J, Wu G, Zhu M-N, Sang Z-P, Cao L, Wei RR (2023) *Hippophae rhamnoides* L.: a comprehensive review on the botany, traditional uses, phytonutrients, health benefits, quality markers, and applications. J Agric Food Chem 71(12):4769–4788

Michel T, Destandau E, Le Floch G, Lucchesi ME, Elfakir C (2012) Antimicrobial, antioxidant and phytochemical investigations of sea buckthorn (*Hippophae rhamnoides* L.) leaf, stem, root and seed. Food Chem 131(3):754–760

Mishra KP, Chanda S, Karan D, Ganju L, Sawhney RC (2008) Effect of Seabuckthorn (*Hippophae rhamnoides*) flavone on immune system: an *in-vitro* approach. Phytother Res 22(11):1490–1495

Naksuriya O, Daowtak K, Tima S, Okonogi S, Mueller M, Toegel S, Khonkarn R (2022) Hydrolyzed flavonoids from Cyrtosperma johnstonii with superior antioxidant, antiproliferative, and anti-inflammatory potential for cancer prevention. Molecules 27(10):3226

Novaes P, Torres PB, Cornu TA, de Carvalho Lopes J, Ferreira MJP, dos Santos DYAC (2019) Comparing antioxidant activities of flavonols from Annona coriacea by four approaches. S Afr J Bot 123:253–258

Olas B, Zuchowski J, Lis B, Skalski B, Kontek B, Grabarczyk L, Stochmal A (2018) Comparative chemical composition, antioxidant and anticoagulant properties of phenolic fraction (a rich in non-acylated and acylated flavonoids and non-polar compounds) and non-polar fraction from *Elaeagnus rhamnoides* (L.) A. Nelson fruits. Food Chem 247:39–45

Padwad Y, Ganju L, Jain M, Chanda S, Karan D, Kumar Banerjee P, Chand Sawhney R (2006) Effect of leaf extract of Seabuckthorn on lipopolysaccharide induced inflammatory response in murine macrophages. Int Immunopharmacol 6(1):46–52

Puganen A, Kallio HP, Schaich KM, Suomela J-P, Yang B (2018) Red/green currant and sea buckthorn berry press residues as potential sources of antioxidants for food use. J Agric Food Chem 66(13):3426–3434

Qi F, Sun JH, Yan JQ, Li CM, Lv XC (2018) Anti-inflammatory effects of isorhamnetin on LPS-stimulated human gingival fibroblasts by activating Nrf2 signaling pathway. Microb Pathog 120:37–41

Ren R, Li N, Su C, Wang Y, Zhao X, Yang L, Li Y, Zhang B, Chen J, Ma X (2020) The bioactive components as well as the nutritional and health effects of sea buckthorn. RSC Adv 10(73):44654–44671

Saeedi-Boroujeni A, Mahmoudian-Sani MR (2021) Anti-inflammatory potential of Quercetin in COVID-19 treatment. J Inflamm Lond 18(1):3

Sharma P, Suryakumar G, Singh V, Misra K, Singh SB (2015) *In vitro* antioxidant profiling of seabuckthorn varieties and their adaptogenic response to high altitude-induced stress. Int J Biometeorol 59(8):1115–1126

Shi H, He J, Li X, Han J, Wu R, Wang D, Yang F, Sun E (2018) Isorhamnetin, the active constituent of a Chinese herb Hippophae rhamnoides L, is a potent suppressor of dendritic-cell maturation and trafficking. Int Immunopharmacol 55:216–222

Skalski B, Kontek B, Lis B, Olas B, Grabarczyk L, Stochmal A, Zuchowski J (2019a) Biological properties of *Elaeagnus rhamnoides* (L.) A. Nelson twig and leaf extracts. BMC Complement Altern Med 19:148

Skalski B, Lis B, Pecio L, Kontek B, Olas B, Zuchowski J, Stochmal A (2019b) Isorhamnetin and its new derivatives isolated from sea buckthorn berries prevent H_2O_2/Fe – induced oxidative stress and changes in hemostasis. Food Chem Toxicol 125:614–620

Slawinska N, Olas B (2023) Selected seeds as sources of bioactive compounds with diverse biological activities. Nutrients 15(1):187

Slawinska N, Zuchowski J, Stochmal A, Olas B (2023) Extract from sea buckthorn seeds-a phytochemical, antioxidant, and hemostasis study; effect of thermal processing on its chemical content and biological activity *in vitro*. Nutrients 15(3):686

Sul OJ, Ra SW (2021) Quercetin prevents LPS-induced oxidative stress and inflammation by modulating NOX2/ROS/NF-kB in lung epithelial cells. Molecules 26(22):6949

Sun J, Sun G, Meng X, Wang H, Luo Y, Qin M, Ma B, Wang M, Cai D, Guo P, Sun X (2013) Isorhamnetin protects against doxorubicin-induced cardiotoxicity *in vivo* and *in vitro*. PLoS One 8(5):e64526

Tang XL, Liu JX, Dong W, Li P, Li L, Hou JC, Zheng YQ, Lin CR, Ren JG (2015) Protective effect of kaempferol on LPS plus ATP-induced inflammatory response in cardiac fibroblasts. Inflammation 38(1):94–101

Tian Y, Puganen A, Alakomi HL, Uusitupa A, Saarela M, Yang B (2018) Antioxidative and antibacterial activities of aqueous ethanol extracts of berries, leaves, and branches of berry plants. Food Res Int 106:291–303

Tian C, Liu X, Chang Y, Wang R, Lv T, Cui C, Liu M (2021) Investigation of the anti-inflammatory and antioxidant activities of luteolin, kaempferol, apigenin and quercetin. S Afr J Bot 137:257–264

Tkacz K, Wojdylo A, Turkiewicz IP, Ferreres F, Moreno DA, Nowicka P (2020) UPLC-PDA-Q/TOF-MS profiling of phenolic and carotenoid compounds and their influence on anticholinergic potential for AChE and BuChE inhibition and on-line antioxidant activity of selected *Hippophae rhamnoides* L. cultivars. Food Chem 309:125766

Upadhyay NK, Kumar MS, Gupta A (2010) Antioxidant, cytoprotective and antibacterial effects of Sea buckthorn (*Hippophae rhamnoides* L.) leaves. Food Chem Toxicol 48(12):3443–3448

Wang L, Chen C, Su A, Zhang Y, Yuan J, Ju X (2016a) Structural characterization of phenolic compounds and antioxidant activity of the phenolic-rich fraction from defatted adlay (Coix lachryma-jobi L. var. ma-yuen Stapf) seed meal. Food Chem 196:509–517

Wang XF, Li H, Liu MP, Jiao HS (2016b) Study on antioxidant activity of different tissue parts of *Hippophae rhamnoides*. China Tradit Pat Med 38:437–440

Wang Y, Zhao L, Huo Y, Zhou F, Wu W, Lu F, Yang X, Guo X, Chen P, Deng Q, Ji B (2016c) Protective effect of proanthocyanidins from sea buckthorn (Hippophae rhamnoides L.) seed against visible light-induced retinal degeneration in vivo. Nutrients 8(5):245

Wang B, Hao D, Zhang Z, Gao W, Pan H, Xiao Y, He B, Kong L (2018) Inhibition effects of a natural inhibitor on RANKL downstream cellular signalling cascades cross-talking. J Cell Mol Med 22(9):4236–4242

Wang X, Cao Y, Chen S, Lin J, Bian J, Huang D (2021a) Anti-inflammation activity of flavones and their structure-activity relationship. J Agric Food Chem 69(26):7285–7302

Wang MY, Zhao H, Wen X, Ho CT, Li SM (2021b) Citrus flavonoids and the intestinal barrier: interactions and effects. Compr Rev Food Sci Food Saf 20(1):225–251

Wang G, Wang Y, Yao L, Gu W, Zhao S, Shen Z, Lin Z, Liu W, Yan T (2022a) Pharmacological activity of quercetin: an updated review. Evid Based Complement Alternat Med:3997190

Wang N, Wen X, Gao Y, Lu S, Li Y, Shi Y, Yang Z (2022b) Identification and characterization of the bioactive polyphenols and volatile compounds in sea buckthorn leaves tea together with antioxidant and alpha-glucosidase inhibitory activities. Front Nutr 9:890486

Wei J, Zhao J, Su T, Li S, Sheng W, Feng L, Bi Y (2023) Flavonoid extract from seed residues of *Hippophae rhamnoides* ssp. sinensis protects against alcohol-induced intestinal barrier dysfunction by regulating the Nrf2 pathway. Antioxidants 12(3):562

Wu W, Yang B, Qiao Y, Zhou Q, He H, He M (2020) Kaempferol protects mitochondria and alleviates damages against endotheliotoxicity induced by doxorubicin. Biomed Pharmacother 126:110040

Xiao Y, Yu Y, Yu X (2012) Study on antioxidant activity of isorhamnosine and quercetin. Lishizhen Med Mater Med Res 23(05):1118–1120

Xie J, Zheng Y (2017) Myricetin protects keratinocyte damage induced by UV through I kappa B/NF kappa B signaling pathway. J Cosmet Dermatol 16(4):444–449

Xu D, Hu MJ, Wang YQ, Cui Y-L (2019) Antioxidant activities of quercetin and its complexes for medicinal application. Molecules 24(6):1123

Yang ZG, Wen XF, Li YH, Matsuzaki K, Kitanaka S (2013) Inhibitory effects of the constituents of *Hippophae rhamnoides* on 3T3-L1 cell differentiation and nitric oxide production in RAW264.7 cells. Chem Pharm Bull 61(3):279–285

Yang JH, Shin BY, Han JY, Kim MG, Wi JE, Kim YW, Cho IJ, Kim SC, Shin SM, Ki SH (2014) Isorhamnetin protects against oxidative stress by activating Nrf2 and inducing the expression of its target genes. Toxicol Appl Pharmacol 274(2):293–301

Yang X, Wang Q, Pang ZR, Pan MR, Zhang W (2017) Flavonoid-enriched extract from Hippophae rhamnoides seed reduces high fat diet induced obesity, hypertriglyceridemia, and hepatic triglyceride accumulation in C57BL/6 mice. Pharm Biol 55(1):1207–1214

Yang Y, Xu X, Liu Q, Huang H, Huang X, Lv H (2020) Myricetin prevents cataract formation by inhibiting the apoptotic cell death mediated cataractogenesis. Med Sci Monit 26:e922519

Yang L, Gao Y, Bajpai VK, El-Kammar HA, Simal-Gandara J, Cao H, Cheng KW, Wang M, Arroo RRJ, Zou L, Farag MA, Zhao Y, Xiao J (2021) Advance toward isolation, extraction, metabolism and health benefits of kaempferol, a major dietary flavonoid with future perspectives. Crit Rev Food Sci Nutr:1–17

Ye X, Zhang C, Dai T, Chen J, Hu P, Li X, Li T (2019) Influence of flavonols with different structure on their antioxidant activity using multiple regression. Food Sci Technol 44(08):279–284

Zhang A, Sun H, Wang X (2013) Recent advances in natural products from plants for treatment of liver diseases. Eur J Med Chem 63:570–577

Chapter 8
Hypolipidemic Effects of Flavonoids from Sea Buckthorn Leaves

Xiaobo Wei, Yaoran Li, and Boyu Huang

Abstract The chemical composition of flavonoids in sea buckthorn leaves is summarized, and different extraction methods (organic solvent extraction, ultrasonic-assisted extraction, microwave extraction, and supercritical fluid extraction) and purification technologies (decolorization of macro-porous resin and decolorization of magnetic multi-walled carbon nanotubes) of flavonoids in sea buckthorn leaves are also introduced. In addition, the hypolipidemic effect of flavonoids from sea buckthorn leaves is summarized. Finally, the regulatory mechanism of flavonoids in mouse hypolipidemic effects is also explained.

Keywords Sea buckthorn leaves · Chemical composition · Flavonoids · Hypolipidemic effect · Regulatory mechanism

Abbreviations

HDL-C	High-density lipoprotein-cholesterol
LDL-C	Low-density lipoprotein-cholesterol
TC	Total cholesterol
TG	Triglycerides

X. Wei
Food Science and Technology, Ningxia University, Yinchuan, Ningxia, China

Y. Li (✉)
Food College, Ningxia University, Yinchuan, China

B. Huang
Department of Food Science and Nutrition, Zhejiang University, Hangzhou, China

J. Tian et al. (eds.), *Sea Buckthorn*, https://doi.org/10.1007/978-981-97-9865-0_8

8.1 Flavonoids of Sea Buckthorn Leaves

Sea buckthorn leaves are rich in flavonoids, polyphenols, polysaccharides, vitamins, trace elements, amino acids, and other substances. Among them, the flavonoids of sea buckthorn leaves have unique functions such as antioxidant (Criste et al. 2020; Wang et al. 2022), anticancer (Jia et al. 2020), hypoglycemic and hypolipidemic functions, and metabolism regulation. Therefore, the development and use of flavonoids in sea buckthorn leaves and the promotion of their industrialization have extensive application prospects.

8.2 Chemical Composition of Flavonoids in Sea Buckthorn Leaves

Both the fruits and leaves of sea buckthorn are rich in flavonoids, but the flavonoid content of sea buckthorn leaves is higher than that of sea buckthorn fruits (Ciesarova et al. 2020; Sytarova et al. 2020). Studies have shown that the total flavonoid content was 854 mg/100 g in fresh fruits and 3888 mg/100 g in dried leaves. As shown in Table 8.1, the flavonoids of the leaves include anthocyanins, quercetin, isorhamnetin, epicatechin, and flavonols (Bal et al. 2011; Ma et al. 2023).

8.3 Extraction Technology of Flavonoids from Sea Buckthorn Leaves

Flavonoids are important bioactive components in sea buckthorn, and the total flavonoid content in sea buckthorn leaves is the highest among all parts of sea buckthorn, it is important to investigate the extraction and purification process of total flavonoids from sea buckthorn leaves. Currently, the extraction methods of flavonoids mainly include organic solvent extraction, ultrasonic-assisted extraction, microwave extraction, and supercritical fluid extraction.

8.3.1 Organic Solvent Extraction

The extraction of flavonoids from sea buckthorn leaves using organic solvents is a more widely used traditional method based on the principle of similar solubility. The flavonoid of sea buckthorn leaves contains many multi-phenolic hydroxyl groups that have a more stable structure due to the existence of hydrogen and hydrophobic bonds. However, hydrophilic organic solvents can cause their hydrogen bonds to break so that they can be better soluble, and therefore, organic

Table 8.1 Main flavonoids in sea buckthorn leaves

Number	Compound name	Molecular formula	References
1	Isorhamnosin	$C_{16}H_{12}O_7$	Hibasami et al. (2005)
2	Isorhamn-3-O-rutin	$C_{28}H_{32}O_{16}$	Ji et al. (2020)
3	Isorhamn-3-O-glucoside	$C_{22}H_{22}O_{12}$	
4	Isorhamnin-3-O-galactosyl-7-O-ramnoside	$C_{28}H_{32}O_{16}$	
5	Quercetin 3-O-galactoside	$C_{21}H_{20}O_{11}$	
6	Isorhamn-3-O-rhamnoside	$C_{21}H_{18}O_{12}$	Ran and Li (2019)
7	Kaempferol-3-O-glucoside	$C_{21}H_{20}O_{11}$	
8	Myricetin	$C_{15}H_{10}O_8$	
9	Isorhamnosin-3-hexoside	$C_{22}H_{22}O_{12}$	Pop et al. (2013)
10	Isorhamnetin-3-O-neneneba neohesperidin	$C_{28}H_{32}O_{16}$	
11	Isorhamn-3-O-rhamnosylglucoside	$C_{28}H_{32}O_{16}$	
12	Isorhamnosin-3-O-rutin-7-O-glucoside	$C_{34}H_{42}O_{21}$	
13	Isorhamnosin-3-O-sophoroside 7-O-rhamnoside	$C_{34}H_{42}O_{21}$	
14	Isorhamnosin-3-O-glucoside 7-O-rhamnoside	$C_{28}H_{32}O_{16}$	
15	Isorhamnosin-3-O-mustard oil-based glucose-glucose-7-O-rhamnoside	$C_{45}H_{52}O_{25}$	
16	Isorhamnosin-3-O-neohesperidoside-7-O-glucoside	$C_{34}H_{42}O_{21}$	
17	Isorhamnosin-3,7-O-dihexoside	$C_{28}H_{32}O_{17}$	
18	Quercetin 3-O-pentoside	$C_{20}H_{18}O_{11}$	
19	Quercetin 3-O-hexoside	$C_{21}H_{20}O_{12}$	
20	Quercetin 3-O-rhamnoside	$C_{27}H_{30}O_{16}$	
21	Quercetin 3-O-sophoroside 7-O-rhamnoside	$C_{33}H_{40}O_{21}$	
22	Quercetin 3-O-glucoside 7-O-rhamnoside	$C_{27}H_{30}O_{16}$	
23	Quercetin 3-O-(2-rutinyl) glucoside	$C_{33}H_{40}O_{21}$	
24	Kaempferol 3-O-rutin	$C_{27}H_{30}O_{15}$	
25	Kaempferol-3-O-neohesperidin	$C_{27}H_{30}O_{15}$	
26	Rutin	$C_{27}H_{30}O_{16}$	
27	Isorhamnosin-3-O-arabinoside-7-O-rammnoside	$C_{27}H_{30}O_{15}$	Qin et al. (2016)
28	Isorhamn-3-O-sophoroside O-glucuronic acid 7-O-rhamnoside-7-O-rhamnoside	$C_{40}H_{50}O_{27}$	
29	Quercetin	$C_{15}H_{10}O_7$	
30	Quercetin 3-O-rhamnosylrhamnoside-7-O-glucoside	$C_{33}H_{40}O_{20}$	
31	Quercetin 7-O-glucoside	$C_{21}H_{20}O_{12}$	
32	Kaempferol-3-O-β-D-glucose-(1 → 2)-β-D-rhamnoside	$C_{27}H_{30}O_{15}$	
33	Kaempferol-3-O-rutin-7-O-rhamnoside	$C_{33}H_{40}O_{19}$	
34	Myricetin-3-O-α-L-rhamnose-(1 → 6)-β-D-glucoside	$C_{27}H_{30}O_7$	
35	Quercetin 3-O-glucoside	$C_{21}H_{20}O_{12}$	
36	Kaempferol	$C_{15}H_{10}O_6$	Cai et al. (2019)
37	Luteolin	$C_{15}H_{10}O_6$	

Drawn by the author

solvents are used to extract the flavonoids from sea buckthorn leaves. Yang et al. (2006) used ethanol, methanol, ethyl acetate, and acetone solvents to extract flavonoids from sea buckthorn leaves, and their research found that ethanol is the best solvent for extracting total flavonoids from sea buckthorn leaves, with the highest extraction rate of 1.96%. Moreover, Hui et al. (2017) studied the ethanol reflux extraction process of flavonoids from sea buckthorn leaves. The optimal process conditions optimized by the orthogonal design experiment were ethanol concentration of 70%, solid–liquid ratio of 1:16 (g/mL), extraction time of 2 h, and extraction thrice. Under these process conditions, the extraction rate of flavonoids from sea buckthorn leaves was 5.64%. The solvent extraction process is simple, convenient, and inexpensive. However, there are problems such as long-time consumption, large material loss, low extraction efficiency, and poor purity of flavonoid products.

8.3.2 Ultrasonic-Assisted Extraction

Ultrasonic treatment is a strengthening technique and is widely used to extract bioactive compounds from natural products. The enhancement process is based on the phenomenon of ultrasonic cavitation, including the formation of stable or transient bubbles caused by the compression and expander cycle of ultrasonic through the liquid, and the subsequent rupture, leading to the release of bioactive compounds. Ultrasonic can also cause the liquid to move from the shear force and turbulence caused by bubble implosion, thereby increasing the solubility of sea buckthorn leaf flavonoids in solvents (Yusoff et al. 2022).

Li et al. (2012) studied ultrasonic-assisted ethanol extraction of flavonoids from sea buckthorn leaves. The extraction rate of flavonoids from sea buckthorn leaves was 5.03% at 40% ethanol concentration, 20 kHz ultrasonic frequency, and 15 min at room temperature. Sun and Yang (2012b) also studied the ultrasonic-assisted ethanol extraction process of flavonoids from sea buckthorn leaves. Under the conditions of ethanol volume fraction of 70.64%, ultrasound power of 561.5 W, solid–liquid ratio of 1:35 g/mL, and ultrasound time of 30.7 min, the extraction rate of flavonoids from sea buckthorn leaves was 10.51 mg/g. In addition, Li et al. (2015) used ethanol hydrochloric acid solution as the extraction agent and determined the optimal extraction process conditions as follows: ethanol concentration of 50%, hydrochloric acid concentration of 3 mol/L, solid–liquid ratio of 1:20 g/mL, ultrasonic power of 80 W, and ultrasonic time of 30 min. Under these conditions, the flavonoid extraction rate reached 6.34%. The ultrasound-assisted extraction of flavonoids from sea buckthorn leaves has the advantages of low extraction temperature, less time, saving solvent, less impurity content of the proposed flavonoids, and a high extraction rate. However, the investment cost of ultrasonic equipment is huge, and the environmental pollution caused by ultrasonic noise exists. Thus, it is only used for laboratory research at present and industrial production is not yet available.

8.3.3 Microwave-Assisted Extraction

Microwave-assisted extraction is an advanced extraction technique that has become a common method for the isolation of bioactive compounds from plant matrixes. Microwave is a non-ionizing electromagnetic wave, between the frequency band of 300 MHz and 300 GHz, which more easily penetrates into sea buckthorn leaves and interacts with polar substances to generate heat, leading to changes in cell structure, thus improving the permeability of the solvent in the sample matrix. Jia et al. (2021) found that under the extraction conditions of microwave power at 325 W, microwave time at 3.5 min, ionic liquid concentration with 0.78 mol/L, material–liquid ratio with 1:30 g/mL, and aqueous ethanol concentration at 80%, the extraction rate of total flavonoids of sea buckthorn leaves was 8.11%. Cui et al. (2018) conducted the experimental design and process optimization of microwave-assisted low-eutectic solvent extraction of rutin, quercetin-3-0-glucose, quercetin, kaempferol, and isorhamnetin from sea buckthorn leaves. They determined the optimal extraction process parameters as follows: low-eutectic solvent of choline chloride/1,4-butanediol of 1:3, extraction temperature of 64 °C, extraction time of 18 min, and liquid-to-solid ratio of 20:1. The average extraction rates of rutin, quercetin-3-0-glucose, scutellarin, kaempferol, and isorhamnetin were 8.967, 1.809, 9.107, 0.448, and 0.489 mg/g, respectively, when the sea buckthorn leaves were extracted under the above-optimized conditions. Moreover, the combination of ultrasound and microwave-assisted extraction is more conducive to the extraction of flavonoids from sea buckthorn leaves. Under the optimized conditions of extraction (ionic liquid of 1.0 M 1-butyl-3methyl imidazole bromine salt, liquid/solid ratio of 12 ml/g, extraction time of 34 min, microwave power of 540 W and a fixed ultrasonic power of 50 W), the extraction yields of rutin, quercetin, kaempferol, and isorhamnetin were 9.18, 5.52, 3.03, 5.64 mg/g, respectively (Li et al. 2018). Microwave-assisted extraction has the advantages of easy processing, significantly shortening extraction time, and effectively reducing economic and environmental costs, which is a new technology for flavonoid extraction with great potential for development. However, because the research and development of microwave extraction equipment suitable for industrial production is relatively late, the microwave-assisted extraction of flavonoids from sea buckthorn leaves is only carried out in the laboratory, and yet to reach the scale of industrial production.

8.3.4 Supercritical Fluid Extraction

The process of supercritical fluid extraction of flavonoids from sea buckthorn leaves consists of two parts: extraction and separation. The solubility of supercritical fluid varies under different pressures and temperatures. The supercritical fluid meets sea buckthorn leaves to extract flavonoids by controlling the pressure and temperature.

Next, the supercritical fluid is transformed into a common gas by changing the pressure and temperature, ultimately resulting in rapid precipitation of sea buckthorn leaf flavonoids. There are no studies reported about the application of supercritical technology in the extraction of sea buckthorn leaf flavonoids, but it has been reported in other plants including *Passiflora* leaves (Moraes et al. 1997) and *Odontonema striatum* leaves (Ouédraogo et al. 2018). Therefore, the application of supercritical extraction technology in the extraction of flavonoids from sea buckthorn leaves has great prospects for application.

8.3.5 Subcritical-Assisted Extraction

Subcritical water extraction technology involves controlling the temperature and pressure of subcritical water, changing the polarity, surface tension, and viscosity of water, and extracting solutes from high to low according to their polarity. Extracting antioxidant components from sea buckthorn leaves using subcritical water extraction technology proves that this technology has a high flavonoid extraction rate, short extraction time, high efficiency, and good environment without damaging the antioxidant components in the extract. Kumar et al. (2011) used subcritical water extraction to extract antioxidant components from sea buckthorn leaves with a total flavonoid content of 47.06–66.03 mg/g rutin.

8.3.6 Enzyme-Assisted Extraction

The principle of enzymatic digestion is to use suitable enzymes to break down parts of the cell wall and intercellular matrix, thus causing damage to the cell wall and accelerating the leaching of active ingredients. The cell wall of sea buckthorn leaves is composed of cellulose and other substances, and the treatment of sea buckthorn leaves using enzymes related to cell wall degradation causes the cell wall structure to be disrupted, resulting in faster leaching of flavonoids. Zhu et al. (2008) studied the extraction process of flavonoids from sea buckthorn leaves using cellulase assistance. The optimal extraction process conditions for flavonoids from sea buckthorn leaves were determined by orthogonal design experiments, with cellulase addition of 4%, solid–liquid ratio of 1:50 g/mL, temperature of 40 °C, and time of 2 h. Enzyme-assisted extraction has the advantages of mild extraction conditions, high extraction rate, and favorable protection of flavonoid activity, but there are few relevant studies in the extraction of sea buckthorn leaf flavonoids. Thus, enzymatic-assisted extraction of flavonoids from sea buckthorn leaves will be a future research hotspot and receive attention from researchers.

8.4 Purification Technology of Flavonoids from Sea Buckthorn Leaves

During the extraction process, the sea buckthorn leaf extract appeared dark green due to the presence of impurities such as chlorophyll and carotenoids. Meanwhile, further application of sea buckthorn leaf flavonoids is also affected by the pigmentation which seriously affects the quality of the finished product and thus the biological activity is greatly affected. Therefore, decolorization has become an important part of the flavonoid extraction, separation, and purification process. At present, the common decolorization methods of flavonoids from sea buckthorn leaves include decolorization of macro-porous resin and the decolorization of magnetic multi-walled carbon nanotubes.

8.4.1 Decolorization of Macro-porous Resin

The macro-porous adsorption resin is mainly made of methyl phenyl ether, phenyl ethylene, methyl propionate methacrylate, and propionitrile, which is a synthetic non-gel-type polymer with a porous steric structure. The macro-porous resin has the advantages of high recovery, good selectivity, mild adsorption conditions, low operating cost, and easy regeneration. Thus, it is widely used for the separation and purification of flavonoids.

Kang (2019) found that AB-8-type macro-porous resin has better adsorption and desorption properties for flavonoids in sea buckthorn leaves compared to D101, X-5, DA201, HPD600, DM-301, DM-130, and NKA-9-type macro-porous resins. The dynamic adsorption conditions were as follows: the loading liquid concentration was 0.8 mg/ml, the flow velocity of the loading liquid was 1.8 BV/h, and the pH value of the loading liquid was 4.8. The dynamic desorption conditions were as follows: the ethanol concentration was 70% in elution, the elution dosage was 4.0 BV, and the flow velocity of elution was 1 BV/h. Under these conditions, the content of flavonoids in the adsorption–desorption experiment of the macro-porous resin is increased by 2.4 times. In contrast, Sun (2012a) found that comparing the static adsorption and elution rates of total flavonoids from sea buckthorn leaves with resins D3520, X-5, and AB-8, the best static adsorption and elution conditions were determined for the X-5 adsorption resin with the optimal adsorption time of 3 h, adsorption temperature of 25 °C, optimal elution time of 2.5 h, resolution temperature of 25 °C, and ethanol solvent concentration of 65%. The optimum adsorption and elution conditions of the resin X-5 were: the concentration of the upper sample was 2.5 mg/mL and the adsorption rate was 2BV/h. The optimum desorption conditions were: the concentration of ethanol was 75%, the desorption rate was 2BV/h, and the elution volume was 3.0BV. The purity of total flavonoids of sea buckthorn leaves obtained under these conditions was 89.72%. Because of the wide variety of macro-porous adsorption resin, it provides a wide space for the screening of flavonoids, coupled with its large adsorption capacity, simple operation, easy regeneration, stable product quality, and easy to achieve industrial production, so macroporous adsorption resin technology has become a key technology for the purification of flavonoids.

8.4.2 Decolorization of Magnetic Multi-walled Carbon Nanotubes

The utilization of nanoparticles in adsorption processes is becoming increasingly important as they exhibit better adsorption properties due to their high porosity. Carbon nanotubes are ideal adsorbents for the removal of natural product pigments due to their rich porous structure and large surface area. Carbon nanotubes are available in various forms, namely single-walled (SWCNT), double-walled (DWCNT), and multi-walled (MWCNT). MWCNT consists of multiple carbon nanotubes nested in each other. Meanwhile, multi-walled carbon nanotubes are tubular structures formed by multiple concentric graphene layers through interlayer van der Waals forces. Compared with single-walled and double-walled carbon nanotubes, multi-walled carbon nanotubes can be fabricated in large quantities and are easier to purify. Yang (2022) synthesized three multi-walled carbon nanotubes with different contents of $MnFe_2O_4$, and the results showed that the best comprehensive performance was achieved at 50% of $MnFe_2O_4$ addition. Then the decolorization process of MWCNT was further optimized, and the best decolorization conditions were employed as follows: decolorization temperature of 60 °C, decolorization time of 30 min, and MWCNT dosage of 1.0%. Magnetic multi-walled carbon nanotubes are a new technology developed in recent years, which will have broader development space in the purification of flavonoids from sea buckthorn leaves.

8.5 Hypolipidemic Effect of Flavonoids from Sea Buckthorn Leaves

Hyperlipidemia is a global health risk factor whose development is closely related not only to lipid absorption and metabolism in the intestine but also to intestinal damage and microbiota changes caused by a high-fat diet (Bai et al. 2023). Flavonoids can significantly reduce triglycerides (TG) and low-density lipoprotein-cholesterol (LDL-C) levels and effectively correct lipid metabolic homeostasis. In addition, it can prevent and treat hyperlipidemia by inhibiting endogenous lipid biosynthesis and promoting lipid redistribution. Thus, flavonoids from sea buckthorn leaves are a potential functional factor for hypolipidemia.

8.5.1 *Effects of Flavonoids from Sea Buckthorn Leaves on Body Weight in Hyperlipidemic Mice*

A typical feature of hyperlipidemic symptoms of obesity in mice is induced by increased body weight. Sun (2012) evaluated the hypolipidemic pharmacological activity of flavonoids from sea buckthorn leaves using a high-fat mouse model. A model group was established using a high-fat diet and a positive control group was established with a 5 mg/kg/d vincristine solution. With the extension of modeling time, the body weight of the mice in the model group increased substantially and was significantly different from that of the normal control group, while the body weights of the mice in the positive control, low-dose, medium-dose, and high-dose groups were not significantly different from that of the control group (Table 8.2). These results indicated that flavonoids from sea buckthorn leaves can effectively alleviate weight gain in hyperlipidemic mice.

8.5.2 *Effects of Flavonoids from Sea Buckthorn Leaves on Serum Biochemical Indicators in Hyperlipidemic Mice*

Lipids are fatty substances in the blood, including total cholesterol (TC), TG, LDL-C, and high-density lipoprotein-cholesterol (HDL-C). They are important indicators of blood lipids, which exist in the blood in the form of lipoproteins, such as non-free fatty acids, combined with different proteins. High lipid content in the body will cause lipid deposits in the lining of the arteries, which will result in reduced vascular elasticity and increased blood viscosity, prompting platelet aggregation and resulting in atherosclerosis, leading to the occurrence of many diseases. Sun (2012) used the method of preventive hyperlipidemia model mice to study the effect of sea buckthorn leaf flavonoids on blood lipids in mice. The study showed that the mice with different doses of sea buckthorn leaf flavonoids by gavage had lower levels of TC, TG, and LDL-C in serum, while high-density lipoprotein (HDL-C) was significantly increased, especially in the middle- and high-dose groups, compared with the

Table 8.2 The effect on rat body weight of flavonoids of sea buckthorn leaves

Time (week)	Control group	Model group	Positive group	Low-dose group	Middle-dose group	High-dose group
0	21.08 ± 0.84	22.74 ± 0.93	20.79 ± 0.90	21.05 ± 0.77	20.33 ± 0.53	20.79 ± 0.84
1	23.52 ± 1.20	30.07 ± 0.76	25.45 + 1.20	28.86 ± 1.01	25.22 ± 0.89	23.58 ± 0.63
2	27.36 ± 0.86	36.52 ± 0.51	26.56 + 0.97	35.72 ± 1.00	29.86 ± 0.53	27.43 ± 0.62
3	30.59 ± 0.71	41.17 ± 0.34	29.92 + 0.69	39.63 ± 0.63	32.32 ± 0.55	29.78 ± 0.41
4	33.39 ± 0.69	50.85 ± 0.52	32.19 ± 0.47	40.50 ± 0.52	34.29 ± 0.62	34.87 ± 0.79

Drawn by the author

hyperlipidemia model mice. Besides, Wei et al. (2016) also studied the effect of sea buckthorn leaf flavonoids on blood lipid levels in experimental hyperlipidemic rats and found that the levels of TC, TG, and LDL-C in the serum of rats in each dose group of flavonoids from sea buckthorn leaves were significantly lower than those in the hyperlipidemic model control group and the level of HDL-C was significantly higher than that of the model control group. These results indicate that flavonoids from sea buckthorn leaves exhibit a significant hypolipidemic effect.

8.5.3 *Effects of Flavonoids from Sea Buckthorn Leaves on Adipogenic and Lipogenic Gene Expression in Hyperlipidemic Mice*

A high-fat diet induces adipocyte hypertrophy and proliferation, leading to weight gain and ultimately obesity. Acetyl-CoA carboxylase (ACC) is the rate-limiting enzyme for fatty acid synthesis. At the same time, carnitine palmitoyltransferase-1 (CPT-1) and peroxisome proliferator-activated receptor-α (PPAR-α) are involved in mitochondrial fatty acid oxidation. Pichiah et al. (2012) found that sea buckthorn leaf extract significantly decreased the expression levels of ACC genes in the liver compared with the hyperlipidemia model group, while sea buckthorn leaf extract significantly increased the expression levels of CPT-1 and PPAR-α genes compared with the control group. Thus, the extract of sea buckthorn leaves prevents high-fat diet-induced obesity in mice through the downregulation of adipogenic and lipogenic gene expression.

8.5.4 *Possible Pathways of Flavonoids Sea Buckthorn Leaves in Alleviating Hyperlipidemia*

Although there is insufficient literature to confirm the pathway of action of sea buckthorn leaf flavonoids in the treatment of hyperlipidemia, we hypothesize, based on the reported literature, that the treatment of hyperlipidemia with sea buckthorn leaf flavonoids may be related to the PPAR signaling pathway associated with lipid metabolism. Fatty acid binding protein (FABP) acts on liver skeletal muscle and adipocytes within it, which in turn stimulates peroxisome proliferator-activated receptor family proteins (PPAR-α, PPAR-γ, PPAR-β), which further activate various pathways such as ketogenesis, lipid transport, lipogenesis, and cholesterol production, thus performing various biological functions related to lipid metabolisms such as fatty acid transport, bile acid biosynthesis, and glucose transport. These proteins further activate various pathways such as ketogenesis, lipid transport, lipogenesis, and cholesterol production, thus performing various biological functions related to lipid metabolism such as fatty acid transport, bile acid biosynthesis, and glucose transport. Xiao et al. (2021) used network pharmacology analysis and in vitro experiments to identify the hypolipidemic mechanism of sea buckthorn flavonoids. The results showed that sea buckthorn flavonoids increase the expression levels of PPAR-γ, PPAR-α, CPT-1, liver X receptor-α (LXR-α), and ATP binding cassette subfamily A member 1 (ABCA1) and reduce the expression levels of sterol regulatory element-binding protein-2 (SREBP-2) and its target gene low-density lipoprotein receptor (LDLR), promote cholesterol conversion to bile acid and cholesterol outflow, inhibit cholesterol de novo synthesis, accelerate fatty acid oxidation, and then treat hyperlipidemia. Therefore, further experiments need to be done later to confirm this speculation.

References

Bai Y, Yue ZL, Wang Y, Li Y, Li C, Liu X, Shi R, Huo N, Li D, Gao S, Han X (2023) Synergistic effect of polysaccharides and flavonoids on lipid and gut microbiota in hyperlipidemic rats. Food Funct 14:921–933

Bal LM, Meda V, Naik SN, Satya S (2011) Sea buckthorn berries: a potential source of valuable nutrients for nutraceuticals and cosmoceuticals. Food Res Int 44:1718–1727

Cai S, Ruan C, Du W, Ding J, Han P, Wang H (2019) Difference in flavonoid constituents in leaf, pulp, and seed of Hippopha rhamnoides. J Plant Resour Environ 28:58–67

Ciesarova Z, Murkovic M, Cejpek K, Kreps F, Tobolkova B, Koplik R, Belajova E, Kukurova K, Dasko L, Panovska Z, Revenco D, Burcova Z (2020) Why is sea buckthorn (*Hippophae rhamnoides* L.) so exceptional? A review. Food Res Int 133:109170

Criste A, Urcan AC, Bunea A, Pripon Furtuna FR, Olah NK, Madden RH, Corcionivoschi N (2020) Phytochemical composition and biological activity of berries and leaves from four romanian sea buckthorn (*Hippophae rhamnoides* L.) varieties. Molecules 25:1170

Cui Q, Liu JZ, Wang LT, Kang YF, Meng Y, Jiao J, Fu YJ (2018) Sustainable deep eutectic solvents preparation and their efficiency in extraction and enrichment of main bioactive flavonoids from sea buckthorn leaves. J Clean Prod 184:826–835

Hibasami H, Mitani A, Katsuzaki H, Imai K, Yoshioka K, Komiya T (2005) Isolation of five types of flavonol from seabuckthorn (Hippophae rhamnoides) and induction of apoptosis by some of the flavonols in human promyelotic leukemia HL-60 cells. Int J Mol Med 15:805–809

Hui R, Feng J, Lin M, Feng B (2017) Optimization of extraction technology for flavonoids in leaves and seeds of *Hippophae rhamnoides* by multiindex-orthogonal test. China Pharm 28:4856–4859

Ji M, Gong X, Li X, Wang C, Li M (2020) Advanced research on the antioxidant activity and mechanism of polyphenols from hippophae species-a review. Molecules 25:917

Jia CO, Du YA, Sun K (2020) Study on effect and mechanism of total flavonoids of Hippophae in inhibiting proliferation and migration of lung cancer A549 cells. Nat Prod Res Dev 32:937–945

Jia X, Zhu M, Zhao S (2021) Microwave-assisted extraction of total flavoniods from seabuckthorn leaves using ternary solvents. J Shanxi Uni 44:771–780

Kang Y (2019) Optimization of the extraction and refinement process of Chinese sea buckthorn leaf flavonoids and preliminary evaluation of their activity. Thesis for M.S., Beijing University of Chinese Medicine, pp 67–83

Kumar MSY, Dutta R, Prasad D, Misra K (2011) Subcritical water extraction of antioxidant compounds from seabuckthorn (*Hippophae rhamnoides*) leaves for the comparative evaluation of antioxidant activity. Food Chem 127:1309–1316

Li F, Liu H, Zhong Y, Liu Y (2012) Technology of total polyphenols and total flavones extracted from sea buckthorn leaves. Food Mach 28:128–130

Li S, Wu F, Chen Y, Yang Y (2015) The optimization of purification technics of total flavonoids from leaves of *Hippophae rhamnoides* L. J Anhui Norm Uni 38:567–571

Li C, Zhang J, Zhao C, Yang L, Zhao W, Jiang H, Ren X, Su W, Li Y, Guan J (2018) Separation of the main flavonoids and essential oil from seabuckthorn leaves by ultrasonic/microwave-assisted simultaneous distillation extraction. R Soc Open Sci 5:180133

Ma QG, He NX, Huang HL, Fu XM, Zhang ZL, Shu JC, Wang QY, Chen J, Wu G, Zhu MN, Sang ZP, Cao L, Wei RR (2023) *Hippophae rhamnoides* L.: A comprehensive review on the botany, traditional uses, phytonutrients, health benefits, quality markers, and applications. J Agric Food Chem 71:4769–4788

Moraes MDLL, Vilegas JHY, Lanças FM (1997) Supercritical fluid extraction of glycosylated flavonoids from *Passiflora* leaves. Phytochem Anal 8:257–260

Ouédraogo JCW, Dicko C, Kini FB, Bonzi-Coulibaly YL, Dey ES (2018) Enhanced extraction of flavonoids from Odontonema strictum leaves with antioxidant activity using supercritical carbon dioxide fluid combined with ethanol. J Supercrit Fluids 131:66–71

Pichiah PB, Moon HJ, Park JE, Moon YJ, Cha YS (2012) Ethanolic extract of seabuckthorn (*Hippophae rhamnoides* L) prevents high-fat diet-induced obesity in mice through down-regulation of adipogenic and lipogenic gene expression. Nutr Res 32:856–864

Pop RM, Socaciu C, Pintea A, Buzoianu AD, Sanders MG, Gruppen H, Vincken JP (2013) UHPLC/PDA-ESI/MS analysis of the main berry and leaf flavonol glycosides from different Carpathian *Hippophae rhamnoides* L. varieties. Phytochem Anal 24:484–492

Qin Z, Zhang Y, Qi M, Zhang Q, Liu S, Li Y, Liu Y, Liu Y (2016) Rapid analysis of compounds in leaves of Chinese seabuckthorn and Tibetan seabuckthorn by UPLC/Q-TOF-MS. China J Chin Mater Med 41:1461–1468

Ran BB, Li WD (2019) Research progress on chemical constituents and their differences between sea buckthorn berries and leaves. China J Chin Mater Med 44:1767–1773

Sun T (2012a) Extraction and separation of flavonoids from Hippophae rhamnoides leaves and hypolidemic effect of mouse experiments. Thesis for M.S., Heilongjiang Bayi Agricultural University, pp 25–35

Sun T, Yang H (2012b) Process of optimal extraction of *Hippophae rhamnoides* leaves flavonoids by ultrasoundassisted. J Heilongjiang Bayi Agric Uni 24:45–50

Sytarova I, Orsavova J, Snopek L, Mlcek J, Byczynski L, Misurcova L (2020) Impact of phenolic compounds and vitamins C and E on antioxidant activity of sea buckthorn (*Hippophae rhamnoides* L.) berries and leaves of diverse ripening times. Food Chem 310:125784

Wang Z, Wang W, Zhu C, Gao X, Chu W (2022) Evaluation of antioxidative and neuroprotective activities of total flavonoids from sea buckthorn (*Hippophae rhamnoides* L.). Front Nutr 9:861097

Wei Z, Zuo F, Tong D, Zeng Y, Meng X (2016) Effect of sea buckthorn leaf flavonoids capsule on experimental hyperlipidemia rats. Pharm Clin Chin Mater Med 7:33–34

Xiao PT, Liu SY, Kuang YJ, Jiang ZM, Lin Y, Xie ZS, Liu EH (2021) Network pharmacology analysis and experimental validation to explore the mechanism of sea buckthorn flavonoids on hyperlipidemia. J Ethnopharmacol 264:113380

Yang J (2022) Study on extraction and decolorization of flavonoids from sea buckthorn leaves. Thesis for M. S., Shihezi University, pp 49–63

Yang X, Chen M, Zhu L, Li H, Yin L (2006) Study on ultrasonic circulation extraction of total flavonoids from sea buckthorn leaves. Trans Chin Soc Agric Mach 37:166–168

Yusoff IM, Mat Taher Z, Rahmat Z, Chua LS (2022) A review of ultrasound-assisted extraction for plant bioactive compounds: phenolics, flavonoids, thymols, saponins and proteins. Food Res Int 157:111268

Zhu H, Zhao M, Wang W, Wang F (2008) Study on enzymatic extraction of flavonoids from seabuckthorn leaves. Agric Technol 28:30–32

Chapter 9
Gut Microbe Regulation of Sea Buckthorn for the Metabolic Syndrome

Rina Wu, Weihe Cang, Shanshan Yang, Danli Yan, Xuwen Ma, Jingjing Li, Jing Li, and Wentao Xing

Abstract Metabolic syndrome, a major health issue, is closely associated with gut microbiota dysbiosis, defined by an imbalance of beneficial and harmful bacteria. This imbalance contributes to various gastrointestinal and metabolic disorders. The current research highlights the effectiveness of natural plant components, especially dietary fibers and polyphenols, in positively influencing gut microbiota composition. Sea buckthorn, rich in polyphenols, has emerged as a key player in regulating gut microbiota and mitigating metabolic inflammation. This chapter examines how sea buckthorn and other treatments targeting gut microbiota impact metabolic diseases like obesity, non-alcoholic fatty liver disease, diabetes, and cardiovascular disease. It also explores the mechanisms of sea buckthorn in addressing metabolic disorders, focusing on lipid metabolism proteins and oxidative stress markers. This comprehensive review underscores the potential of sea buckthorn as a natural intervention strategy for various chronic metabolic diseases. By influencing gut microbiota composition and diversity, sea buckthorn holds promise in maintaining intestinal barrier integrity, reducing inflammation, and ultimately contributing to metabolic health.

Keywords Sea buckthorn · Metabolic syndrome · Gut microbiota · Gut microbiota dysbiosis

Abbreviations

AMPK	AMP-activated protein kinase
CAT	Catalase
CC1	Acetyl-CoA carboxylase
CD	Crohn's disease
CRC	Colorectal cancer
CVD	Cardiovascular Disease

R. Wu (✉) · W. Cang · S. Yang · D. Yan · X. Ma · J. Li · J. Li · W. Xing
Food Science College, Shenyang Agricultural University, Shenyang, China

J. Tian et al. (eds.), *Sea Buckthorn*, https://doi.org/10.1007/978-981-97-9865-0_9

CYP2E1	Cytochrome P450 2E1
FAS	Fatty acid synthase
FXR	Farnesoid X receptor
GSH-Px	Glutathione peroxidase
HDL-C	High-density lipoprotein cholesterol
HFD	High-fat diet
IBD	Inflammatory bowel disease
IECs	Intestinal epithelial cells
LDL-C	Low-density lipoprotein cholesterol
LPS	Lipopolysaccharides
NAFLD	Non-alcoholic fatty liver disease
PPARα	Peroxisome proliferator-activated receptor alpha
SCFAs	Short-chain fatty acids
SIRT1	Sirtuin 1
SOD	Superoxide dismutase
SREBP-1	Sterol regulatory element-binding protein-1
SSP	Sea buckthorn seed protein
T2D	Type 2 diabetes
TC	Total cholesterol
TG	Triglycerides
TGR5	G-protein-coupled bile acid receptor
UC	Ulcerative colitis

9.1 Introduction

Metabolic syndrome is an increasingly concerning health issue, and there is a close relationship between gut microbiota dysbiosis and metabolic abnormalities. Gut microbiota dysbiosis includes a decrease in beneficial bacteria and an increase in potentially harmful bacteria. There is growing evidence that gut dysbiosis is associated with diseases, including gastrointestinal diseases such as inflammatory bowel disease (IBD), ulcerative colitis (UC), Crohn's disease (CD), and colorectal cancer (CRC), as well as metabolic disorders outside the gut such as obesity, diabetes, and cardiovascular diseases.

The proposition of utilizing active ingredients from natural plants to modulate gut microbiota has emerged as a potential intervention strategy for numerous chronic metabolic diseases. The main sources of high-quality nutrients such as dietary fiber and polyphenols have been (Gâtlan and Gutt 2021) shown to have a positive impact on the composition and diversity of gut microbiota, thereby preventing various diseases and metabolic disorders. Sea buckthorn is a plant rich in polyphenols and other active ingredients, known for regulating gut microbiota and anti-inflammatory activity (Chen et al. 2023; Ciesarová et al. 2020). Sea buckthorn has been shown to improve intestinal barrier integrity and reduce inflammation (Liang et al. 2022). Therapies targeting gut microbiota and barrier function, such as prebiotics, probiotics, dietary changes, and anti-inflammatory drugs, have demonstrated significant efficacy in reducing metabolic inflammation in animal models.

This chapter introduced gut microbiota dysbiosis and impaired intestinal barrier function associated with metabolism-related inflammation. Sea buckthorn can have positive effects in reducing metabolic inflammation by improving gut microbiota and barrier function.

9.2 Interactions Between Gut Microbes and Metabolic Diseases

The gut microbiota is integral to managing the intestinal barrier and immune system, both vital for sustaining gut equilibrium. The epithelial layer of the gut is the primary defense separating our body from external elements. Intestinal epithelial cells (IECs) are key in preserving the integrity of this intestinal barrier (Chelakkot et al. 2018). Dysbiosis of gut microbiota has been linked to chronic low-grade inflammation, which is a hallmark of obesity, type 2 diabetes, and non-alcoholic fatty liver disease (NAFLD) (Chen et al. 2019; Frazier and Chang 2020).

9.2.1 Intestinal Barrier

The gut microbiota plays a crucial role in regulating the intestinal barrier and immune system, which are essential for maintaining gut homeostasis. The gut epithelial layer forms the major barrier that separates our body from the external environment. The maintenance of the intestinal epithelial barrier is an essential function of the intestinal epithelial cells (IECs) (Iyer et al. 2018).

9.2.2 Metabolites

Some gut bacteria can produce endotoxins, such as LPS, which can enter the systemic circulation and trigger low-grade inflammation, a known factor in the development of metabolic disorders such as type 2 diabetes and obesity. For example, a study found that mice fed a high-fat diet had increased levels of LPS in their bloodstream, which led to inflammation and insulin resistance. Gut bacteria modify bile acids, which can then influence metabolism through the farnesoid X receptor (FXR) and the G-protein-coupled bile acid receptor (TGR5) (Janssen and Kersten 2015). Bile acids have been implicated in regulating lipid and glucose metabolism, as well as energy expenditure. For example, a study found that mice lacking gut bacteria had altered bile acid metabolism and were more susceptible to diet-induced obesity

(Yu et al. 2023). Gut microbes can influence the efficiency of nutrient absorption and synthesize essential nutrients, including certain vitamins and amino acids, impacting metabolic processes. For example, a study found that gut bacteria can synthesize vitamin B_{12}, which is important for energy metabolism (Fan and Pedersen 2021).

9.2.3 Immune System

The gut microbiota metabolizes proteins and complex carbohydrates, synthesizes vitamins, and produces an enormous number of metabolic products that can mediate cross-talk between gut epithelial and immune cells (Yoo et al. 2020). The gut microbiota also produces short-chain fatty acids (SCFAs), which have been shown to play a role in metabolic and inflammatory pathways. The gut has a direct neural connection with the brain through the vagus nerve, and bacteria can stimulate the afferent neurons of the enteric nervous system (ENS). The gut microbiota also plays a role in bile acid metabolism and affects the biotransformation of bile acids.

9.3 Sea Buckthorn Modulates Gut Microbes to Influence Metabolic Diseases

Sea buckthorn has numerous health benefits, especially its cardioprotective effect. It enhances cardiovascular health by suppressing platelet aggregation, lowering cholesterol and blood pressure, and offering powerful antioxidant activities. The plant's leaves, fruits, and oil are abundant in bioactive compounds such as vitamins A, C, and E, unsaturated fatty acids, flavonoids, and plant sterols, which are beneficial to heart health. These components contribute to sea buckthorn's therapeutic potential in preventing and treating cardiovascular diseases by regulating platelet function and providing antioxidant benefits (Visan et al. 2023). In essence, the diverse bioactive compounds of sea buckthorn hold promise for supporting cardiovascular well-being.

9.3.1 Obesity

Obesity is considered as an imbalance in the body's energy metabolism, stemming from a complex interplay of genetic and environmental factors. Studies in animal models and populations have shown that the development of obesity is associated with a marked imbalance in the gut microbiota, especially in the predominant proportion of gut bacteria (Casariu et al. 2013). There is growing evidence that changes in the gut microbiota affect host metabolism, including energy regulation, nutrient digestion and absorption, and lipid metabolism. Conventional flora or specific commensal flora also affect energy homeostasis and the expression of genes that regulate nutrient metabolic pathways such as lipids (Cha et al. 2012). Studies have shown that the percentage of lactobacilli decreases in obese individuals compared to healthy individuals. As a result, the ratio of sterol/lactobacilli in the gut or feces increases (Mortensen et al. 2018). In summary, the gut microbiota is essential for maintaining gut homeostasis and, in turn, prevents the development of obesity.

High-fat dietary patterns and unhealthy lifestyles contribute to obesity and obesity-related chronic metabolic diseases. Increasing evidence confirms that the pathogenesis of obesity and related complications is related to gut microbial dysbiosis. Thus the gut microbiota may be considered a potential target for the prevention or treatment of obesity and related chronic diseases. Numerous studies have reported that sea buckthorn extract has positive effects such as lowering plasma cholesterol, increasing intestinal probiotics, enhancing lipid-metabolizing enzyme activities, strengthening antioxidant capacity, and reducing the incidence of chronic diseases. Notably, sea buckthorn has been observed to inhibit hepatic lipogenesis by downregulating the high-fat diet-induced overexpression of SREBP-1c, PPAR-γ, and ACC (Mohamed et al. 2020). Furthermore, sea buckthorn appears to play a role in fatty acid catabolism dominated by fatty acid β-oxidation. PPAR-α is a ligand-activated nuclear hormone receptor that stimulates fatty acid oxidation by upregulating downstream genes (e.g., ACOX and CPT1) (Montaigne et al. 2021). ACOX is the rate-limiting enzyme for fatty acid β-oxidation by peroxisomal enzymes, and CPT1 is the rate-limiting enzyme for fatty acid β-oxidation, which determines the rate at which fatty acid oxidation occurs. Regarding research on preventing or ameliorating obesity and obesity-related diseases, it has been suggested that supplementation with plants rich in polyphenols, polysaccharides, or flavonoids may improve obesity symptoms by activating fatty acid β-oxidation or increasing lipid metabolism. Sea buckthorn's capacity to mitigate lipid metabolism disorders associated with obesity is likely attributed to its rich content of biologically active compounds, such as polyphenols, flavonoids, and polysaccharides. It counters the effects of a high-fat diet by downregulating genes involved in fat synthesis and storage, such as PPAR-γ, ACC, and SCD1 (Mulati et al. 2021). Additionally, sea buckthorn may promote fat breakdown and support weight loss by inhibiting fat accumulation genes and activating genes responsible for fat catabolism.

9.3.2 Non-alcoholic Fatty Liver Disease

Recent studies have provided evidence that non-alcoholic fatty liver disease (NAFLD) can be mitigated through dietary and lifestyle modifications that modulate the gut–liver axis and attenuate dysbiosis of the gut microbiota. Sea buckthorn can reduce hepatic oxidative stress to alleviate lipid peroxidation and improve the pathophysiology of the liver in obese mice. Sea buckthorn polysaccharide has a hepatoprotective effect and can prevent chemical liver injury in mice (Wang et al. 2022). Recent research indicates that sea buckthorn oils, especially sea buckthorn pulp oil (SBPO), can alleviate symptoms of NAFLD in mice. These oils help reduce body fat, improve liver function, and increase beneficial gut bacteria, potentially offering a dietary means to combat NAFLD and related metabolic issues like obesity and high cholesterol (Smida et al. 2019). These findings highlight the promise of SBPO as a dietary intervention for the amelioration of NAFLD and its related metabolic complications through the modulation of the gut microbiota (Guo et al. 2021).

9.3.3 Diabetes

Diabetes mellitus, characterized by pancreatic defects in insulin secretion or action, is a hyperglycemic metabolic disorder often precipitated by a high glycemic index diet. The management of diabetes predominantly aims at reducing glucose levels. In this context, the sorbitol pathway becomes particularly relevant; in hyperglycemic states, its activity is heightened, leading to increased sorbitol production. Dupak et al. (2022) documented that sea buckthorn polyphenols exert a hypoglycemic effect by inhibiting aldose reductase activity within this pathway, thus curtailing sorbitol synthesis and the accumulation of advanced glycation end-products.

Sea buckthorn, with its rich composition of proteins, essential amino acids, and sea buckthorn seed protein (SSP), has been identified as a high-quality plant protein source that mitigates lipid disorders and hyperglycemia. The impact of SSP extends to a reduction in serum glutamate-oxaloacetate transaminase activity, indicative of its protective effect against cellular damage. The gut microbiota, increasingly recognized for its role in type 2 diabetes (T2D) development, is positively influenced by SSP. Yuan et al. (2018) found that SSP supplementation altered the intestinal microbiota, favoring beneficial strains such as *Bifidobacterium* and *Lactobacillus* while reducing pathogenic ones like *Clostridium globigii* This modulation was associated with improved glucose tolerance, increased insulin sensitivity, and reduced insulin resistance. The underlying mechanisms involved the upregulation of the AMPK/SIRT1 pathway and the downregulation of hepatic glucose metabolism genes.

In parallel, sea buckthorn fruit oil extracts have shown efficacy in mitigating diabetes through the PI3K/Akt signaling pathway, enhancing glucose uptake and insulin sensitivity, and modulating the expression of key enzymes and proteins involved in glucose and lipid metabolism (Gao et al. 2017).

Insulin resistance and metabolic syndrome, key features of T2D, are also mitigated by the influence of sea buckthorn on gut microbiota (Mazhar et al. 2023; Zhang et al. 2021). It promotes a favorable shift in the Firmicutes to Bacteroidetes ratio (Guo et al. 2020) and enriches *Akkermansia* populations, a genus linked to improved metabolic profiles. *Akkermansia* not only optimizes glucose and lipid metabolism but also stimulates the secretion of glucagon-like peptide-1 (GLP-1), a hormone crucial for enhancing insulin secretion and reducing appetite. Propionate, a metabolite produced by *Akkermansia,* serves as a GLP-1 stimulant (Li et al. 2023). Moreover, the interaction between the *Akkermansia*-derived protein P9 and integrin ICAM-2 on immune cells has been identified to trigger GLP-1 release, offering therapeutic potential for hyperglycemia management (An et al. 2023; Yuan et al. 2018).

9.3.4 Cardiovascular Disease

Cardiovascular disease (CVD) is the principal cause of mortality worldwide, with a considerable prevalence in Europe. The key risk factors for CVD include hypertension, lipid imbalances such as high LDL cholesterol and triglycerides, along with low HDL cholesterol, and increased platelet aggregation. Notably, diets rich in fruits, particularly berries, and vegetables correlate with lower CVD risk (van den Munckhof et al. 2018).

Sea buckthorn stands out for its cardioprotective properties, attributed to its diverse biological effects, including anti-hypertensive, lipid-lowering, anti-aggregatory, and antioxidant actions. Despite robust in vitro and in vivo findings, further human-based studies are needed to elucidate its role in modulating oxidative stress linked to CVD (Olas 2016; Wang et al. 2023).

The antioxidant potential of sea buckthorn is largely due to its phenolic content, notably flavonols (Olas 2016). Additionally, sea buckthorn promotes cardiovascular health by enhancing gut microbiota balance and short-chain fatty acid (SCFA) production, which is crucial for reducing inflammation, maintaining intestinal health, and managing blood pressure. Sea buckthorn polysaccharides improve gut dysbiosis and support SCFA-producing bacteria, thereby lowering coronary artery disease risk. Its seed oil further aids in cholesterol management and supports beneficial gut bacteria (Kobyliak et al. 2016).

Isorhamnetin and quercetin, flavonoids derived from sea buckthorn, have shown promise in cardiovascular protection, with isorhamnetin notably preventing myocardial hypertrophy by targeting the PI3K-Akt pathway (Lin et al. 2022). These compounds have also been effective in models of myocardial ischemia, suggesting therapeutic potential for cardiac conditions.

Sea buckthorn's comprehensive cardiovascular benefits stem from its ability to modulate lipid profiles, enhance endothelial function, support gut health, and alleviate inflammation. These attributes underscore its potential inclusion in CVD prevention and treatment strategies.

9.4 Potential Mechanisms by Which Sea Buckthorn Components Affect Metabolic Diseases

9.4.1 Effect of Sea Buckthorn on Proteins Related to Lipid Metabolism

AMP-activated protein kinase (AMPK) is a cellular energy regulator that maintains energy homeostasis by modulating metabolic processes. It influences lipid metabolism by phosphorylating acetyl-CoA carboxylase (ACC1) and by regulating the expression and activity of the fatty acid synthase (FAS) gene through sterol regulatory element-binding protein-1 (SREBP-1). Sea buckthorn polysaccharides (SP) have been shown to activate AMPK, enhancing the phosphorylation of AMPKα in high-fat diet (HFD) mice livers, which implies a stimulation of the AMPK pathway by SP. Moreover, AMPK activation affects fatty acid oxidation through peroxisome proliferator-activated receptor alpha (PPARα) and carnitine palmitoyltransferase-1 (CPT-1), as well as through phosphorylation of PGC-1α. SP has also been observed to increase mRNA expression of genes related to fat oxidation, like CPT-1α, CPT-2, and sirtuin 1 (SIRT1), and to reduce the expression of genes involved in fat synthesis, suggesting its regulatory role on proteins associated with lipid metabolism via the AMPK pathway. Additionally, studies have shown that SP reduces body, liver, and abdominal fat weight, serum triglycerides (TG), total cholesterol (TC), low-density lipoprotein cholesterol (LDL-C), and liver TG levels, while increasing high-density lipoprotein cholesterol (HDL-C) in HFD-induced mice, indicating SP's potential in combating obesity-related symptoms (Lan et al. 2022).

Clinical studies have established a link between visceral/ectopic fat and the onset of metabolic syndromes, including dyslipidemia, poor glucose tolerance, hypertension, atherosclerosis, and heart failure (Neeland et al. 2018). Activating brown (BAT) and beige (WAT) adipocytes, which increase energy expenditure, is a potential strategy to mitigate obesity and its associated risks. Key markers of the adipocyte browning process include PRDM16 and PGC1α (Ma et al. 2022), which regulate the development and function of BAT. However, pharmacological approaches to activate this browning process can have side effects, prompting the search for natural alternatives like sea buckthorn. SP's role in reducing lipid accumulation in HFD-induced mice may be attributed to the activation of BAT, enhancing thermogenesis. Sea buckthorn also affects liver enzymes involved in lipid regulation, thereby reducing hepatic triglycerides and cholesterol, as well as decreasing hepatic lipid accumulation. This indicates the potential of sea buckthorn as a natural therapeutic agent for obesity and its comorbidities.

9.4.2 *Effect of Sea Buckthorn on Biomarkers of Oxidative Stress in Liver and Erythrocytes*

Oxidative stress in obesity may be an important pathogenic mechanism of obesity-related metabolic syndrome. Possible mechanisms contributing to obesity-associated oxidative stress include increased oxygen consumption, decreased antioxidant capacity of free radicals produced through mitochondrial respiration, increased fat deposition, and cellular damage leading to increased rates of free radical formation (e.g., superoxide anion and hydroxyl radical). Biomarkers of oxidative stress in liver and erythrocytes, including lipid peroxidation, CYP2E1 activity, and antioxidant enzyme activities (SOD, GSH-Px, and CAT) (Rong et al. 2022) (Taskinen et al. 2019).

CYP2E1 is an enzyme ubiquitous in various organs of animals and is increased in obesity, diabetes, fasting, and high-fat diets. Moreover, its increase in obesity-related diseases in the liver is well documented as one of the most important P450s in terms of toxicology. In the livers of high-fat-fed mice, CYP2E1 activity was significantly higher than in normal diet-fed mice. However, supplementation with sea buckthorn reduced CYP2E1 activity compared to high-fat-fed control mice. In the current study, reduced SOD and CAT activities were observed in the liver and erythrocytes of high-fat-fed mice. Erythrocyte GSH-Px activity was also reduced in high-fat-fed control mice compared to normal diet-fed mice. Interestingly, supplementation with sea buckthorn extract reversed SOD and CAT activities in the liver and erythrocytes. The results of Lee et al. (2011) verified this claim by showing that supplementation with sea buckthorn leaf tea increased SOD and CAT activities in liver and erythrocytes and reduced lipid peroxide levels, thus protecting against superoxide anion formation from the high-fat diet and providing significant protection against high-fat-induced oxidative stress.

The gut microbiota (GM) is considered as a specialized organ involved in the pathophysiology of NAFLD. Dysregulation of the gut microbiota affects intestinal permeability, leading to the arrival of some pathogens or bacterial-derived factors in the gut to the liver via the hepatic-intestinal circulation, leading to liver injury, steatohepatitis, and fibrosis progression. The pathways involved are diverse and include changes in bile acid metabolism, intestinal ethanol production, short-chain fatty acid production, and other byproducts (Arab et al. 2020). GM in patients with NAFLD is dominated by *Bacteroides thickettsiensis, Bacteroides anomalies, Aspergillus* spp., and *Actinobacteria* at the portal level. In addition, *Fusobacterium rectum, Anaplasma unusualum,* and *Escherichia coli* were the most abundant at the species level. Lifestyle, dietary, pharmacological, and environmental factors can cause abnormal transgenic components. These components can directly affect the immune system and the intestinal mucosa's integrity. This, in turn, can change the levels of inflammatory cytokines and fatty acid oxidation, leading to liver disease (Ni et al. 2022).

Of course, a rational diet and a healthy lifestyle can also improve gut microbiota dysbiosis through the gut–liver axis. This, in turn, reduces the risk of

NAFLD. Palmitoleic acid has been recognized as a potential nutraceutical for modulating NAFLD because of its ability to alleviate hepatic steatosis and inflammation in HFD-fed mice. Sea buckthorn fruit oil extract containing 81% concentrated palmitoleic acid may reduce body weight, decrease visceral fat accumulation and adipocyte size, improve lipid indices (Ma et al. 2024), and attenuate hepatocellular injury in HFD-induced hyperlipidemic hamsters via AMPK and Akt pathways. For example, Wang et al. (2022) showed that sea buckthorn pulp oil and sea buckthorn seed oil reduced visceral fat accumulation and adipocyte size, suppressed serum TG, TC, and LDL-C levels, lowered hepatic TG and LDL-C levels, decreased hepatocellular injury scores, and altered the intestinal microbiota by enhancing the abundance and diversity of the microbial community.

9.5 Conclusion

In this chapter, we focus on the role of sea buckthorn in managing metabolic syndromes, emphasizing its effects on gut microbiota, obesity, non-alcoholic fatty liver disease, diabetes, and cardiovascular disease. Sea buckthorn, rich in polyphenols and other active ingredients, positively impacts gut microbiota composition and diversity, helping maintain intestinal barrier integrity and reducing inflammation. It offers the potential to treat metabolic inflammations through therapies targeting gut microbiota and barrier function. The gut microbiota, crucial for intestinal barrier and immune system regulation, plays a significant role in metabolic health. Dysbiosis is linked to chronic inflammation, a feature of various metabolic disorders. In summary, sea buckthorn is a promising natural intervention strategy for many chronic metabolic diseases.

References

An Y, Dai H, Duan Y et al (2023) The relationship between gut microbiota and susceptibility to type 2 diabetes mellitus in rats. Chin Med 18:49

Arab JP, Arrese M, Shah VH (2020) Gut microbiota in non-alcoholic fatty liver disease and alcohol-related liver disease: current concepts and perspectives. Hepatol Res 50:407–418

Casariu ED, Virgolici B, Lixandru D et al (2013) Sea buckthorn pulp oil treatment can prevent metabolic syndrome in hypertriglyceridemic waist phenotype obese children. Farmacia 61:1043–1053

Cha YS, Moon HJ, Pichiah PBT et al (2012) Anti-obesity effect of sea buckthorn (Hippohae rhamnoides L) on diet induced obese C57BL/6N mice. Proc Nutr Soc 71:E119

Chelakkot C, Ghim J, Ryu SH (2018) Mechanisms regulating intestinal barrier integrity and its pathological implications. Exp Mol Med 50:1–9

Chen MT, Hui SC, Lang HD et al (2019) SIRT3 deficiency promotes high-fat diet-induced non-alcoholic fatty liver disease in correlation with impaired intestinal permeability through gut microbial dysbiosis. Mol Nutr Food Res 63:1800612

Chen Y, Cai Y, Wang K et al (2023) Bioactive compounds in sea buckthorn and their efficacy in preventing and treating metabolic syndrome. Foods 12:1985

Ciesarová Z, Murkovic M, Cejpek K et al (2020) Why is sea buckthorn (Hippophae rhamnoides L.) so exceptional? A review. Food Res Int 133:109170

Dupak R, Hrnkova J, Simonova N et al (2022) The consumption of sea buckthorn (Hippophae rhamnoides L.) effectively alleviates type 2 diabetes symptoms in spontaneous diabetic rats. Res Vet Sci 152:261–269

Fan Y, Pedersen O (2021) Gut microbiota in human metabolic health and disease. Nat Rev Microbiol 19:55–71

Frazier K, Chang EB (2020) Intersection of the gut microbiome and circadian rhythms in metabolism. Trends Endocrinol Metab 31:25–36

Gao S, Guo Q, Qin C et al (2017) Sea buckthorn fruit oil extract alleviates insulin resistance through the PI3K/Akt signaling pathway in type 2 diabetes mellitus cells and rats. J Agric Food Chem 65:1328–1336

Gâtlan AM, Gutt G (2021) Sea buckthorn in plant based diets. An analytical approach of sea buckthorn fruits composition: nutritional value, applications, and health benefits. Int J Environ Res Public Health 18:8986

Guo C, Han L, Li M et al (2020) Seabuckthorn (Hippophaë rhamnoides) freeze-dried powder protects against high-fat diet-induced obesity, lipid metabolism disorders by modulating the gut microbiota of mice. Nutrients 12:265

Guo ZL, Cheng JY, Zheng L et al (2021) Mechanochemical-assisted extraction and Hepatoprotective activity research of flavonoids from sea buckthorn (Hippophae rhamnoides L.) pomaces. Molecules 26:7615

Iyer SS, Gensollen T, Gandhi A et al (2018) Dietary and microbial Oxazoles induce intestinal inflammation by modulating aryl hydrocarbon receptor responses. Cell 173:1123–1134 e1111

Janssen AW, Kersten S (2015) The role of the gut microbiota in metabolic health. FASEB J 29:3111–3123

Kobyliak N, Virchenko O, Falalyeyeva T (2016) Pathophysiological role of host microbiota in the development of obesity. Nutr J 15:1–12

Lan Y, Sun Q, Ma Z et al (2022) Seabuckthorn polysaccharide ameliorates high-fat diet-induced obesity by gut microbiota-SCFAs-liver axis. Food Funct 13:2925–2937

Lee HI, Kim MS, Lee KM et al (2011) Anti-visceral obesity and antioxidant effects of powdered sea buckthorn (Hippophae rhamnoides L.) leaf tea in diet-induced obese mice. Food Chem Toxicol 49:2370–2376

Li J, Yang G, Zhang Q et al (2023) Function of Akkermansia muciniphila in type 2 diabetes and related diseases. Front Microbiol 14:1172400

Liang J, Zhang G, Song Y et al (2022) Targeted metabolome and transcriptome analyses reveal the pigmentation mechanism of Hippophae (sea buckthorn) fruit. Foods 11:3278

Lin X, Yuen M, Yuen T et al (2022) Regulatory effect of sea-buckthorn procyanidins on oxidative injury HUVECs. Front Nutr 9:850076

Ma Z, Sun Q, Chang L et al (2022) A natural anti-obesity reagent derived from sea buckthorn polysaccharides: structure characterization and anti-obesity evaluation in vivo. Food Chem 375:131884

Ma Y, Yao J, Zhou L et al (2024) Comprehensive untargeted lipidomic analysis of sea buckthorn using UHPLC-HR-AM/MS/MS combined with principal component analysis. Food Chem 430:136964

Mazhar M, Zhu Y, Qin L (2023) The interplay of dietary fibers and intestinal microbiota affects type 2 diabetes by generating short-chain fatty acids. Foods 12:1023

Mohamed EA, Tulcan C, Alexa E et al (2020) Sea buckthorn and grape extract might be helpful and sustainable phyto-resources as associated hypolipidemic agents-preliminary study. Sustainability 12:9297

Montaigne D, Butruille L, Staels B (2021) PPAR control of metabolism and cardiovascular functions. Nat Rev Cardiol 18:809–823

Mortensen MW, Spagner C, Cuparencu C et al (2018) Sea buckthorn decreases and delays insulin response and improves glycaemic profile following a sucrose-containing berry meal: a randomised, controlled, crossover study of Danish sea buckthorn and strawberries in overweight and obese male subjects. Eur J Nutr 57:2827–2837

Mulati A, Zhang X, Zhao T et al (2021) Isorhamnetin attenuates high-fat and high-fructose diet induced cognitive impairments and neuroinflammation by mediating MAPK and NF kappa B signaling pathways. Food Funct 12:9261–9272

Neeland IJ, Poirier P, Despres JP (2018) Cardiovascular and metabolic heterogeneity of obesity: clinical challenges and implications for management. Circulation 137:1391–1406

Ni Y, Lu M, Xu Y et al (2022) The role of gut microbiota-bile acids axis in the progression of non-alcoholic fatty liver disease. Front Microbiol 13:908011

Olas B (2016) Sea buckthorn as a source of important bioactive compounds in cardiovascular diseases. Food Chem Toxicol 97:199–204

Rong L, Zou J, Ran W et al (2022) Advancements in the treatment of non-alcoholic fatty liver disease (NAFLD). Front Endocrinol (Lausanne) 13:1087260

Smida I, Pentelescu C, Pentelescu O et al (2019) Benefits of sea buckthorn (Hippophae rhamnoides) pulp oil-based mouthwash on oral health. J Appl Microbiol 126:1594–1605

Taskinen MR, Packard CJ, Boren J (2019) Dietary fructose and the metabolic syndrome. Nutrients 11:1987

van den Munckhof ICL, Kurilshikov A, ter Horst R et al (2018) Role of gut microbiota in chronic low-grade inflammation as potential driver for atherosclerotic cardiovascular disease: a systematic review of human studies. Obes Rev 19:1719–1734

Visan S, Soritau O, Tatomir C et al (2023) The bioactive properties of carotenoids from Lipophilic Sea buckthorn extract (Hippophae rhamnoides L.) in breast cancer cell lines. Molecules 28:4486

Wang Z, Zhou S, Jiang Y (2022) Sea buckthorn pulp and seed oils ameliorate lipid metabolism disorders and modulate gut microbiota in C57BL/6J mice on high-fat diet. Front Nutr 9:1067813

Wang J, Ren C, Jin L et al (2023) Seabuckthorn Wuwei Pulvis attenuates chronic obstructive pulmonary disease in rat through gut microbiota-short chain fatty acids axis. J Ethnopharmacol 314:116591

Yoo JY, Groer M, Dutra SVO et al (2020) Gut microbiota and immune system interactions. Microorganisms 8:1587

Yu H, Nie R, Shen C (2023) The role of bile acids in regulating glucose and lipid metabolism. Endocr J 70:359–374

Yuan H, Shi F, Meng L et al (2018) Effect of sea buckthorn protein on the intestinal microbial community in streptozotocin-induced diabetic mice. Int J Biol Macromol 107:1168–1174

Zhang J, Ni Y, Qian L et al (2021) Decreased abundance of Akkermansia muciniphila leads to the impairment of insulin secretion and glucose homeostasis in lean type 2 diabetes. Adv Sci (Weinh) 8:e2100536

Chapter 10
Bioactivities of Sea Buckthorn Phenolics and Their Anti-cancer Mechanism Toward Human Breast Cancer Cells

Xiaobo Wei, Yaoran Li, Peiyi Zhu, and Jinhu Tian

Abstract A general introduction of sea buckthorn phenolics is summarized, highlighting their various bioactivities including antioxidant, antimicrobial, anti-inflammatory, anti-cancer, cardiovascular protective, and hypoglycemic effects. Additionally, the mechanisms underlying the anti-cancer properties of sea buckthorn phenolics, particularly their regulation of the cell cycle, apoptosis pathways, and the PI3K/Akt/mTOR pathway in human breast cancer cells, are discussed both in vitro and in vivo.

Keywords Sea buckthorn phenolics · Bioactivities · Breast cancer · Anti-cancer mechanism · Regulatory mechanism

Abbreviations

ADH	Atypical ductal hyperplasia
AG	α-glucosidase
AMVN	2,2-azobis-2,4-dimethylvaleronitrile
CRC	Colorectal carcinoma
ICAM	Intercellular cell adhesion molecule
LDL	Low-density lipoprotein
MCP	Monocyte chemoattractant protein
NBE	Normal breast epithelium
NOS	Arginine metabolism

X. Wei
Food Science and Technology, Ningxia University, Yinchuan, Ningxia, China

Y. Li (✉)
Food College, Ningxia University, Yinchuan, China

P. Zhu · J. Tian
Department of Food Science and Nutrition, Zhejiang University, Hangzhou, China

J. Tian et al. (eds.), *Sea Buckthorn*, https://doi.org/10.1007/978-981-97-9865-0_10

ROS	Reactive oxygen species
STZ	Streptozocin
TCS	Triclosan

10.1 Introduction

Polyphenols are a complex class of phenolic secondary metabolites that cannot be synthesized by human body and need to be obtained through diet. Polyphenols are general terms for substances with benzene rings and various hydroxyl structures. There are many kinds of polyphenols in sea buckthorn, mainly including phenolic acids and flavonoids, among which phenolic acids mainly consist of derivatives of hydroxybenzoic acid and hydroxycinnamic acid, such as chlorogenic acid, gallic acid, protocatechuic acid, *p*-hydroxybenzoic acid, coumaric acid, eugenol, cinnamic acid, *p*-coumaric acid, ferulic acid, and caffeic acid. Flavonoids mainly include flavanols, isoflavones, and anthocyanins, such as isorhamnetin, kaempferol, quercetin, catechin, and epicatechin. Many studies have shown that sea buckthorn polyphenols play an important role in antioxidant, anti-inflammatory, cardiovascular protection, and anti-cancer.

10.2 Antioxidant Activity of Sea Buckthorn Phenolics

The antioxidant activity of polyphenols depends on the structure of their functional groups, and the number of hydroxyl groups in polyphenols greatly affects several mechanisms of antioxidant activity, such as scavenging free radicals and metal ion chelating ability. The mechanisms involved in the antioxidant capacity of sea buckthorn polyphenols include the inhibition of reactive oxygen species (ROS) formation by impeding enzymatic pathways involved in their synthesis, scavenging ROS, and upregulating or protecting antioxidant defenses. The phenolic fraction of sea buckthorn fruits is the main contributor to its overall antioxidant activity. Olsson et al. (2004) conducted a comparative analysis of the total antioxidant potential among diverse phytochemicals present in sea buckthorn, revealing a significant positive correlation between the antioxidant capacity and the levels of phenolics and ascorbic acid. The hydrophobic phenolics demonstrated heightened efficacy in inhibiting the lipid peroxidation process induced by 2,2-azobis-2,4-dimethylvaleronitrile (AMVN) and iron ascorbate, while the water-soluble total phenolic extracts were slightly less effective in inhibiting this oxidative process. Kim et al. (2011) found that butanol-extracted phenolic fractions from sea buckthorn had good scavenging ability against DPPH radicals. Olas et al. (2016) found that sea buckthorn berries phenolics (dominant compounds in this fraction—flavonoids: 214.04 mg/g) exhibited antioxidant activity at concentrations of 0.5–50 mg/ml, which was mainly reflected in the inhibition of reactive oxygen species production, lipid peroxidation, and protein carbonylation. Sea buckthorn polyphenols extracts (gallic acid, quercetin, kaempferol, and isorhamnetin) protected liver tissues of carbon tetrachloride-induced oxidative stress-injured SD rats through in vivo antioxidant effects (Maheshwari et al. 2011).

10.3 Antimicrobial Activity of Sea Buckthorn Phenolics

Plant polyphenols prevent food spoilage by inhibiting the development and growth of harmful microorganisms. Therefore, they are known as natural antibacterial agents. At present, there are relatively few studies on the antimicrobial activity of sea buckthorn polyphenols. Jiao et al. (2007) studied the antibacterial effect of sea buckthorn leaf polyphenols using 13 common food-contaminating bacteria (*Staphylococcus aureus*, *Escherichia coli*, *Bacillus subtilis*, *Bacillus cereus*, *Aspergillus cereus*, *Beer yeast*, *Saccharomyces cerevisiae*, *Fusarium oxysporum*, *Streptomyces interrogans*, *Aspergillus niger*, *Rhizopus oligosporus*, and *Trichoderma elegans*) as test organisms. The results showed that the ethanol-extracted polyphenols of sea buckthorn leaves were effective against six common food-contaminating bacteria (*Escherichia coli*, *Salmonella* spp., *Bacillus subtilis*, and *Bacillus cereus*). The results showed that sea buckthorn leaf polyphenols had good inhibitory effects on six common food-contaminating bacteria (*Escherichia coli*, *Salmonella*, *Proteus mirabilis*, *Staphylococcus aureus*, *Bacillus subtilis*, and *Bacillus cereus*) and showed an obvious quantitative–effect relationship, with the strongest inhibitory effects on *Proteus mirabilis* and *Bacillus subtilis*. Tang (2022) also found that sea buckthorn leaf polyphenols had significant inhibitory effects on *Staphylococcus aureus*, *Escherichia coli*, *Bacillus cereus*, *Listeria monocytogenes*, and *Salmonella* spp. It was hypothesized that inhibition was achieved by inhibiting the growth of bacterial proteins and enzymes or by disrupting bacterial membranes and impeding the normal expression of bacterial proteins. It was also found that the inhibited bacteria included both Gram-negative and Gram-positive bacteria, indicating that sea buckthorn leaf polyphenols have broad-spectrum antibacterial activity.

10.4 Anti-inflammatory Activity of Sea Buckthorn Phenolics

Sea buckthorn polyphenols exert anti-inflammatory effects mainly through free radical scavenging activity, modulation of cellular activity in inflammatory cells, regulation of the activity of enzymes involved in arachidonic acid metabolism (phospholipase A2, COX) and arginine metabolism (NOS), and modulation of the production of other pro-inflammatory molecules. Gu et al. (2022) found that sea buckthorn flavonoids reduced the expression levels of inflammatory factors, including IL-1α, IL-1β, IL-6, monocyte chemoattractant protein-1 (MCP-1), and MCP-3 in the supernatant of the HaCaT cells treated by IFN-gamma/TNF-alpha. In addition, sea buckthorn flavonoids administration in a dose-dependent manner decreased the expression of *p*-NF-κB, *p*-ERK, and *p*-P38 in IFN-γ/TNFα-treated HaCaT cells. These results suggest that sea buckthorn flavonoids exert anti-inflammatory effects by regulating the Th1/Th2 balance, which in turn reduce MC903-induced atopic dermatitis (AD)-like lesions in mice.

In addition, isorhamnetin, a representative phenolic compound in sea buckthorn, has been widely reported to alleviate inflammation. Qiu et al. (2016) treated type II diabetes mice with 50–150 mg/kg isorhamnetin by gavage for 12 weeks and found that it can reduce the expression levels of tumor necrosis factor-α (TNF-α), inflammatory factors interleukin-1β (IL-1β), IL-6, ICAM-1, and transforming growth factor-β1 (TGF-β1), reduce the production of inflammatory mediators, thereby mitigating renal function damage caused by diabetes. Similarly, isorhamnetin markedly decreased TNF-α, IL-1β, and IL-6 concentrations and suppressed the activation of NF-κB signaling in LPS-induced inflammatory mice (Li et al. 2016).

10.5 Anti-cancer Activity of Sea Buckthorn Phenolics

The phenolics in sea buckthorn have several inhibitory effects on the growth of various cancer cells, including colon and liver cancer. Colorectal carcinoma (CRC) is the most common gastrointestinal cancer, and miRNA is a new therapeutic target. Wang et al. found that the main component of sea buckthorn polyphenols (HPs60) is kaempferol and its derivatives, which have significant anti-cancer activity. HPs60 significantly inhibited the expression of cyclinD1, cyclinD2, cyclinD3, cyclinE1, and Bcl-2 as targets of miR-195-5p and miR-497-5p, and promoted the expression of Caspase 2, the target gene of miR1247-3p. These results indicate that HPs60 can be explored as a natural bioactive ingredient and reveal potential targets for HPs60 in the treatment of colorectal cancer.

Liver cancer is one of the most common malignancies worldwide, and an increase in the initial synthesis of fatty acids in tumor cells is an important feature of cancer progression. Meanwhile, glucose metabolism is a major source of cellular energy, and disorders of glucose metabolism are associated with many diseases such as diabetes and cancer, implying that metabolic reprogramming from oxidative phosphorylation to glycolysis plays an integral role in the development of liver cancer. Accumulating studies have shown that plant-derived polyphenols inhibit liver cancer development by regulating lipid and glucose metabolism (Li et al. 2023).

After the action of sea buckthorn flavonoids on hepatocellular carcinoma cell line BEL-7402 cells for 48 h, the cancer cells gradually became rounded, with solidified nuclei and bud-like protrusions on the cell surface, showing typical morphological changes of apoptosis, and these results indicated that sea buckthorn flavonoids could induce apoptosis in hepatocellular carcinoma cells (Sun et al. 2003). Similarly, Grey et al. (2010) compared the effects of sea buckthorn extracts obtained by ethyl acetate extraction and ethanol extraction on the proliferation of liver cancer cells HepG2 and colon cancer cells Caco-2. Their study revealed significant inhibition of proliferation in both types of cancer cells following exposure to extracts derived from sea buckthorn. Interestingly, the antiproliferative activity of the extract of ethyl acetate was higher than that of the extract obtained by the ethanol method, and the reason for these differences may be related to the differences in the types of compounds in the extracts. The main products of the ethanol method were quercetin rutinoside, isorhamnetin rutinoside, kaempferol glucoside, and proanthocyanidins, while the components in the ethyl acetate extract were mainly proanthocyanidins and corosolic acid, suggesting that the inhibitory effect of sea buckthorn flavonoid rutinoside substances on cancer cell activity pit can be lower than that of proanthocyanidins and corosolic acid. Another high-content flavonol in sea buckthorn, kaempferol, has a good effect in inhibiting liver cancer. When kaempferol acts on liver cancer cell SK-HEP-1, it can upregulate the expression of tumor suppressor proteins such as p-MAPK, LC3-II, Atg 5, Atg 7, Atg 12, and Beclin 1, and inhibit the overexpression of CDK1, Cyclin B, p-Akt, and p-mTOR. Kaempferol may block the cell cycle of SK-HEP-1 at G2/M through the CDK1/Cyclin B cycle regulation pathway and AMPK, Akt signaling pathway, inducing autophagy of cancer cells, thereby exerting long-term cancer prevention (Huang et al. 2013). Recently, Guo et al. (2017) studied the total phenolic content of different subspecies of sea buckthorn berries and investigated their antiproliferative properties on HepG2 human liver cancer cells through in vitro experiments. The results indicate that *H. rhamnoides* L. subsp. *sinensis* berries have the highest total phenolic content, with *H. rhamnoides* L. subsp. *yunnanensis* berry polyphenols having the greatest anti-liver cancer cell proliferation properties.

10.6 Protective Effect of Sea Buckthorn Phenolics on Cardiovascular System

Sea buckthorn flavonoids are clinically important raw materials for the treatment of cardiovascular diseases, and their preventive effects on heart disease have been unanimously recognized by many researchers (Suomela et al. 2006; Xu et al. 2011). Tang et al. (2016) found that sea buckthorn extract reduced the cholesterol content in the body, improved blood rheology, and prevented cholesterol deposition in the arterial wall, which in turn prevented the development of atherosclerosis. In addition, sea buckthorn flavonoids have a good ability to scavenge reactive oxygen radicals in living organisms, which in turn acts as anti-myocardial ischemia, anti-arrhythmia, reduces the area of myocardial infarction, relieves angina pectoris, and improves cardiac function (Seeram 2008). Eccleston et al. (2002) studied the changes in plasma cholesterol, LDL-C, platelet aggregation and plasma intercellular cell adhesion molecule 1 (ICAM-1) in patients after taking sea buckthorn juice (355 mg/kg) for 4 weeks. The results showed that dietary supplementation with sea buckthorn phenolic extracts could relatively reduce the susceptibility of patients to low-density lipoprotein (LDL) oxidation in plasma and prevent further deterioration of atherosclerotic patients. Isorhamnetin could prevent the development of myocardial hypertrophy by attenuating the effect of angiotensin II-induced cardiac hypertrophy and blocking the activation of PI3K/Akt signaling pathway. Moreover, Xu et al. (2011) provided a comprehensive overview of current research endeavors concerning sea buckthorn extracts in both preclinical and clinical contexts for the prevention and management of cardiovascular diseases. Their synthesis highlighted that flavonoids and other components with high antioxidant activity in sea buckthorn extracts are the main cause of their protective effects on the cardiovascular system.

10.7 Hypoglycemic Effect of Sea Buckthorn Phenolics

Sea buckthorn polyphenols are mainly composed of phenolic acids and flavonoids, which exert hypoglycemic effects by modulating α-amylase and glucosidase activities and inhibiting postprandial blood glucose elevation levels. The possible mechanism of sea buckthorn polyphenols in controlling blood glucose elevation to prevent diabetes is that sea buckthorn leaf polyphenols form a complex with α-amylase or glucosidase to slow down the hydrolysis of carbohydrates and glucose absorption and reduce the diffusion of glucose from the active site, thus reducing the postprandial blood glucose elevation (Zhao et al. 2020). Yan et al. (2023) found that the main components of sea buckthorn leaf alcohol extract are flavonoids. It reduced fasting blood glucose and increased oral glucose tolerance in streptozocin (STZ)-induced hyperglycemic mice and effectively inhibited the elevation of blood glucose induced by starch intake in normal mice. The alcoholic extract of sea buckthorn leaves showed good inhibitory activity (IC50 = 16.94 μg/mL) against yeast-derived activity of α-glucosidase (AG). Sea buckthorn leaf alcohol extract (25–100 μg/mL) inhibited glucose uptake by Caco-2 cells in a dose-dependent manner. These results suggest that the alcohol extract derived from sea buckthorn leaves exerts hypoglycemic effects by inhibiting alpha-glucosidase (AG) activity, thereby modulating glucose uptake by intestinal epithelial cells. The phenolic substances in sea buckthorn are rich and diverse. In addition to the reported biological activities of these components, the activities and pathways of action of other functional components require more extensive and in-depth research.

10.8 Anti-cancer Mechanism of Sea Buckthorn Polyphenols on Human Breast Cancer Cells In Vitro and In Vivo

Breast cancer is a highly heterogeneous chronic disease with a biological phenotype that varies among individuals. Current breast cancer subtypes include ductal epithelial A subtype (ER+/PR+, HER2−, Ki-67 < 20%), ductal epithelial B subtype (ER+/

PR+, Ki-67 > 20%), HER-2 overexpression subtype (ER−, PR−, HER2+), and basal cell-like/triple-negative subtype (ER−, PR−, HER2−) and normal mammary gland-like subtype of breast. Breast carcinogenesis is a complex multi-step process ranging from atypical ductal hyperplasia (ADH) to metastatic cancer. The normal breast epithelium (NBE) includes the ductal lumen, the outer myoepithelial cells, and the basement membrane. In the initiation phase, NBE is exposed to a range of stresses (environmental chemicals, diet, lifestyle, and hormones), which lead to the accumulation of epigenetic and genetic changes, as well as oxidative stress and chronic inflammation. Endocrine, autocrine, and paracrine signaling cascades influence all stages of breast carcinomatosis development and create a tumor microenvironment through the secretion of growth factors, cytokines, chemokines, and protein hydrolases that promote breast tumor promotion and progression. Epigenetic regulation may also promote breast tumorigenesis through the upregulation of oncogenes. Molecular mechanisms of chemoprevention by natural dietary phenolic compounds include (a) promotion of phase II detoxification and antioxidant activity, (b) induction of apoptosis, autophagy, cell cycle arrest, and senescence, (c) modulation of endocrine signaling, (d) inhibition of inflammation, proliferation, transformation, angiogenesis, EMT, migration, invasion and adhesion processes, and (e) inhibition of ABC and glucose transporter proteins.

10.8.1 In Vitro

(a) Cell Cycle Regulation and Cell Apoptosis Pathway

Cell cycle refers to the process in which cells become two daughter cells through orderly mitosis. It usually includes G0/G1 (prophase of DNA synthesis), S (anaphase of DNA synthesis), G2 (anaphase of DNA synthesis), other intercellular phases and Mitosis phase M. The process of cell cycle requires the participation of many regulatory factors, mainly including cell cyclin, cyclin-dependent kinase (CDK), and its inhibitors (CKI) (Vermeulen et al. 2003). Apoptosis is a strictly controlled cell suicide pattern characterized by nuclear pyknosis, cell shrinkage, cell membrane foaming, and DNA fragmentation. Caspase, a member of cysteine protease, is the central regulator of apoptosis. Promoters of caspase (including caspase-2, -8, -9, -10, -11, and -12) are tightly bound to pro-apoptotic signals. Once activated, these caspases cleave and activate downstream effector caspases (including caspase-3, -6, and -7), which in turn cleave cell proteins according to specific Asp residues, thereby executing apoptosis (Fuchs and Steller 2011).

Sea buckthorn flavonoids induced characteristic changes in growth inhibition and apoptosis in human estrogen-dependent breast cancer cells Bcap-37 at concentrations ranging from 10 to 1000 μg/mL. Further studies revealed that sea buckthorn flavonoids could down-regulate the expression of DNA replication-related kinases and cyclin-dependent kinase CDK1 and cancer-related genes including c-myc and c-fos. They could upregulate the expression of pro-apoptotic genes including Caspase-9, AFDD, and BAI in Bcap-37 cells. Therefore, it is hypothesized that sea buckthorn flavonoids inhibit breast cancer proliferation through the regulation of cell cycle-related gene expression, thereby impeding interphase progression, suppressing DNA replication, and curtailing oncogene-initiated DNA replication and transcription. Moreover, these flavonoids are believed to enhance the sensitivity of breast cancer cells and indirectly inducing cells to enter the apoptotic cascade response (Zhang et al. 2005). Breast carcinogenesis often exhibits intracellular lipid synthase overexpression. Wang et al. (2014) found that sea buckthorn proanthocyanidins could effectively inhibit lipid synthase expression after acting on estrogen receptor-negative breast cancer MDA-MB-231 cells, suggesting that proanthocyanidins promote apoptosis in breast cancer cells through a lipid metabolism regulatory pathway of lipid synthase inhibition. In addition, Kim et al. (2016) studied the anti-proliferation effect of kaempferol in MCF-7 breast cancer cells induced by triclosan (TCS). The results showed that kaempferol significantly inhibited TCS-induced cell growth. TCS treatment up-regulated the protein expression of Cyclin D1, Cyclin E and Cathepsin D, and down-regulated the expression of p21 and bax, and kaempferol treatment reversed gene expression induced by TCS. These results indicate that kaempferol has an anti-cancer effect on the proto-oncogenic activity of TCS in breast cancer.

(b) PI3K/Akt/mTOR Pathway

Phosphoinositide 3-kinase/protein kinase B/mammalian target of Sirolimus (PI3K/Akt/mTOR) signal transduction pathway is an important intracellular transmission pathway, which is abnormally activated in the occurrence and development of breast cancer, and is one of the most important pathways in the pathogenesis of breast cancer (Kirkegaard et al. 2005). As a main monomer of sea buckthorn polyphenols, isorhamnetin has a significant effect in the treatment of breast cancer. After treatment with isorhamnetin, the expression of Akt and cytochrome protein in breast cancer cells decreased, the proliferation of cancer cells was inhibited, and the apoptosis effect was intensified (Hu et al. 2015). In addition, Guo (2019) studied the effects of sea buckthorn polyphenols (isorhamnetin, quercetin, and kaempferol) on breast cancer cells MCF-7 and MDA-MB-231. Isorhamnetin has a stronger inhibitory effect on breast cancer cells MCF-7 and MDA-MB-231 than quercetin and kaempferol, and quercetin is better than kaempferol. The combination of isorhamnetin and quercetin showed strong synergistic effects in inhibiting the proliferation of MCF-7 and MDA-MB-231 cells. In addition, the effects of isorhamnetin and kaempferol on the expression levels of related proteins in NF-κB, PI3K/Akt/mTOR, and p38MAPK signaling pathways in MDA-MB-231 cells were further investigated. The results showed that isorhamnetin and kaempferol dose-dependently down-regulated the expression levels of cyclin D1, PCNA, Bcl-2 and NF-κB, and simultaneously up-regulated the expression of Bax and increased the phosphorylation level of p38MAPK, while differential regulation of p-Akt expression levels was observed. However, upon exposure to isorhamnetin at concentrations of 15, 25, and 35 μM for 12 h, a dose-dependent reduction in p-Akt expression was observed in cancer cells. Respectively, treatment with kaempferol at concentrations of 40, 60, and 80 μM initially led to a reduction in p-Akt expression followed by upregulation. These contrasting responses suggest that inhibition of Akt activation may not be the main mode of action of kaempferol against breast cancer.

10.8.2 In Vivo

In the field of breast cancer research, it is crucial to systematically model the observation of the cancer process and treatment. In preclinical cancer treatment assessment, researchers commonly utilize human tumor cell lines or patient-derived tumors transplanted into immunodeficient mice to establish in situ or ectopic tumor animal models. These models serve as platforms for observing and evaluating tumor growth and metastasis, enabling in-depth exploration of the mechanisms underlying the action of clinical drugs. Guo (2019) established a xenograft tumor animal model of breast cancer cell MDA-MB-231 in nude mice and evaluated the effects of isorhamnetin and kaempferol on the growth of breast cancer in the form of single- and synergistic action. The results showed that after 4 weeks of drug administration, the tumor volume and weight were effectively suppressed in experimental animals, and the anti-tumor effect of isorhamnetin was better than that of kaempferol. When administered in combination with half a concentration of isorhamnetin and kaempferol, the high-dose group had the best inhibitory effect on tumor growth. Isorhamnetin and kaempferol inhibit the growth of transplanted tumors by inhibiting the activation of PI3K/Akt signaling pathway and p38MAPK signaling pathway, further activating the endogenous mitochondrial apoptosis pathway, making tumor cells enter the apoptosis process and preventing the metastasis and invasion of breast cancer cells between tissues by downregulating the expression levels of MMP-2 and MMP-9. In addition, isorhamnetin and kaempferol showed differences in the regulation of PI3K/Akt pathway, indicating that sea buckthorn flavonols may synergistically inhibit the growth of breast cancer cells in vitro and in vivo through the common regulation of the same signal pathway, or different signal transduction pathways. Similar results were reported by Kim et al. (2016). These in vitro effects of triclosan (TCS) and kaempferol were also confirmed in a xenogeneic mouse model transplanted with MCF-7 breast cancer cells. Compared with the control group, tumor volume increased after TCS treatment. However, in several analyses of tumor tissue, TCS showed the potential to induce tumor growth by increasing the induction of BrdU incorporation into the nucleus, increasing the expression of PCNA and Cyclin D1, and reducing the expression of p21. TCS significantly reduced Bax expression in breast cancer tissue and increase Cathepsin D expression, which is related to apoptosis and metastasis of breast cancer. In contrast, the intervention of kaempferol significantly reduced the tumor growth effect in vivo and the anti-apoptotic and metastatic potential of TCS, as expected from in vitro experimental results.

References

Eccleston C, Baoru Y, Tahvonen R, Kallio H, Rimbach GH, Minihane AM (2002) Effects of an antioxidant-rich juice (sea buckthorn) on risk factors for coronary heart disease in humans. J Nutr Biochem 13:346–354

Fuchs Y, Steller H (2011) Programmed cell death in animal development and disease. Cell 147:742–758

Grey C, Widén C, Adlercreutz P, Rumpunen K, Duan RD (2010) Antiproliferative effects of sea buckthorn (*Hippophae rhamnoides* L.) extracts on human colon and liver cancer cell lines. Food Chem 120:1004–1010

Gu Y, Wang X, Liu F, Zhang J, Zhang X, Liu J, Li S, Wang D, Guan H, Hou D (2022) Total flavonoids of sea buckthorn (*Hippophae rhamnoides* L.) improve MC903-induced atopic dermatitis-like lesions. J Ethnopharmacol 292:115195

Guo R (2019) Bioactivities and bioavailability of *Hippophaë rhamnoids* L. phenolics and their anticancer mechanism towards human breast cancer cells both in vitro and in vivo. Thesis for M.S. South China University of Technology, pp 74–119

Guo R, Guo X, Li T, Fu X, Liu RH (2017) Comparative assessment of phytochemical profiles, antioxidant and antiproliferative activities of Sea buckthorn (*Hippophaë rhamnoides* L.) berries. Food Chem 221:997–1003

Hu S, Huang L, Meng L, Sun H, Zhang W, Xu Y (2015) Isorhamnetin inhibits cell proliferation and induces apoptosis in breast cancer via Akt and mitogen-activated protein kinase kinase signaling pathways. Mol Med Rep 12:6745–6751

Huang WW, Tsai SC, Peng SF, Lin MW, Chiang JH, Chiu YJ, Fushiya S, Tseng MT, Yang JS (2013) Kaempferol induces autophagy through AMPK and AKT signaling molecules and causes G2/M arrest via downregulation of CDK1/cyclin B in SK-HEP-1 human hepatic cancer cells. Int J Oncol 42:2069–2077

Jiao X, Yin L, Cheng Y (2007) Extraction and antibacterial properties of flavonoids of sea buckthorn leaves. Food Sci 28:124–129

Kim JS, Kwon YS, Sa YJ, Kim MJ (2011) Isolation and identification of sea buckthorn (*Hippophae rhamnoides*) phenolics with antioxidant activity and alpha-glucosidase inhibitory effect. J Agric Food Chem 59:138–144

Kim SH, Hwang KA, Choi KC (2016) Treatment with kaempferol suppresses breast cancer cell growth caused by estrogen and triclosan in cellular and xenograft breast cancer models. J Nutr Biochem 28:70–82

Kirkegaard T, Witton CJ, McGlynn LM, Tovey SM, Dunne B, Lyon A, Bartlett JMS (2005) AKT activation predicts outcome in breast cancer patients treated with tamoxifen. J Pathol 207:139–146

Li Y, Chi G, Shen B, Tian Y, Feng H (2016) Isorhamnetin ameliorates LPS-induced inflammatory response through downregulation of NF-kappaB signaling. Inflammation 39:1291–1301

Li S, Yin S, Ding H, Shao Y, Zhou S, Pu W, Han L, Wang T, Yu H (2023) Polyphenols as potential metabolism mechanisms regulators in liver protection and liver cancer prevention. Cell Prolif 56:13346

Maheshwari DT, Yogendra Kumar MS, Verma SK, Singh VK, Singh SN (2011) Antioxidant and hepatoprotective activities of phenolic rich fraction of Seabuckthorn (*Hippophae rhamnoides* L.) leaves. Food Chem Toxicol 49:2422–2428

Olas B, Kontek B, Malinowska P, Zuchowski J, Stochmal A (2016) *Hippophae rhamnoides* L. fruits reduce the oxidative stress in human blood platelets and plasma. Oxidative Med Cell Longev 2016:4692486

Olsson ME, Gustavsson K-E, Andersson S, Nilsson Å, Duan RD (2004) Inhibition of cancer cell proliferation in *vitro* by fruit and berry extracts and correlations with antioxidant levels. J Agric Food Chem 52:7264–7271

Qiu S, Sun G, Zhang Y, Li X, Wang R (2016) Involvement of the NF-kappaB signaling pathway in the renoprotective effects of isorhamnetin in a type 2 diabetic rat model. Biomed Rep 4:628–634

Seeram NP (2008) Berry fruits: compositional elements, biochemical activities, and the impact of their intake on human health, performance, and disease. J Agric Food Chem 56:627–629

Sun B, Zhang P, Qu W, Zhang X, Zhuang X, Yang H (2003) Study on effect of flavonoids from oil-removed seeds of *Hippophaerhamnoides* on inducing apoptosis of human hepatoma cell. J Chin Med Mater 26:875–878

Suomela J-P, Ahotupa M, Yang B, Vasankari T, Kallio H (2006) Absorption of flavonols derived from sea buckthorn (*Hippophaë rhamnoides* L.) and their effect on emerging risk factors for cardiovascular disease in humans. J Agric Food Chem 54:7364–7369

Tang M (2022) Application of seabuckthorn leaf polyphenols in apple juice preservation. Food Sci Technol:93–95

Tang L, Lv H, Li S, Bi H, Gao X, Zhou J (2016) Protective effects of polyphenol extracts from sea buckthorn (*Hippophaë rhamnoides* L) on rat hearts. Open J Mol Integr Physiol 6:10–18

Vermeulen K, Van Bockstaele DR, Berneman ZN (2003) The cell cycle: a review of regulation, deregulation and therapeutic targets in cancer. Cell Prolif 36:131–149

Wang Y, Nie F, Ouyang J, Wang X, Ma X (2014) Inhibitory effects of sea buckthorn procyanidins on fatty acid synthase and MDA-MB-231 cells. Tumour Biol 35:9563–9569

Xu YJ, Kaur M, Dhillon RS, Tappia PS, Dhalla NS (2011) Health benefits of sea buckthorn for the prevention of cardiovascular diseases. J Funct Foods 3:2–12

Yan C, Ding Z, Li X, Mao X, Zongsheng Y, Zhifang W, Ye J, Kurihara H, Li Y-F, Liang L, He R (2023) The chemical constituents and hypoglycemic activity of alcoholic extract of sea buckthorn leaves. Acta Pharm Sin 58:396–404

Zhang P, Mao Y, Sun B, Qian M, Qu W (2005) Changes in apoptosis-related genes expression profile in human breast carcinoma cell line Bcap-37 induced by flavonoids from seed residues of *Hippophae Rhamnoides* L. Chin J Cancer 24:454–460

Zhao C, Wan X, Zhou S, Cao H (2020) Natural polyphenols: A potential therapeutic approach to hypoglycemia. eFood 1:107–118

Chapter 11
Immunological Enhancement Effect of Sea Buckthorn Powder

Fangyuan Song, Pin Chen, and Jinhu Tian

Abstract Besides the general introduction of sea buckthorn, the processing status of sea buckthorn powder, including processing technology and application status of sea buckthorn powder, is described, which emphatically involves the factors that affect the processing quality of sea buckthorn. The physical basis of immune regulation in sea buckthorn is also discussed, and the nutritional components affecting immunity in sea buckthorn are explained. The effects of sea buckthorn on immune system are summarized. Finally, the product development and the problems and prospects of sea buckthorn powder processing are provided in detail.

Keywords Sea buckthorn power · Processing status · Immune regulation · Product development

Abbreviations

ANAE	α-naphthyl acetate esterase
HPLC	High-performance liquid chromatography
LPS	Lipopolysaccharide
PHA	Phytohemagglutinin
RP-HPLC	Reversed phase-high performance liquid chromatography
VC	Vitamin C
VP	Vitamin P

F. Song (✉)
Institute of Agricultural Products Processing, Xinjiang Academy of Agricultural Reclamation Sciences, Shihezi, China

P. Chen · J. Tian
Department of Food Science and Nutrition, Zhejiang University, Hangzhou, China

J. Tian et al. (eds.), *Sea Buckthorn*, https://doi.org/10.1007/978-981-97-9865-0_11

11.1 General Introduction Of Sea Buckthorn

Sea buckthorn (*Hippophae rhamnoides L., Elaeagnaceae*) is a perennial shrub or tree of the genus *Hippophae* in the family Elaeagnus, also known as sour vinegar willow, black thorn, sour thorn, Qizarigana (Mongolian name), Darbu (Tibetan name), Jihan et al. (Yildiz et al. 2012; Ursache et al. 2017). Sea buckthorn is widely distributed in temperate regions of Europe and Asia (Ma 2020), China, Britain, and Mongolia (Gao et al. 2000). Sea buckthorn is abundant in China, with a relatively wide distribution area that includes four kinds and five subspecies of sea buckthorns (He et al. 2008; Liu et al. 2014), distributed in Xinjiang, Shanxi, Liaoning, and Gansu. Sea buckthorn is a kind of tree species with both ecological and economic uses, which can prevent damage by wind and sand and maintain soil and water, which has great value in food and pharmaceutical fields (Kumar et al. 2013; Priyanka et al. 2015). Sea buckthorn is full of treasures and is known as "natural nutrient bank," and its roots, stems, leaves, and fruits are rich in 428 kinds of bioactive substances such as multiple vitamins, amino acids, and trace elements. Total flavonoids, esters, and phenolic acids of sea buckthorn can assist in reducing blood sugar and blood fat and increasing vascular properties (Aleksandra and Izabela 2017; Zhang et al. 2004), and reduce blood cholesterol content (Yang and Kallio 2001). Sea buckthorn plays an obvious role in protecting respiratory tract and gastrointestinal digestive system (Himanshi et al. 2018).

Sea buckthorn is processed into many products, including foods, health care products, cosmetics, pharmaceuticals, and so on, in eight categories made up of more than 200 products. Sea buckthorn juice, sea buckthorn fruit powder, sea buckthorn vinegar, sea buckthorn tea, sea buckthorn fruit wine, and so on (Yao et al. 2020) are relatively popular in the market. Sea buckthorn powder has low water content and is difficult for microorganisms to grow and reproduce, thus extending the shelf life of products (Wu et al. 2022), reducing the cost of storage and transportation, and making it convenient to carry and prepare, which is popular among consumers. At the same time, the fruit powder can be prepared into drinks, fruit wine, fruit vinegar, and other products through compounding, brewing, fermentation, and other processes, greatly improving the application of raw materials (Li et al. 2017).

11.2 Processing Status of Sea Buckthorn Fruit Powder

Sea buckthorn fruit powder is a kind of powder product that is dried by spray drying or vacuum freeze-drying technology and has the advantages of easy preservation, transportation, and use. At present, studies on sea buckthorn fruit powder are mostly focused on component detection, in vitro antioxidant capacity, and application, while those on product processing technology are few and not in-depth. For example, a previous study, which simultaneously determined quercetin, kaempferol, rutin, and isorhamnonose in sea buckthorn fruit powder from three different producing areas by high-performance capillary zone electrophoresis, found that the sea buckthorn fruit powder with the highest active component came from Urumqi, Xinjiang (Yang et al. 2017). Zhang et al. (2010) analyzed and detected seven water-soluble vitamins, including VC, VB1, VB2, VB6, VB12, folic acid, and niacin in sea buckthorn fruit powder, wolfberry, and prickly thorn on Qinghai-Tibet Plateau by high-performance liquid chromatography (HPLC) and found that sea buckthorn fruit powder had the highest VC content and rich VB1 content. No other water-soluble vitamins were detected. Dai et al. (2015) used ion chromatography to simultaneously detect the contents of seven organic acids, namely citric acid, malic acid, oxalic acid, quinic acid, lactic acid, acetic acid, and formic acid in sea buckthorn fruit powder and the established method successfully achieved a good separation within 25 min. Wang et al. (2013) also established a method for detecting lead, arsenic, and mercury heavy metal in sea buckthorn by using microwave digestion technology double channel atomic fluorescence method. Previous research studied the ability of sea buckthorn fruit powder to remove free radicals and found that sea buckthorn fruit powder had better antioxidant ability in vitro (Qiang et al. 2014; Ma et al. 2014). Cui et al. (2016) mixed sea buckthorn fruit powder with wheat flour in different proportions to study its influence on flour characteristics and found that adding sea buckthorn fruit powder could hinder the formation of gluten network and change the gelatinous properties, saltiness, gelatinization, and gluten ultrastructure of flour.

11.2.1 Processing Technology of Sea Buckthorn

Drying is the key technology to determine the quality of sea buckthorn powder. The commonly used drying techniques include vacuum freeze-drying, spraying drying, hot air drying, and so on. The processing principle, temperature, time, and energy consumption of different drying technologies are different. Therefore, the quality of fruit powder prepared by different drying methods is different. At present, there are many studies focusing on the effect of different drying methods on the quality of fruit power.

11.2.2 Processing Technology of Sea Buckthorn Fruit Powder

A previous study showed that used five drying methods found that vacuum freeze-drying could better preserve color and nutrition in terms of product quality (Lin et al. 2022). The study contrasted the effect of vacuum freeze-drying, low-temperature hot air drying, and high-temperature hot air drying on the content of active components of sea buckthorn extract, and the results demonstrated that the higher the degree, the more the loss of flavonoids, vacuum freeze-drying, and high-temperature drying technology can prevent oxidation, which is conducive to protection-holding flavonoids in the extract (Li et al. 2020). Vacuum freeze-drying enables materials to remove water from frozen materials in a state of low temperature and vacuum by using the principle of ice crystal sublimation (Chu 2020) to achieve the purpose of freeze-drying. Moreover, porous structure is left on the tissue of ice crystals after sublimation, which enables good rehydration. Freeze-drying can retain the nutritional composition, color, and flavor of the raw materials to the maximum extent. At present, most enterprises apply this technology in the production of high-quality fruit powder (Caparino et al. 2012; Kim et al. 2012; Reyes et al. 2011).

11.2.3 Suitability of Variety for Processing Sea Buckthorn Powder

At present, the research on sea buckthorn powder at home and abroad mainly focuses on processing technology. The comprehensive suitability of sea buckthorn varieties to the nutrient composition and processing characteristics of sea buckthorn powder is rarely studied. Some varieties have multiple processing suitability, but it is difficult to ensure that the quality of different types of products processed is good. Generally, the suitability of varieties is closely related to the content of quality characteristic components of raw materials. The physiological metabolism of varieties with different origins and different genetic conditions is different, and the content of accumulated material components is also different (Wang et al. 2020, 2021). Therefore, understanding the chemical composition and content of raw materials of different varieties can provide certain theoretical support for the suitability of varieties.

Wu et al. (2021) determined the contents of seven fatty acids in sea buckthorn seed oil at the same time, screened oleic acid and linoleic acid according to principal component analysis and comprehensive score, and concluded that oleic acid and linoleic acid were markers causing quality differences in sea buckthorn seed oil. At the same time, under drought stress, the changes of multiple physiological and biochemical indexes of four sea buckthorn varieties were analyzed, and through principal component analysis and comprehensive evaluation, it was found that local Chinese sea buckthorn varieties had the strongest drought-resistant ability, which provided certain theoretical basis for the selection of drought-resistant tree species (He et al. 2015).

11.2.4 Study on Quality of Sea Buckthorn Powder

Quality characteristics of high-quality sea buckthorn powder can be summarized as rich nutrition, bright color, fine and even particles, good flavor and taste, rich flavor, good solubility during brewing, clear soup color without impurities, and so on (Li et al. 2017; Gao et al. 2020).

At present, vitamin C, total phenols, flavonoids, and carotenoids are used as indicators in the nutritional studies of sea buckthorn fruit powder. Luo et al. (2021) analyzed the nutritional components of five kinds of sea-buckthorn with large fruit in Altay, Xinjiang, and determined the contents of vitamin C, total flavonoids, and fruit oil. Cui et al. (2021) believed that the chemical composition of sea buckthorn mainly consists of flavonoids and vitamins, as well as saponins, amino acids, and trace elements. High and diversified nutrient content is an important factor for product quality. Therefore, raw materials with high nutrient content should be selected in raw material selection, processing technology, storage, and transportation. At the same time, nutrients should be retained as far as possible.

Color is an important index to evaluate fruit powder products, which affects consumers' preference, and color of sea buckthorn fruit powder is related to the content of pigment substances (carotenoids) in sea buckthorn fruits, which is also affected by fruit variety, climate, processing technology, and other aspects.

The silty properties of sea buckthorn powder affect the rehydration, pungency, and taste of the product. The product characteristics of sea buckthorn powder require it to have good impact and fluidity, but in the process of processing and storage, the fruit powder often undergoes moisture absorption and caking, reducing the quality of the product. Luo added sea buckthorn fruit residue and silica to sea buckthorn fruit powder in proportion, and the crude fiber in sea buckthorn fruit residue not only increased the nutrition of the product but also played a certain role in anti-caking. This study provides a new idea for the comprehensive utilization of sea buckthorn (Luo et al. 2020).

11.3 The Physical Basis of Immune Regulation in Sea Buckthorn

Sea buckthorn fruit is the source of many bioactive ingredients. At present, nearly 200 kinds of active ingredients have been found in sea buckthorn fruit, including flavonoids (isorhamnosin, quercetin, kaempferol, etc.), phenolic acids (gallic acid, vanillic acid, p-coumaric acid, etc.), vitamin C, vitamin E, carotenoids, amino acids and minerals (iron, calcium, phosphorus, and potassium). These active ingredients play an increasingly important role because of their special physiological functions in different aspects.

11.3.1 Flavonoids

The flavonoids of sea buckthorn mainly exist in roots, stems, leaves, flowers, and fruits. At present, more than 30 flavonoid components have been detected in sea buckthorn. The main compounds include isorhamnetin, quercitrin, and kaempferol as parent nuclei and glucose, rhamnose, and sophora as glycogen (Chen et al. 2007; Zu et al. 2006). These substances can reflect the chemical properties and biological activities of sea buckthorn fruit to a certain extent. The flavonoid content of sea buckthorn fruit was different among different subspecies. There is a significant difference between the Chinese subspecies sinensis and the Mongolian subspecies. The content of total flavonoids and main compounds in the former is higher than that in the latter. However, contrary to this result, Guo used reversed-phase high-performance liquid chromatography (RP-HPLC) system to compare the composition and content of phenolic compounds in seabeaches fruit of Chinese subspecies, Mongolian subspecies, Yunnan subspecies (*Yunnanensis*), and *Turkestanica* subspecies. The total flavone content was ranked as Yunnan subspecies > Mongolian subspecies > Turkey subspecies > China subspecies. It was also found that flavonoids of sea buckans fruit generally appeared in free state and the contents of iso-rhamnoside 3-*O*-glucose-7-*O*-rhamnoside were the highest, whose values ranged from 112 to 187 mg/100 g (Guo et al. 2017).

11.3.2 Organic Acids

The fruits of sea buckthorn contain many organic acids, such as malic acid, citric acid, succinic acid, tartaric acid, and oxalic acid. The total amount of acid is 3.86–4.52% (Li et al. 2007b), and a small amount of tannin is about 0.02–0.06%, but the tannin content in the leaves is more than 10.00%. These compounds can prevent coronary atherosclerosis, significantly increase the flexibility and plasticity of blood vessels, and have significant antihypertensive effect. The organic acids in sea buckthorn can relieve the toxicity of barbiturates, antibiotics, and other drugs and prevent the physiological damage caused by teratoid factors, X-ray, hyperbaric oxygen delivery, and stress response.

11.3.3 Vitamins

It has been proven that sea buckthorn contains vitamins C, E, A, B1, B2, B6, B12, and vitamin K, P, and other vitamins (Li et al. 2007a). Vitamin C is the most abundant vitamin in the fruits of sea buckthorn. The content of vitamin C in fresh fruits from different producing areas is 1.6–17.16 mg/g, 20 times that of hawthorn, four times that of kiwi, six times that of citrus, 200 times that of apple, and 80 times that of tomato (Li et al. 2007a). Because sea buckthorn fruit does not contain an ascorbate oxidase component, vitamin C is very stable. Vitamin A is extremely abundant in sea buckthorn. The content of β-carotene in seed oil, fruit residue oil, and juice oil in China were 484.5 mg, 187 mg, and 67.2 mg per 100 g as determined by HPLC. Carotenoids with provitamin A activity include β-carotene, γ-carotene, cryptoxanthin, and β-zeaxanthin. The content of vitamin E in sea buckthorn is the highest in fruit trees. Vitamin E content per 100 g of sea buckthorn oil was 206.9 mg, 2.9 mg in fresh fruit, and 28.0 mg in pomace.

11.3.4 Polysaccharide

Polysaccharides are natural macromolecular substances, which are the main immunoactivity components of most herbal medicines. It used to be discarded as an invalid component, but now it has attracted widespread attention because of its prominent role in promoting and restoring the immune function of the body. In recent years, it has been found that the sugar chain of polysaccharides plays a decisive role in molecular biology (Lu 1995). In addition, none of the immunologically active polysaccharides have been found to have mutagenic and carcinogenic effects. Therefore, polysaccharides are the most important immunologically active ingredients that can be expected to play an important role in the future immune modulators, and some people even believe that polysaccharides may be the material basis for nourishing traditional Chinese medicine and strengthening the body's immune function. The polysaccharides isolated from sea buckthorn were mainly neutral polysaccharides composed of glucose, fructose, galactose, arabinose, mannose, and rhamnose, among which glucose and fructose accounted for 80% of the total sugars, and the polysaccharide content of sea buckthorn fruits was higher than that of sea buckthorn leaves (Wang et al. 2017)

11.3.5 Proteins and Amino Acids

The protein contents in flesh, juice, and seed of sea-buckthorn were 2.89%, 0.90%–1.20%, and 24.38%, respectively. Compared with other fresh fruits of plants, the protein content of sea buckthorn fruit is quite high. The amino acids in the leaves and fruits of sea buckthorn were studied abroad. There were 18 kinds of amino acids in the pulp and juice, including eight kinds of amino acids that could not be synthesized by the human body.

11.3.6 Terpenoids and Steroids

The fruits of sea buckthorn contain ursolic acid, oleanolic acid, glutaric acid, stigmasterol, digitalin, and balsamine (Ma and Cui 1987). There are 52 such ingredients in fruit, mainly cyclolanosterol, ergosterol, β-sitosterol, 24-methylene-cycloabutenol, and libanol. β-sitosterol is the highest content in fruit (about 50%) and ursuline, oleanolic acid, α- and β-vanillin in fruit and leaves (Liu 1998).

11.4 Immune Regulation of Sea Buckthorn

11.4.1 The Effect of Immune Enhancers on Immune Organs

Immune organs mainly consist of central immune organs such as bone marrow, thymus, and supraliminal sac and peripheral immune organs such as spleen, lymphatic tissue, and so on. The central immune organ controls the immune response of the body and endows lymphocytes with immune function, making them T and B lymphocytes. Peripheral immune organs are places where immunoactivity cells settle, proliferate, and respond to antigens. Therefore, the development of immune organs directly affects the immunity of the body.

The weight of infected organs is one of the indicators of immune function. One general immune organ weight reduction for the immunosuppression and immune organ weight increase immune enhancement performance. The effects of sea buckthorn on immune organs were mainly studied in spleen, thymus, and Fassi capsule.

11.4.2 Effect of Total Flavonoids of Sea Buckthorn Powder on Immune Function

Studies have shown that sea buckthorn powder can regulate blood lipids, reduce oxidative damage in liver, and prevent lipid metabolism disorders (Xu et al. 2016), which may be related to the high content of flavonoids in sea buckthorn (Zhuang et al. 2012). Bai studied the effects of sea-buckthorn flavonoids on the immune function of aging rats induced by D-galactose. The results showed that compared with the control group, sea-buckthorn flavonoids treated groups improved the immune organ index of rats to varying degrees, increased the number of white blood cells, and enhanced the phagocytosis activity of abdominal macrophages. The positive lymphocyte rate and serum hemolysin content labeled by acid α-naphthyl acetate esterase (ANAE) staining were improved, indicating that flavonoids of sea buckass could enhance the immune function of D-galactose-induced aging in rats (Bai 2009).

A previous study found that the extract of sea buckthorn fruit could inhibit the proliferation of chromium on lymphocytes and speculated that the ability of its extract to stimulate the proliferation of lymphocytes was due to the existence of cytokinome in sea buckthorn, which endowed it with manipulatory properties (Geetha et al. 2002a, b). Then, they reported in their later studies that when the concentration of flavonoids in sea buckthorn fruit was 100 μg/mL, the apoptosis of lymphocytes induced by tert-butanol was inhibited and the caspase activity was decreased by reducing the intracellular calcium content. In addition, sea buckthorn fruit can improve the immune response to T-2 toxin reaction of 15-day-old chicks, and the combination of sea buckthorn fruit and glucan can be used as an anti-T-2 toxin additive for immune protection of the body (Geetha et al. 2009). The immune effect of sea buckthorn fruit is generally through its biologically active components to stimulate the proliferation of lymphocytes in vivo and activate specific cell-mediated immune responses. The total flavonoids of sea buckthorn can enhance specific and non-specific immune function of mice, increase leukocyte lysozyme, phagocytosis, and total complement content, promote antibody production, increase serum antibody level, increase blood cell proportion and spleen-specific rose-forming cells, and enhance the proliferation of activated lymphocytes. The anti-ovine hemolysin production caused by cyclophthylphthalamine was significantly protected, and the decrease caused by antagonistic cyclophthylphthalamine was significantly reduced, and the lymphocyte transformation was promoted at low concentrations. These results indicated that total flavonoids of sea buckthorn could significantly enhance the immune function.

11.4.3 Immune Effect of Vp of Sea Buckthorn

Vitamin P is a kind of bioflavonoid, which has good biological activity, such as cardiovascular protection function, and widely exists in various fruits and vegetables and wild berries. Sea-buckthorn contains a large amount of sea-buckthorn Vp, which has immune-promoting effects, antioxidant damage effects, and anti-lipid peroxidation effects. Studies have found that sea-buckthorn Vp has a certain protective effect on the hematopoietic system of X-ray-radiated mice, can protect the internal organs of radiated mice, and improve the body immunity (Li et al. 2014). Vp of sea buckthorn has a certain protective effect on the hematopoietic system of X-ray-radiated mice, can protect the internal organs of radiated mice, and improve the body immunity (Yang et al. 2016).

11.4.4 Effect of Vp Powder of Sea Buckthorn

Sea-buckthorn Vp powder is prepared by spraying and drying the raw juice of sea buckthorn after instantaneous sterilization at high temperatures, which is convenient for transportation and storage, and can be used as raw materials and auxiliary materials of medicine carrier, food, health food, and cosmetics. Studies have shown that sea-buckthorn Vp powder is rich in vitamins, polyphenols, and flavonoids (Lu et al. 2008), has good antioxidant properties, and can effectively remove 1,1-diphe nyl-2-trinitrophenylhydraine free radicals, hydroxyl free radicals, superoxide anion free radicals and nitrite (Qiang et al. 2014).

11.4.5 Effects of Sea Buckthorn Polysaccharide on the Immune System

Polysaccharide compounds play an immunomodulatory role by activating immune cells through various mechanisms and improving the body's specific and non-specific immune functions (Hirazumi and Furusawa 1999). Through the determination of immune organ index, macrophage phagocytosis function, lymphocyte proliferation ability, and hemolysin level in mice, studies have shown that polysaccharide of sea buckspine has anti-oxidation, anti-aging, and anti-fatigue biological activities and can enhance non-specific immunity, humoral immunity, and cellular immunity to some extent (Ren 2019). Polysaccharides from sea buckthorns can increase the weight of immune organs in mice, significantly transform spleen T lymphocytes induced by ConA, enhance the phagocytosis ability of peritoneal macrophages, and increase the serum hemolysin level of mice. A previous study showed that polysaccharide from sea buckthorn can significantly promote the proliferation of macrophages and the release of cytokines in vitro. In vivo, polysaccharides of sea buckthorn can enhance the immune organ index of immunocompromised mice, enhance the phagocytosis function of macrophages, and increase the level of cytokines in spleen tissue (Wang et al. 2018).

11.5 The Effect of Sea Buckthorn on the Immune System

11.5.1 Effects of Sea Buckthorn on Cellular Immunity

Cellular immunity is an important immunity mediated by T cells. Reaction lymphocyte transformation is commonly used for the determination of cellular immune function test and E knot formation test. Xu et al. (1997) gavaged the sea buckthorn seed oil injected into mice stomach, and found that sea buckthorn seed oil could promote the formation of the mice E knot promote phytohemagglutinin (PHA) stimulation of lymphocyte transformation function. Zhong et al., (1989) found that sea buckthorn total flavonoids can make the T cell percentage, the thymus index, and lymphocyte acid esterase tag number higher positive rate SRFC, mice spleen cells increase, SRFC caused can adversely affect cyclophosphamide reduce enhances the lymphocyte proliferation of ConA stimulation. Geetha et al. (2002a, b) reported that with lymphocytes of hippophae rhamnoides as a model, antioxidant and immune regulation in vitro studies have shown that *Hippohaerhamnoides* leaves and fruit of the ethanol extract (500 g/ml) can inhibit the chromium by potassium dichromate (Kr^{6+}), lead to cell apoptosis and DNA fragmentation, stop Kr^{6+} on lipopolysaccharide (LPS), and withstand sword bean globulin caused by the inhibition of lymphocyte proliferation. To protect the cells, sea buckthorn showed that it can regulate cellular immunity (Geetha et al. 2002a, b). Prakash et al. (2010) have shown that sea buckthorn recoverable by the radiation of ethanol extract (CO^{60}-γ) results in the decrease of the $CD4^+/CD8^+$ ratio of recovery from 1.5 to 2.1. Thus, it showed that it performs a special regulation of cellular immune function by regulating T cell subsets (Prakash et al. 2010).

11.5.2 The Effect of Sea Buckthorn on Humoral Immune Function

Humoral immunity is another important immune response mediated by B cells. Humoral immunity protects against extracellular microorganisms and their toxins. Humoral immunity is often measured by direct antibody level test and hemolytic plaque test. Plaque test is a cell-mediated hemolysis test, which is a method to determine the number of antibody-forming cells in vitro. The number of hemolysis plaque formation can roughly reflect the number of antibody-forming cells (Xu 1994). A previous study showed that sea bucks pine powder (10 g/kg) could significantly increase the median hemolytic value (HC50) of SRBC-immunized mice (Li et al. 1994). Studies have shown that sea buckthorn powder can effectively stimulate the development of central immune organs (thymus) in mice and enhance the immune effect of peripheral mucosal organs (digestive tract mucosa and respiratory tract mucosa) (Li et al. 2021).

11.6 Development of Sea Buckthorn Powder Product

As a solid beverage, effervescent tablets have the advantages of small size, easy to carry and preserve, simple eating method, fast disintegration rate, and high bioavailability (Hong et al. 2018, Wang 2017). After the effervescent tablets dissolve in drinking water, under the action of effervescent disintegrating agents, a large amount of carbon dioxide will be produced, making the mouth feel like soda water and sometimes the bubbles generated by disintegration will make the effervescent tablets in the water roll up and down, speeding up the disintegration and melting, which can also relieve pressure while working (Naji et al. 2021).

Sea buckthorn, as a nutrient-rich characteristic fruit, has not been fully developed yet. In the continuous pursuit of health, the research on sea buckthorn products has gradually attracted people's attention. Sea buckthorn fruit powder is a kind of new healthy product, which is easy to carry and ingest. At present, sea buckthorn fruit powder has been applied to various food types such as bread, milk powder, tablet candy, and so on to improve senses, flavor, and enrich nutrition (Ivanišová and Kačániová 2020). Liu developed a sea-buckthorn and vegetable meal replacement powder by combining sea-buckthorn fruit powder, red adzuki bean powder, and pearl seed powder (Liu et al. 2019). The product has rich nutrition and unique taste and simple to process and is easily scalable for enterprise production. Sea buckthorn fruit powder was mixed with red dates powder (Liu et al. 2020), wheat flour (Hong et al. 2018), and milk powder (Zhao and Hao 2012) to make sea buckthorn chewable tablets, bread and milk powder, and other products.

Sea buckthorn fruit powder is rich in nutrition, and it can be prepared into a novel effervescent tablet product, which is convenient to carry and brew and has good taste and can retain its nutritional components. RCO, was used to optimize the formulation of sea buckthorn fruit powder effervescent tablets, and the sea buckthorn fruit powder effervescent tablets prepared under the best conditions in this study (Wang et al. 2019; Lian 2022) in line with the requirements of the Pharmacopoeia of the People's Republic of China on the difference in tablet weight, disintegration time and hardness, and the nutritional composition is significantly higher than that of commercially available sea buckthorn effervescent tablets. The preparation of sea buckthorn effervescent tablets has enriched the product variety, promoted the high-value utilization of sea buckthorn, and provided a new idea for the comprehensive utilization of sea buckthorn fruit powder.

11.7 Problems and Prospects of Sea Buckthorn Fruit Powder Processing

At present, much international research has been done on the processing technology of fruit and vegetable powder and the quality change during storage, but the basic research and technical practice are not complete. So there is still a broad space for the development of scientific research and technological innovation on fruit and vegetable powder. Spray drying powder, due to the high inlet air temperature, destroys the color, flavor, and nutritional components of the product and even produces a burnt taste due to the addition of a large number of drying agents, making the purity of the product low, thereby limiting its market application range (Lu et al. 2015). When vacuum drying powder is made, the water separated from fruits and vegetables is not easily excluded from the system, which affects the performance of vacuum chestnut, resulting in rough fruit and vegetable powder products, poor flavor, serious loss of functional factors and nutrients, and poor taste and application performance (Tang et al. 2013). Freeze-drying can retain the nutritional content of fruit and vegetable products, but its high energy consumption limits the industrialization and promotion (Cang 2018). Therefore, the research on the drying technology of fruits and vegetables and the optimization of drying process have always been the subject of research. In the processing of fruit and vegetable powder, many by-products such as skin residue, core shell, and seeds are produced, which have not been effectively utilized, and a comprehensive utilization of fruits and vegetables has become a hot issue in the international fruit and vegetable processing industry (Zhu and Lu 2013). With the advent health consciousness era, the fruit and vegetable processing industry now has new opportunities and also faces new challenges, and functional fruit and vegetable powder is getting increasing attention from the public. In this situation, researchers and producers should pay special attention to the development of fruit and vegetable powder processing technology, improve the quality of fruit and vegetable powder, and strive to diversify and globalize its use.

References

Aleksandra Z, Izabela N (2017) Abundance of active ingredients in sea-buckthorn oil. Lipids Health Dis 16(1):95–96

Bai HM (2009) Effects of sea buckthorn flavonoids on antioxidant and immune function in rats. Northeast Agricultural University, Harbin

Cang H (2018) Application of vacuum freeze-drying technology in food industry. Food Saf Guide 21:143

Caparino OA, Tang J, Nindo CI et al (2012) Effect of drying methods on the physical properties and microstructures of mango (Philippine 'Carabao' var.) powder. J Food Eng 111(1):135–148

Chen C, Zhang H, Xiao W et al (2007) High-performance liquid chromato graphic fingerprint analysis for different origins of sea buckthorn berries. J Chromatogr A 1154(1):250–259

Chu Y (2020) Overview of vacuum freeze drying technology for food. Anhui Agric Sci Bull 26(18):193–194

Cui Y, Jian D, Mingming Y, Wanchun Z, Jianfang LI, Xiang G (2016) Effects of sea buckthorn fruit powder and jujube powder on Flour Characteristics. Food Sci 37(11):24–29

Cui L, Yiyi F, Shiwei L et al (2021) Nutritional value and industry development of Sea-buckthorn. Food Res Dev 42(11):218–224

Dai D, Yinggui Y, Wanxin H (2015) Simultaneous determination of 7 organic acids in sea buckthorn fruit powder by ion chromatography. Physicochem Insp (Chemistry Branch) 51(03):304–306

Gao X, Ohlander M, Jeppsson N, BjRk L, Trajkovski V (2000) Changes in antioxidant effects and their relationship to phytonutrients in fruits of sea buckthorn (hippophae rhamnoides L.) during maturation. J Agric Food Chem 48(5):1485–1490

Gao W, Chen F, Wang X et al (2020) Recent advances in processing food powders by using superfine grinding techniques: a review. Compr Rev Food Sci Food Saf 19(4):2220–2224

Geetha S, Ram MS, Singh V et al (2002a) Anti-oxidant and immunomodulatory properties of seabuckthorn (*Hippophae rhamnoides*) – an in vitro study. J Ethnopharmacol 79(3):373–378

Geetha S, Sai Ram M, Singh V, Ilavazhagan G, Sawhney RC (2002b) Effect of sea buckthorn on sodium nitroprusside-induced cytotoxicity in murine macrophages. Biomed Pharmacother 56(9):463–467

Geetha S, Ram MS, Sharma SK et al (2009) Cytoprotective and antioxidant activity of seabuckthorn (*Hippophae rhamnoides L.*) flavones against tert-butyl hydroperoxide-induced cytotoxicity in lymphocytes. J Med Food 12(1):151–158

Guo R, Guo X, Li T et al (2017) Comparative assessment of phytochemical profiles, antioxidant and antiproliferative activities of sea buckthorn (*Hippophae rhamnoides L.*) berries. Food Chem 221:997–1003

He S, Yuan X, Wang J (2008) Development and utilization of Seabuckthorn resources in China. Sci Agric Sin 11:87–92

He C, Li M, Luo H et al (2015) Comparison of drought resistance of different Seabuckthorn varieties. Res For Sci 28(5):634–639

Himanshi T, Divya S, Divya S et al (2018) Anti-inflammatory activity of the functional groups present in *Hippophae rhamnoides* (Seabuckthorn) leaf extract. Inflammopharmacology 26(1):802–811

Hirazumi A, Furusawa E (1999) An immunomodulatory polysaccharide rich substance from the fruit juice. Morinda centifolia with antitumor activity. Phytother Res 13(5):380–387

Hong W, Zhai M, Li S et al (2018) Optimization of Seabuckthorn French Bread formulation by response surface method. Food Feed Ind (11):25–28+45

Ivanišová E, Kačániová M (2020) Biological properties of sea buckthorn (*Hippophae rhamnoides L.*) derived products. Acta Sci Pol Technol Aliment 19(2):195–205

Kim S, Choi Y, Lee H et al (2012) Physicochemical properties of jujube powder from Air, vacuum, and freeze drying and their correlations. J Korean Soc Appl Biol Chem 55(2):555–563

Kumar S, Tirpude R, Maheshwari D et al (2013) Antioxidant and antimicrobial properties of phenolic rich fraction of Seabuckthorn (Hippophae rhamnoides L.) leaves in vitro. Food Chem 141(4):982–995

Li L, Shi K, Bai J et al (1994) Effects of sea-buckthorn powder on immune function and cholesterol. J Northwest Pharm 9(5):218–221

Li X, Kong L, Liu H (2007a) Research progress on effective components of Sea buckthorn. J Jilin Agric Univ 29(2):162–167

Li Y, Zhang H, Wang P (2007b) Research progress in nutrient components and effects of sea buckthorn. Prim Health Care China 21(3):73

Li G, He Y, Ding X et al (2014) Study on preparation and immune function enhancement of Vp powder of sea buckthorn. Food Sci 21:229–233

Li S, Yan Z, Zhang Z (2017) Research on processing technology of fruit and vegetable powder. Henan Agric 8(6):53–55

Li Z, Wen K, Zhao J et al (2020) Effects of different drying methods on the content of active components in Seabuckthorn extracts. J Anhui Agric Sci 48(4):175–177+186

Li S, Chen X, Yang M et al (2021) Effects of Sea-buckthorn meal and Sea-buckthorn protoplasm on immune enhancement in mice. Feed Ind 42(22):37–43

Lian Y (2022) Characteristics analysis of lyophilized powder of different varieties of Xinjiang Big fruit sea Buckthorn and product development of effervescent tablets. Xinjiang Agricultural University

Lin P, Yao N, Che F et al (2022) Effects of drying methods on drying characteristics and quality of Sea Buckthorn. Food Ind Sci Technol 43(4):41–48

Liu Y (1998) Analysis of phytosterols in sea buckthorn oil. Sichuan Grain Oil Sci Technol 2:51–52

Liu Y, Lian Y, Wang Y, Li M, Xiao P (2014) Review on the research and development of Seabuckthorn and its significance. Chin J Tradit Chin Med 39(9):1547

Liu J, Cai Y, Wang X et al (2019) Development and rheological properties of Seabuckthorn nutritive meal replacement powder. Sci Technol Food Ind 40(8):163–169

Liu J, Cai Y, Zhou T et al (2020) Formulation optimization and quality analysis of Sea-buckthorn powder chewable tablets. Sci Technol Food Ind 5(2):1–11

Lu Y (1995) Chinese medicine chemistry. Academy Press, pp 5–3

Lu C, Shan Y, Liu H et al (2008) King of natural vitamins: application of sea buckthorn in food ingredients. China Food Addit (Supp. 1):229–295

Lu Y, Wang M, Li H, Yan Y (2015) Application status of spray drying technology in food industry. Anhui Agric Sci 43(11):276–278

Luo L, Cai Y, Sun J et al (2020) Analysis on the caking resistance of Seabuckthorn fruit residue to Seabuckthorn fruit powder. Mod Food Sci Technol 36(4):220–227+234

Luo F, Li J, Wang F (2021) Study on nutrients of Sea Buckthorn in Altay Area. Guide Food Saf 21(2):122–123

Ma Y (2020) Cultivation technology and resource development and utilization of sea-buckthorn. Agric Dev Equip 218(02):171

Ma Z, Cui Y (1987) Analysis of chemical constituents of Sea Buckthorn from China. Study on Botany of Wuhan 5(4):397–403

Ma N, He M, Lu P, Geng C, Liu M (2014) Study on antioxidant activity of sea buckthorn fruit powder. Int Res Dev 12(02):25–28

Naji T, Emadzadeh B, Shahidi N et al (2021) Physicochemical and antioxidant properties of barberry juice powder and its effervescent tablets. Chem Biol Technol Agric 8(23):1–11

Prakash H, Bala M, Ali A, Goel HC (2010) Modification of gamma radiation induced response of peritoneal macrophages and splenocytes by hippophae rhamnoides (rh-3) in mice. J Pharm Pharmacol 57(8):1065–1072

Priyanka S, Geetha S, Virendra S et al (2015) In vitro antioxidant profiling of sea buckthorn varieties and their adaptogenic response to high altitude-induced stress. Int J Biometeorol 59(8):392–401

Qiang W, Julina SJ, Suo Y (2014) Study on antioxidant activity of sea Buckthorn fruit powder from Qinghai. Chin Wild Plant Resour 33(01):24–27

Ren W (2019) Study on antioxidant, anti-aging, anti-fatigue and immune activities of Sea buckthorn polysaccharide. Xinjiang Agricultural University

Reyes A, Evseev A, Mahn A et al (2011) Effect of operating conditions in freeze-drying on the nutritional properties of blueberries. Int J Food Sci Nutr 62(3):358–369

Tang R, Guo X, Wang Y, Zou Q, Li X (2013) Study on preparation of ripe sweet potato whole powder by vacuum drying. Food Sci Technol 38(06):94–97

Ursache F, Ghinea I, Turturica M, Aprodu I, Rapeanu G, Stanciuc N (2017) Phytochemicals content and antioxidant properties of sea buckthorn (*Hippophae rhamnoides*) as affected by heat treatment quantitative spectroscopic and kinetic approaches. Food Chem 233(OCT.15):442

Wang C (2017) Study on compound hawthorn effervescent tablets. Jilin University, Changchun

Wang Z, Duan J, Shi C (2013) Determination of Lead, Arsenic and Mercury in Sea buckthorn and Sea buckthorn fruit powder by microwave digestion and double channel atomic fluorescence spectrophotometer. Contemp Chem Ind 42(11):1624–1626

Wang R, Yang X, Xu W et al (2017) Research progress of Sea buckthorn Polysaccharide. Food Res Dev 38(13):221

Wang R, Li S, Chen C et al (2018) Study on the immunomodulatory effects of Sea buckthorn polysaccharide on macrophages and immunosuppressed mice. Central South Pharm 18(3):384–388

Wang Y, Hao J, Yan S, Zhang T, Dang L, Wang X (2019) Preparation process optimization and quality analysis of sea-buckthorn freeze-dried powder effervescent tablets. Sci Technol Food Ind 44(10):235–241

Wang C, Wang H, Wang M et al (2020) Study on chemical constituents and antioxidant activities of Sea buckthorn fruit from different habitats. West China Pharmaceut J 35(5):513–517

Wang F, Chen H, Wang H et al (2021) Organic composition and tea suitability analysis of Seabuckans leaves in four regions of Shanxi Province. J Shanxi Agric Univ 41(3):17–24

Wu X, Xie Q, Li Q et al (2021) Study on quality evaluation of Sea buckthorn seed oil based on content determination of 7 fatty acids and stoichiometry. J Food Saf Qual Insp 12(20):8128–8135

Wu X, Liu J, Zhao Y et al (2022) Research progress of new drying techniques for fruits and vegetables. Chin Fruit Veg 42(1):2–7+15

Xu Q (1994) Pharmacology of Chinese medicine. Guizhou Science and Technology Press, Guiyang, pp 280–331

Xu X, Che H et al (1997) Effect of Sea buckthorn seed oil on Immune function of mice. Sea Buckthorn 10(3):2932

Xu B, Song C, Du J et al (2016) Effect of sea buckthorn powder on blood lipid and liver in rats with high fat diet protective effect of viscera. Chin J Vet Med 52(9):111–113

Yang B, Kallio H (2001) Fatty acid composition of lipids in sea buckthorn (*Hippophae rhamnoides L.*) berries of different origins. J Agric Food Chem 49(4):1011–1020

Yang F, Suo Y, Hu N (2016) Effects of Vp of sea-buckthorn on protection and immune function of radiation-damaged mice. Nat Prod Res Dev 28(12):1947–1951

Yang X, Feng F, Liu L, Chen Z (2017) Simultaneous determination of active substances in Seabuckthorn fruit powder from 3 habitats by high performance capillary zone electrophoresis. Food Sci 38(14):144–150

Yao N, Che F, Li Y et al (2020) Overview of nutritional value and comprehensive development and utilization of Sea-buckthorn. Preserv Process 20(2):226–232

Yildiz H, Sengul M, Celik F et al (2012) Bioactive content of sea buckthorn (*Hippophae rhamnoides* L.) berries from turkey. Agric Conspec Sci

Zhang M, Zhang CJ, Shrestha S (2004) Study on the preparation technology of superfine ground powder of Agrocybe chaxingu Huang. J Food Eng 67(3):333–337

Zhang F, Zhao Y, Liu Y, Suo Y (2010) Comparative analysis of Water-soluble vitamins in the fruit powder of Prickly thorn, Lycium barbarum and sea buckthorn on Qinghai-Tibet Plateau. Food Sci 31(02):179–182

Zhao H, Hao W (2012) Development of sea-buckthorn nutritional formula. Dairy Sci Technol 35(4):29–32

Zhong F, Jiang Y, Wu F et al (1989) Effect of total flavonoids of Sea-buckthorn on cellular immune function in mice. Chin Herb Med 20(3):43

Zhu L, Lu Y (2013) Production of single cell protein feed from citrus peel residue waste by fermentation. Food Ind Sci Technol 34(22):131–134

Zhuang X, Zhang W, Pang X et al (2012) Combined effect of total flavonoids from seed residues of *Hippophae rhamnoides L.* and zinc on advanced glycation end products-induced endo-thelial cell dysfunction. Food Chem 133(3):905–911

Zu Y, Li C, Fu Y et al (2006) Simultaneous determination of catechin, rutin, quercetin kaempferol and isorhamnetin in the extract of sea buckthorn (*Hippophae rhamnoides L.*) leaves by RP-HPLC with DAD. J Pharm Biomed Anal 41(3):714–719

Chapter 12
Research Progress on Extraction and Functional Activity of Sea Buckthorn Proteins and Polypeptides

Yue Zhao, Jinhu Tian, and Changqing Wei

Abstract Sea buckthorn proteins and peptides have received increasing attention as functional ingredients in food and pharmaceuticals due to their potential health benefits, for example, antioxidant and immunomodulatory activities. Various methods, including aqueous extraction, enzymatic hydrolysis, and membrane separation, have been used to extract these bioactive compounds from plants. This chapter summarizes and organizes the extraction methods of sea buckthorn proteins and peptides and compares their efficiency, yield, and quality. Additionally, this chapter analyzes the reported biological activities of sea buckthorn proteins and peptides, including their antioxidant, anti-inflammatory, and immunomodulatory activities, which are attributed to their mechanisms of action on human health. Finally, this chapter identifies gaps and limitations in current research on sea buckthorn proteins and peptides, such as the lack of human clinical trials and standardization of extraction and analysis methods. Future research should focus on conducting well-designed human clinical trials and standardizing extraction and analytical methods and establish their potential applications in food and medicine.

Keywords Sea buckthorn · Protein · Peptide · Functional activity

Y. Zhao
Food College, Shihezi University, Shihezi, China

J. Tian
Department of Food Science and Nutrition, Zhejiang University, Hangzhou, China

C. Wei (✉)
School of Food Science and Technology, Shihezi University, Shihezi, China

J. Tian et al. (eds.), *Sea Buckthorn*, https://doi.org/10.1007/978-981-97-9865-0_12

Abbreviations

DPPH	2,2-Diphenyl-1-picrylhydrazyl
FTIR	Fourier transform infrared spectrometry
HPLC	High-performance liquid chromatography
SDS-PAGE	Sodium dodecyl sulfate-polyacrylamide gel electrophoresis

12.1 Introduction

Sea buckthorn (*Hippophae rhamnoides* L.) belongs to an exceptionally valuable and ecological plant species with enormous medicinal values. It can control soil erosion and reduce the damage caused by wind and sand and can grow in extreme soil environments such as arid, barren, and saline-alkali land (Yang and Kallio 2002; Ruan and Li 2002). In China, sea buckthorn is mainly distributed in northwest, southwest, and north China. Sea buckthorn is rich in nutrients and contains a variety of bioactive substances, which are not only limited to the berries but also exist in the seeds and leaves of sea buckthorn (Yue et al. 2017; Ranjith et al. 2006; Vincze et al. 2007; Sharma et al. 2008). With the continuous growth of the global population, people's demand for food proteins is also increasing and food processing by-products have received extensive attention. Sea buckthorn fruit is rich in nutrients such as proteins, vitamins, and bioactive substances such as flavonoids, steroids, triterpenes, and tannins. It has the characteristics of homology of medicine and food and has been widely valued in the research and development of health products and food (Ciesarová et al. 2020). Various parts of sea buckthorn contain a number of proteins, and the leaves and fruits also contain a high amount of non-protein nitrogen (Gao et al. 2020). As a by-product of sea buckthorn processing, sea buckthorn seed protein has great potential for development. In recent years, researchers have focused on exploring the potential of sea buckthorn proteins and peptides as functional food ingredients. The protein in sea buckthorn meal is a high-quality protein and the amino acid composition is close to complete protein (Chen et al. 2023).

Proteins and peptides, as essential macronutrients, play pivotal roles in human physiological processes. They support tissue growth, organ maintenance, and metabolic regulation. Sea buckthorn proteins and peptides stand out due to their unique amino acid composition and potential health advantages. Numerous studies have isolated and characterized these compounds, highlighting their versatility for various functional applications (Xiang et al. 2020; Zhu et al. 2021, 2023). These sea buckthorn proteins and peptides hold promise across several sectors, including the food, healthcare, and cosmetics industries. Extracting and purifying these compounds are essential steps for functional characterization and potential utilization. A range of methods, from conventional techniques like solvent extraction, precipitation, and chromatography to advanced ones such as membrane separation, have been employed for their isolation (Lin et al. 2022).

Furthermore, the functional properties of sea buckthorn proteins and peptides have undergone extensive examination. They exhibit robust antioxidant activity, combating oxidative damage caused by free radicals. Additionally, they demonstrate anti-inflammatory and antimicrobial activities, suggesting potential applications in disease prevention and treatment (Yuan et al. 2016, 2018, 2019; Zhu et al. 2023).

Incorporating sea buckthorn proteins and peptides into food products can enhance their nutritional value and functional characteristics. For example, they improve emulsification, texture, sensory attributes, and foaming properties (Khan et al. 2023). These compounds also hold promise in the health industry, contributing to functional foods, dietary supplements, and pharmaceutical products that address various health conditions. Their benefits span antidiabetic, anticancer, and cardioprotective activities. Moreover, the cosmetic industry can harness sea buckthorn proteins and peptides as natural ingredients for skincare, hair care, and cosmetic products. Their applications include promoting skin hydration, enhancing skin elasticity, and reducing the appearance of wrinkles (Krejcarová et al. 2015).

In conclusion, sea buckthorn proteins and peptides have demonstrated versatile potential in the realms of health, cosmetics, and food industries. The subsequent sections of this review will comprehensively cover the progress made in extracting, assessing functional activities, and exploring potential applications of these compounds.

12.2 Extraction Methods of Sea Buckthorn Proteins and Peptides

With the development of protein technology, numerous studies have been done on the extraction of proteins from sea buckthorn seed meal. The extraction and purification of sea buckthorn proteins and peptides are important for the characterization of their functional properties and subsequent applications. Currently, there are many methods to extract sea buckthorn proteins and peptides, including alkali extraction, solvent extraction, and enzymatic extraction, in which alkali extraction is the main method (Uransanaa et al. 2003). This part will provide a detailed introduction and comparison of these methods.

12.2.1 Extraction Method of Sea Buckthorn Protein

12.2.1.1 Alkali Acid Extraction Method

Alkali acid extraction method is a common protein extraction method. The general process is to dissolve the raw material in sodium hydroxide solution and select the appropriate material–liquid ratio, extraction time, extraction temperature, pH, and other conditions for extraction. Then, the pH is adjusted to the isoelectric point of the protein to sink the protein, and then the protein is obtained after drying (Sari et al. 2015). To explore suitable protein extraction methods, three different deep eutectic solvents were applied. The results indicated that alkaline could effectively extract 56.9% protein from sea buckthorn seed meal and its protein content was 73.1% higher than deep eutectic solvents (Lin et al. 2022).

12.2.1.2 Alcohol Extraction

The protein extracted by the alcohol method is mostly alcohol-soluble protein. Cui et al. (2010) used sea buckthorn seed meal as the alcohol method to extract sea buckthorn seed protein. Based on the unifactor test, the process is optimized through orthogonal test and the influence of ethanol concentration, liquid ratio, extraction temperature, and extraction frequency on the protein content in the product were discussed. The best process parameters obtained by the orthogonal test are ethanol concentration 70%, liquid ratio 1:7, extraction temperature 55 °C, and extraction five times. The protein content in the product was 55.18% and the protein yield was 78.65%.

12.2.1.3 Solvent Extraction Method

Solvent extraction has several advantages, including its ability to selectively extract peptides from complex mixtures, its ease of use, and its versatility. Lin et al. (2022) developed three different deep co-crystal solvents to extract proteins from sea buckthorn meal and compared them with alkalinity. The results showed that deep co-crystal solvents resulted in less extracted product β-sheet, more β-turn, more essential amino acids and higher total amino acid content, and choline chloride-urea extraction was better than alkali extraction.

12.2.1.4 Composite Extraction Method

Sea buckthorn seed meal contains many kinds of proteins. Different extraction methods have been used to extract and purify sea buckthorn proteins with different efficiencies, yields, and quality. To improve the yield, various methods can be combined so that most of sea buckthorn proteins are extracted. Ling et al. (2012) used sea buckthorn seed meal as the raw material to extract the seed meal proteins by alkali extraction method and then the proteins are hydrolyzed in the alkali extraction residue with alkaline protease. After two steps of extraction, the total extraction rate of sea buckthorn proteins was 67.24%.

In a nutshell, the efficiency, yield, and quality of sea buckthorn proteins and peptide extraction depend on the specific extraction method and its operating conditions. Further optimization of extraction conditions and exploration of new extraction methods may lead to improved extraction efficiency and yield of sea buckthorn proteins. The choice of extraction method for sea buckthorn proteins depends on the specific characteristics of the target compounds, the desired yield and purity, and the cost-effectiveness of the method. The optimization of extraction conditions is crucial for maximizing the extraction yield and quality of sea buckthorn proteins. The use of multiple extraction methods in combination may also enhance the extraction efficiency and yield of target compounds (Shashirekha et al. 2013; Horax et al. 2011).

12.2.2 Peptides Extraction Methods

12.2.2.1 Enzymatic Hydrolysis

Enzymatic hydrolysis is another commonly used method for the extraction of sea buckthorn peptides. This method uses proteases to hydrolyze protein molecules into smaller peptide fragments and amino acids, which facilitates their extraction. Factors such as enzyme type, pH, temperature, and enzyme-to-substrate ratio can affect the efficiency of this method. Zuo et al. (2019) used sea buckthorn leaf powder as raw material to select alkaline protease as the best enzyme preparation and the best enzymatic process is arrived at through the orthogonal test, that is, 19.88% crude peptides with enzyme addition amount of 14000 U/g, liquid ratio of 1:50, pH 10.0, enzyme temperature of 45 °C, and enzyme lysis time of 2 h.

12.2.2.2 Membrane Separation

A separation technique that uses a semipermeable membrane to separate peptides from other components based on their size, charge, and hydrophobicity is membrane separation. This method can be applied to various types of membranes, including ultrafiltration, nanofiltration, and reverse osmosis membranes. The efficiency of this method is influenced by factors such as membrane pore size, pressure, flow rate, and the presence of co-solvents (Wei et al. 2012).

12.2.2.3 Electrophoresis Method

Since there are many kinds of peptides, it is necessary to purify and identify the desired peptide. Electrophoresis method is widely used for separating chemicals. Polyacrylamide gel electrophoresis is a common electrophoresis technique using polyacrylamide gel as a supporting medium for the separation of proteins and peptides. Kashyap et al. (2020) applied sodium dodecyl-sulfate polyacrylamide gel electrophoresis (SDS-PAGE) and matrix-assisted laser desorption ionization and fluorescence time analyzer for analysis and identification.

12.3 Characterization of Sea Buckthorn Proteins and Peptides

Sea buckthorn berries are regarded as a rich source of proteins and peptides with diverse structures and physical-chemical properties. This part aims to show the structure and properties of sea buckthorn proteins and peptides.

12.3.1 Structure of Sea Buckthorn Proteins

Sea buckthorn proteins are a complex mixture of various types of proteins, including albumins, globulins, prolamins, glutelins, and enzymes. The exact composition and abundance of these proteins can vary depending on factors such as the variety of sea buckthorn, the extraction method used, and the processing conditions (Krejcarová et al. 2015). The molecular weight of sea buckthorn proteins can also vary depending on the specific protein. Xiong et al. (2019) established a two-dimensional electrophoresis mass spectrometry system suitable for the protein extraction of sea buckthorn pulp and seed proteins by TCA–acetone method. The protein expression in the seed of sea buckthorn was lower than those in the pulp of sea buckthorn by two-dimensional electrophoresis and mass spectrometry system. The results of protein separation using different pH strips showed that the amount of acidic protein in the pulp of sea buckthorn was more than that of basic protein, but the seed protein of sea buckthorn contained more basic protein.

12.3.2 Structure of Sea Buckthorn Peptides

Only subtle differences in various amino acid contents were observed in sea buckthorn peptides compared to proteins. The molecular weight distribution of the sea buckthorn peptides is related to the enzymes used due to the different restriction sites in each enzyme. Most alkaline proteases act between the peptide chains, and the restriction sites are C-terminal aromatic or hydrophobic amino acids, such as tyrosine, phenylalanine, and alanine. Neutrophil protease is different, as an endonuclease, with the peptide bond generated by providing amino groups at hydrophobic macromolecular amino acids (Khan et al. 2023).

12.3.3 Methods for the Characterization of Sea Buckthorn Proteins and Peptides

The characterization of sea buckthorn proteins and peptides is important for their identification, purification, and application in various industries. Various analytical techniques are available for the characterization of proteins and peptides, including SDS-PAGE, high-performance liquid chromatography (HPLC), and Fourier transform infrared spectrometry (FTIR) spectroscopy (Sougrakpam et al. 2016; Wang et al. 2013; Li et al. 2020).

Optical analysis is one of the most important means to characterize the internal structure of animal and plant proteins. According to the properties of matter and radiation energy, it can be roughly divided into two categories: spectroscopy and non-spectroscopy. Currently, the spectroscopy commonly used for structural analysis of proteins is ultraviolet absorption light spectroscopy, fluorescence spectroscopy, circular dichroism spectroscopy, and Fourier transform infrared spectroscopy. Non-spectral techniques include dynamic light scattering methods and the small-angle X-ray scattering method (Adal et al. 2017; Tan et al. 2019; Zhang et al. 2019; Liu et al. 2023). These methods can also be applied to the structural characterization of sea buckthorn proteins and peptides. It is common to use SDS-PAGE to separate and characterize proteins grounded in their molecular weight. The proteins are then separated by size using SDS-PAGE. SDS-PAGE has been used to analyze the molecular weight and purity of sea buckthorn proteins (Kashyap et al. 2020; Wang et al. 2022). Another powerful technique for the separation and quantification of proteins and peptides based on their physicochemical properties is HPLC. In this technique, a stationary phase (typically a gel or resin) and a mobile phase (typically a buffer or solvent) are employed to partition analytes according to their size, charge, or hydrophobicity (Dołowy and Pyka 2014; Wang et al. 2022). FTIR spectroscopy is a non-destructive method for the characterization of proteins and peptides based on their vibrational modes. This technique utilizes infrared radiation to stimulate the vibrational modes of a sample, with the resulting spectrum employed to detect the functional groups within the sample (Arumugam et al. 2020). FTIR spectroscopy has been leveraged to determine the secondary structure of proteins and peptides derived from sea buckthorn (Khan et al. 2023).

12.4 Functional Activity of Sea Buckthorn Proteins and Peptides

Sea buckthorn is a nutritious plant whose fruits contain large amounts of vitamins, minerals, and antioxidants. In recent years, researchers have found that the antioxidant and immunomodulatory activities of sea buckthorn have a wide range of potential applications in medicine and food. This chapter will summarize a series of studies on the antioxidant and immunomodulatory activities of sea buckthorn proteins and peptides.

12.4.1 Antioxidant Activity

Antioxidants are substances that protect cells from oxidative damage caused by free radicals (Sultan 2014; Kunwar and Priyadarsini 2011). The antioxidant properties of peptides are related to factors such as molecular weight, amino acid composition, and enzyme type of polypeptides. When the content of amino acids such as tyrosine, tryptophan, histidine, cysteine, and methionine in the peptide chain is high, the antioxidant properties of peptides will also increase. Some studies have found that these amino acids can combine with peroxides to form imines or hydroxyl derivatives (Peng et al. 2021; Chen et al. 2021; Matsui et al. 2018; Nwachukwu and Aluko 2019). In recent years, many studies have focused on exploring the antioxidant components and mechanisms in sea buckthorn and its extracts. In their 2013 study, SaeRom et al. explored the antioxidant properties and cytoprotective mechanism of sea buckthorn fruit extract in mouse embryonic fibroblasts (Lim et al. 2013). The study revealed a cytoprotective effect mediated by medical clearance activity and altered cell cycle regulation, particularly in apoptotic cell death. In the same year, another paper demonstrated the high antioxidant activity of the extraction of sea buckthorn by cocatalytic lecithin liposome oxidation assay (Fan et al. 2013). The high acetone extract showed promising potential for DPPH radical scavenging. One of the main bioactivities of sea buckthorn proteins and peptides is their antioxidant activity. The sea buckthorn peptides were extracted using different methods and their antioxidant activity was evaluated. A significant change was found in the secondary structure and the peak intensity of the amido I, II, and III bands of sea buckthorn proteins by enzymatic hydrolysis. Antioxidant activity increased significantly for all the samples with different enzyme treatments (Khan et al. 2023). The results showed that the enzymatic hydrolysis method was the most effective for extracting sea buckthorn proteins with high antioxidant activity.

12.4.2 Immunomodulatory Activity

Immune function plays an important role in preventing and controlling chronic diseases. Recently, many immunomodulating peptides were prepared from food proteins such as mushrooms (Sheu et al. 2004), egg white (Yang et al. 2001), sheep bone proteins (Yang et al. 2009), green microalga chlorella vulgaris (Morris et al. 2007), and so on. Peptides have also been found to exhibit immunomodulatory activity. This means that they help to regulate the immune system, which plays a key role in defending the body against infection and disease (Hou et al. 2012). However, few immunomodulating proteins or peptides were separated from sea buckthorn. At present, the immunoregulatory activities of sea buckthorn seed peptides with different molecular weights have been studied, but activity has not been investigated systematically. According to animal experiments, the molecular weight of 500–2000 Da peptides can promote the growth of broilers, and the molecular weight of 2000–3500 Da peptides can significantly improve the disease-resistant immune function of broilers (Lian et al. 2012).

12.4.3 Anti-inflammatory Activity

Inflammation is a natural response of the immune system to injury or infection, but chronic inflammation can contribute to the development of many diseases (Newton and Dixit 2012; Poulose et al. 2015; Dvorakova and Landa 2017). Sea buckthorn proteins have been found to exhibit anti-inflammatory activity and shown to reduce inflammation in several animal studies, potentially by inhibiting the production of inflammatory cytokines. Sea buckthorn seed proteins have anti-inflammatory effects on streptozotocin-induced diabetic IRC mice and can reduce the levels of inflammatory factors in diabetic mice (Yuan et al. 2016). Meanwhile, the anti-inflammatory effects of components from leaves of sea buckthorn leaf in mouse peritoneal macrophages were investigated. The results showed that the extracts could significantly inhibit the production of NO and proinflammatory cytokines induced by lipopolysaccharide (Tanwar et al. 2018). These provided a basis for the biomedical applications.

12.4.4 Neuroprotective Activity

Some peptides act as neurotransmitters, which are signaling molecules that transmit signals between nerve cells (Mehan et al. 2022). Natural foods were reported to contain neuroprotective peptides include walnut (Wang et al. 2018), anchovy (Zhao et al. 2019), and so on. Sea buckthorn powder can significantly improve the cell viability induced by Aβ, up to 95%, and the inhibition rate of Aβ is up to 38%. The results indicated that sea buckthorn powder is a potential drug for improving Alzheimer's disease (Dong et al. 2020). Zhu et al. (2023) investigated the effects of sea buckthorn seed proteins and its arginine-rich peptides on memory impairment induced by D-galactose-induced brain aging in mice. Intake of sea buckthorn seed protein and peptide diets may be beneficial for memory impairment and further highlights the potential value of sea buckthorn seed proteins and its arginine-rich peptides in neuroprotection and improving cognitive function (Zhu et al. 2023).

12.4.5 Hypoglycemic Activity

In 2018, Yuan et al. conducted a study on the effect of sea buckthorn seed proteins on the distribution of intestinal microbial community in type 2 diabetic mice. The results of the study showed that the weight and blood sugar levels of the mice supplemented with sea buckthorn seed proteins decreased significantly, while the numbers of certain beneficial bacteria recovered. In addition, through the amplification of the 16S rDNA gene, the researchers found that the intestinal microbial species in the treatment group were adjusted to some extent (Yuan et al. 2018; Shu et al. 2020). The changes of hepatic glucose metabolism and related signaling pathways were also explored sea buckthorn seed proteins can reduce insulin resistance, inhibit the expression of hepatic glucose metabolism genes, and upregulate related signaling pathways to improve the symptoms of diabetic mice (Yuan et al. 2019). Yuan et al. (2016) investigated the hypoglycemic and anti-inflammatory effects of sea buckthorn seed proteins on streptozocin-induced diabetes in a mouse model. They found that sea buckthorn proteins reduced insulin resistance and controlled the effects of related inflammatory factors. These studies provide us with new insights into how sea buckthorn seed proteins affect diabetes and provide valuable references for future drug development and food applications.

12.4.6 Antifreeze Proteins

Cold damage is mitigated by limiting the growth of ice crystals. Sea buckthorn antifreeze proteins help cope with freezing conditions, and ice-binding proteins prevent water molecules from joining the growing crystal plane by noncollinear adsorption (Deswal 2015). The antifreeze proteins of sea buckthorn cope with freezing conditions. These are ice-binding proteins that prevent water molecules from joining the growing crystal plane by means of noncollinear adsorption. The antifreeze proteins in the leaves and berries of sea buckthorn were purified, and it was found that the berry antifreeze proteins had homology with the disease resistance proteins. The ice recrystallization inhibition test confirmed that the purified leaf and berry antifreeze proteins had high ice recrystallization inhibition activity, and the cold stress regulated target interaction, which provided a basis for crop improvement of freeze-sensitive crops or biomedical applications (Sharma et al. 2018).

Overall, the diverse structures and functions of proteins and peptides make them important biomolecules with a wide range of potential biological activities. Peptides have been investigated for their potential therapeutic applications in various diseases, including infections, metabolic disorders, and neuroprotective activity. However, further research is needed to fully understand the mechanisms underlying their effects and to explore their potential clinical applications.

12.5 Challenges and Potential Solutions

Although sea buckthorn proteins and peptides have shown promising potential for various health benefits, there are still some gaps and limitations in current research. One limitation is that most studies have been conducted in vitro or in animal models, and there is a lack of human clinical trials to confirm the health benefits of these compounds. Additionally, the bioavailability of sea buckthorn protein and peptides in humans is not well understood, and their optimal dosages and administration methods have not been established. Another limitation is the lack of standardization in sea buckthorn protein and peptide extraction and analysis methods. Different extraction methods and analytical techniques can result in varying yields and compositions of these compounds, making it difficult to compare results across studies.

To address these limitations, future research should focus on conducting well-designed human clinical trials to confirm the health benefits of sea buckthorn proteins and peptides. Additionally, standardization of extraction and analysis methods should be established to ensure consistency and comparability across studies.

References

Adal E, Sadeghpour A, Connell S et al (2017) Heteroprotein complex formation of bovine lactoferrin and pea protein isolate: a multiscale structural analysis. Biomacromolecules 18(2):625

Arumugam V, Venkatesan M, Ramachandran K et al (2020) Purification, characterization, and antibacterial properties of peptide from marine ascidian Didemnids. Int J Pept Res Ther 26:201–208

Chen M, Ning P, Jiao Y et al (2021) Extraction of antioxidant peptides from rice dreg protein hydrolysate via an angling method. Food Chem 337:128069

Chen A, Feng X, Dorjsuren B et al (2023) Traditional food, modern food, and nutritional value of sea buckthorn (Hippophae rhamnoides L.): a review. J Future Foods 3(3):191–205

Ciesarová Z, Murkovic M, Cejpek K et al (2020) Why is sea buckthorn (*Hippophae rhamnoides L.*) so exceptional? A review. Food Res Int 133:109170

Cui M, Tang N, Guo J et al (2010) Technology study of extraction of sea buckthorn seed protein by alcohol method. Grain Oil Process 10:137–139

Deswal R (2015) Antifreeze proteins from a Himalayan shrub sea buckthorn and their potential biotechnological applications.

Dołowy M, Pyka A (2014) Application of TLC, HPLC and GC methods to the study of amino acid and peptide enantiomers: a review. Biomed Chromatogr 28(1):84–101

Dong K, Fernando W, Durham R et al (2020) A role of sea buckthorn on Alzheimer's disease. Int J Food Sci Technol 55(9):3073–3081

Dvorakova M, Landa P (2017) Anti-inflammatory activity of natural stilbenoids: a review. Pharmacol Res 124:126–145

Fan J, Zhang J, Song H et al (2013) Antioxidant activity and phenolic components of sea buckthorn (*Hippophae rhamnoides*) seed extracts[C]//Proceedings of the 2013 international conference on advanced mechatronic systems. IEEE:96–101

Gao Y, Hou Z, Ren D (2020) Preparation and oxidation stability evaluation of sea buckthorn seed oil microcapsules. Sci Technol Food Ind 41(20):169–176

Horax R, Hettiarachchy N, Kannan A et al (2011) Protein extraction optimisation, characterization, and functionalities of protein isolate from bitter melon (*Momordica charantia*) seed. Food Chem 124(2):545–550

Hou H, Fan Y, Li B et al (2012) Purification and identification of immunomodulating peptides from enzymatic hydrolysates of Alaska pollock frame. Food Chem 134(2):821–828

Kashyap P, Kumar S, Singh D (2020) Performance of antifreeze protein HrCHI4 from *Hippophae rhamnoides* in improving the structure and freshness of green beans upon cryopreservation. Food Chem 320:126599

Khan ZS, Sodhi NS, Fayaz S et al (2023) Valorization of sea buckthorn seed protein to hydrolysates: impact on morphological, structural, functional, and antioxidant properties. JSFA Rep 3(5):222–232

Krejcarová J, Straková E, Suchý P et al (2015) Sea buckthorn (Hippophae rhamnoides L.) as a potential source of nutraceutics and its therapeutic possibilities – a review. Acta Vet Brno 84(3):257–268

Kunwar A, Priyadarsini KI (2011) Free radicals, oxidative stress and importance of antioxidants in human health. J Med Allied Sci 1(2)

Li T, Shi C, Zhou C et al (2020) Purification and characterization of novel antioxidant peptides from duck breast protein hydrolysates. LWT 125:109215

Lian et al (2012) Active peptide of animal feed made by the remnant of seabuckthorn seed. Academic Periodical of Farm Products Processing

Lim S, Go E, Go G, Shin H, Sung J (2013) Antioxidative mechanisms of sea buckthorn fruit extract in mouse embryonic fibroblast cells. Food Sci Biotechnol 22:197–204

Lin J, Xiang H, Sun-Waterhouse D et al (2022) Deep eutectic solvents and alkaline extraction of protein from sea buckthorn seed meal: a comparison study. Food Sci Human Wellness 11(4):1028–1035

Ling M, Cui M, Zhao C et al (2012) Study on the alkali enzyme two-step extraction process and function of sea buckthorn seed meal protein. Food Industry Science and Technology 33(17):240–244

Liu C, Lv N, Song Y et al (2023) Interaction mechanism between zein and β-lactoglobulin: insights from multi-spectroscopy and molecular dynamics simulation methods. Food Hydrocoll 135:108226

Matsui R, Honda R, Kanome M et al (2018) Designing antioxidant peptides based on the antioxidant properties of the amino acid side-chains. Food Chem 245:750–755

Mehan S, Bhalla S, Siddiqui E et al (2022) Potential roles of glucagon-like peptide-1 and its analogues in dementia targeting impaired insulin secretion and neurodegeneration. Degener Neurol Neuromuscular Dis:31–59

Morris H, Carrillo O, Almarales A et al (2007) Immunostimulant activity of an enzymatic protein hydrolysate from green microalga Chlorella vulgaris on undernourished mice. Enzym Microb Technol 40(3):456–460

Newton K, Dixit V (2012) Signaling in innate immunity and inflammation. Cold Spring Harb Perspect Biol 4(3):a006049

Nwachukwu I, Aluko R (2019) Structural and functional properties of food protein-derived antioxidant peptides. J Food Biochem 43(1):e12761

Peng L, Kong X, Wang Z et al (2021) Baijiu vinasse as a new source of bioactive peptides with antioxidant and anti-inflammatory activity. Food Chem 339:128159

Poulose S, Thangthaeng N, Miller M et al (2015) Effects of pterostilbene and resveratrol on brain and behavior. Neurochem Int 89:227–233

Ranjith A, Kumar K, Venugopalan V et al (2006) Fatty acids, tocols, and carotenoids in pulp oil of three sea buckthorn species (*Hippophae rhamnoides*, H. salicifolia, and H. tibetana) grown in the Indian Himalayas. J Am Oil Chem Soc 83(4):359–364

Ruan C, Li D (2002) Community characteristics of Hippophae rhamnoides forest and water and nutrient condition of the woodland in Loess Hilly Region. J Appl Ecol 13(9):1061–1064

Sari Y, Mulder W, Sanders J et al (2015) Towards plant protein refinery: review on protein extraction using alkali and potential enzymatic assistance. Biotechnol J 10(8):1138–1157
Sharma U, Sharma K, Sharma N et al (2008) Microwave-assisted efficient extraction of different parts of Hippophae rhamnoides for the comparative evaluation of antioxidant activity and quantification of its phenolic constituents by reverse-phase high-performance liquid chromatography (RP-HPLC). J Agric Food Chem 56(2):374–379
Sharma B, Sahoo D, Deswal R (2018) Single-step purification and characterization of antifreeze proteins from leaf and berry of a freeze-tolerant shrub sea buckthorn (Hippophae rhamnoides). J Sep Sci 41(20):3938–3945
Shashirekha M, Mallikarjuna S, Rajarathnam S (2013) Status of bioactive compounds in foods, with focus on fruits and vegetables. Crit Rev Food Sci Nutr *55*(10):1324–1339
Sheu F, Chien P, Chien A et al (2004) Isolation and characterization of an immunomodulatory protein (APP) from the Jew's Ear mushroom Auricularia polytricha. Food Chem 87:593–600
Shu, Jian, Liu et al (2020) Hypoglycemic activity and renal protection effect of sea buckthorn seed protein peptide in db/db mice. Technol Food Ind 41(21):317–321
Sougrakpam Y, Deswal R (2016) Hippophae rhamnoides N-glycoproteome analysis: a small step towards sea buckthorn proteome mining. Physiol Mol Biol Plants 22(4):1–12
Sultan S (2014) Reviewing the protective role of antioxidants in oxidative stress caused by free radicals. Asian Pac J Health Sci 1(4):401–406
Tan L, Hong P, Yang P et al (2019) Correlation between the water solubility and secondary structure of tilapia-soybean protein co-precipitates. Molecules 24(23):4337
Tanwar H, Shweta SD et al (2018) Anti-inflammatory activity of the functional groups present in Hippophae rhamnoides (Seabuckthorn) leaf extract. Inflammopharmacology 26:291–301
Uransanaa M, Gerel D, Jamyansan Y et al (2003) Protein and amino acid composition of sea buckthorn seeds (Hippophae rhamnoides mongolica Rouse). Mong J Biol Sci 1(1):85–88
Vincze I, Bányai-Stefanovits É, Vatai G (2007) Concentration of sea buckthorn (*Hippophae rhamnoides L.*) juice with membrane separation. Sep Purif Technol 57(3):455–460
Wang J, Su Y, Jia F et al (2013) Characterization of casein hydrolysates derived from enzymatic hydrolysis. Chem Cent J 7:1–8
Wang S, Su G, Zhang Q et al (2018) Walnut (Juglans regia) peptides reverse sleep deprivation-induced memory impairment in rat via alleviating oxidative stress. J Agric Food Chem 66(40):10617–10627
Wang D, Li W, Yao Y et al (2022) Research progress on extraction and functional activity of sea buckthorn protein and polypeptides. 43(3):447–455
Wei M, Tang L, Yang Q et al (2012) Membrane separation and antioxidant activity research of peanut peptides. Appl Mech Mater 209:2009–2012
Xiang H, Waterhouse D, Liu P et al (2020) Pancreatic lipase-inhibiting protein hydrolysate and peptides from sea buckthorn seed meal: preparation optimization and inhibitory mechanism. LWT 134:109870
Xiong C, Ryan C, Wu B et al (2019) Establishment of two-dimensional electrophoresis-mass spectrometry system for pulp and seeds protein of sea buckthorn. Mol Plant Breed 17(8):2600–2606
Yang B, Kallio H (2002) Effects of harvesting time on triacylglycerols and glycerophospholipids of sea buckthorn (Hippophae rhamnoides L.) berries of different origins. J Food Compos Anal 15(2):143–157
Yang M, Eoum H, Na K, Araki S et al (2001) Enhanced phagocytic activity of neutrophils caused by administration of egg white derivatives (EWD) in cats injected with cyclophosphamide (CPA). J Vet Med Sci 63:269–274
Yang H, Liu Y, Ma L, Kong B (2009) Hydrolyzing condition andimmunocompetence of sheep bone protein enzymatic lysates. Agric Sci China 8:1332–1338
Yuan H, Zhu X, Wang W et al (2016) Hypoglycemic and anti-inflammatory effects of sea buckthorn seed protein in diabetic ICR mice. Food Funct 7(3):1610–1615
Yuan H, Shi F, Meng L et al (2018) Effect of sea buckthorn protein on the intestinal microbial community in streptozotocin-induced diabetic mice. Int J Biol Macromol 107:1168–1174

Yuan H, Meng L, Wang W et al (2019) The effects of sea buckthorn seed protein on glucose metabolism in streptozotocin-induced diabetic ICR mice. Pak J Pharm Sci 32(5):2011–2017

Yue X, Shang X, Zhang Z et al (2017) Phytochemical composition and antibacterial activity of the essential oils from different parts of sea buckthorn (Hippophae rhamnoides L.). J Food Drug Anal 25(2):327–332

Zhang N, Guo Q, Shi Y et al (2019) Preparation of casein non-phosphopeptide–soybean polypeptide complex, its structure and emulsifying properties' evaluation. Eur Food Res Technol 245:355–363

Zhao T, Zheng L, Zhang Q et al (2019) Stability towards the gastrointestinal simulated digestion and bioactivity of PAYCS and its digestive product PAY with cognitive improving properties. Food Funct 10(5):2439–2449

Zhu X, Wang W, Cui C (2021) Hypoglycemic effect of hydrophobic BCAA peptides is associated with altered PI3K/Akt protein expression. J Agric Food Chem 69(15):4446–4452

Zhu X, Cai L, Liu J et al (2023) Effect of sea buckthorn seed protein and its arginine-enriched peptides on combating memory impairment in mice. Int J Biol Macromol 232:123409

Zuo L, Jiao T, Zou L et al (2019) Study on the preparation technology of sea buckthorn leaf polypeptide. Food Ferment Technol 55(1):30–34

Chapter 13
Protein Extraction Process and Anti-fatigue Activity of Sea Buckthorn Seed

Wei Wang

Abstract The proteins from sea buckthorn were briefly introduced, and different extraction processes of proteins were discussed. Then, the properties of protein from sea buckthorn were analyzed, such as digestibility, antioxidant activity, anti-fatigue activity, inflammation, etc. Finally, other functional compounds from sea buckthorn protein were also introduced, and their application in health food processing was examined.

Keywords Sea buckthorn · Protein · Extraction · Function

Abbreviations

SSM	sea buckthorn seed meal
SSP	sea buckthorn seed protein
DES	deep eutectic solvent
FI	florence intensity
SDS	sodium dodecyl sulfates
EHT	electron high tension
5-HT	serotonin
DA	dopamine
NE	noradrenaline
CIF	chemotherapy-induced fatigue
AMPK	adenylate-activated protein kinase
POE	P. Orientale
SOD	superoxide dismutase
GPx	glutathione peroxidase
NO	nitric oxide

W. Wang (✉)
School of Biological & Chemical Engineering, Zhejiang University of Science & Technology, Hangzhou, China

J. Tian et al. (eds.), *Sea Buckthorn*, https://doi.org/10.1007/978-981-97-9865-0_13

5-FU	5-fluorouracil
CAT	catalase
TRF	tocotrienol-rich fraction
PUFA	polyunsaturated fatty acid
BCAAs	branched chain amino acids

13.1 Protein in Sea Buckthorn

The protein content in different types of sea buckthorn berries exhibits significant variation. In wild sea buckthorn found in Mongolia, although the seeds constitute merely 7.2% of the fruit's fresh weight, they contain 37.79% of the protein in the whole fruit. This positions sea buckthorn seeds as a vital source of complete proteins, offering all essential and conditionally essential amino acids. The abundance and variety of protein and amino acid types render sea buckthorn seeds an outstanding dietary protein source (Xiang et al. 2020).

Traditionally, the sea buckthorn industry has focused on processing the fruit's flesh into edible products like jam, puree, and juice. Due to their inedibility, the seeds are separated from the fruit to extract seed oil. Post oil extraction, the protein content in the sea buckthorn seed meal (SSM) can exceed 20%, yet this resource is often relegated to animal feed or discarded. Therefore, the current processing methods do not fully utilize the rich protein resources in sea buckthorn seeds, resulting in a great waste of resources. Advancements in the extraction processes of sea buckthorn seed proteins are crucial as they will not only deepen the development of these proteins but also amplify the nutritional value and enhance the comprehensive utilization of sea buckthorn.

13.2 Extraction Process of Sea Buckthorn Seed Protein

The preparation of sea buckthorn seed protein (SSP) primarily involves three steps: degreasing, removal of flavonoids, and extraction.

13.2.1 Sample Preparation

After homogenizing the sea buckthorn fruits in water with a blender, the fruit flesh and peel are removed. Subsequently, the settled seeds are collected. These isolated seeds are then briefly soaked in a minimal volume of methanol to remove excess water and facilitate the drying process. After draining, the seeds are air-dried to reduce their moisture content to approximately 10%. Finally, the air-dried seeds are stored at room temperature for further use (Mehta et al. 2013).

13.2.2 Degreasing

Sea buckthorn seeds were ground into powder form and sieved through a 20-mesh sieve for subsequent experiments.

13.2.2.1 Cold Pressing

Traditionally, sea buckthorn seed oil has been extracted using a method known as cold pressing. This method employs a hydraulic press to mechanically apply pressure to the sea buckthorn seed powder, which squeezes out the oil and leaves behind the SSM. However, cold pressing might not remove all the oil from the seeds, indicating its potential inefficiency. Despite being a straightforward and environmentally friendly technique, cold pressing does not guarantee complete oil extraction (Li et al. 2007).

13.2.2.2 Solvent Extraction

Solvent extraction is commonly used for industrial-scale defatting of sea buckthorn seeds and utilizes organic solvents, such as hexane or petroleum ether, to isolate the lipid fraction from the seeds. The sea buckthorn seed powder is mixed with the solvent, and the mixture is then shaken or stirred to facilitate the extraction. After extraction, the solvent is evaporated, leaving behind the SSM.

Ultrasound-assisted extraction is an emerging technology in food and bioprocess engineering in recent years. During the extraction process, an ultrasonic system with a probe-type tip is employed to pretreat the sea buckthorn seed powder before the conventional solvent extraction. Compared to conventional extraction techniques, the application of ultrasound-assisted extraction produces higher oil yield with better retention of bioactive in bio-matrices and reduces extraction time while utilizing a lower volume of solvent-to-feed ratio (Sanwal et al. 2021).

13.2.2.3 Supercritical Fluid Extraction

Supercritical fluid extraction has appeared as a cutting-edge technology for extracting essential oils and natural products. Supercritical fluid extraction utilizes a supercritical fluid, such as carbon dioxide, to extract lipids from sea buckthorn seed powder. Under specific temperature and pressure conditions, carbon dioxide reaches a supercritical state, acting as both a liquid and a gas, which enhances its solubility and extraction power. Furthermore, as the solvent can be easily removed from the product by depressurization, no solvent traces remain in the extracted oil products (Zheng et al. 2017).

13.2.3 Protein Extraction

There are many types of proteins, each exhibiting considerable variation in properties. Most of the proteins are water-soluble proteins, which can be dissolved in water, dilute acid, or dilute alkali solution. Conversely, fat-soluble proteins can be dissolved in organic solvents such as ethanol, propanol, and butanol. Additionally, the solubility of proteins in different solvents varies due to differences in their molecular structures. External factors such as temperature, pH value, and extraction time can significantly influence protein solubility. Therefore, the selection of the SSP extraction method should be based on the type of protein in the raw material and the purpose of extraction.

13.2.3.1 Solid-Liquid Extraction

Solid-liquid extraction utilizes the properties of volatile organic solvents to transfer certain components from the raw materials into the solvent. Subsequently, the organic solvents are recovered through processes such as evaporation or distillation, yielding the desired components.

Ethanolic Extraction (Removal of Flavonoids)

Flavonoids are easily soluble in water, ethanol, methanol, and other polar solvents. Ethanol is being less toxic and easily recyclable through evaporation, so the use of ethanol for protein extraction simultaneously serves the purpose of removing flavonoids and other pigments.

An appropriate amount of SSM was taken, and it was mixed with 5 parts of 60% ethanol. The mixture was then extracted by shaking or stirring at approximately 50 °C for a certain period. The precipitate was extracted by repeating the extraction several times as described above until 0% protein was found in the washing liquid. The resulting solid phase was dried in a ventilated place to ensure complete ethanol evaporation, then pulverized, and sieved through an 80-mesh sieve to obtain SSP. The liquid phase was evaporated to recover ethanol. Meanwhile, the utilization of flavonoids in the liquid phase holds significant importance in the comprehensive processing of sea buckthorn seeds.

Deep Eutectic Solvent

Deep eutectic solvents (DES) are combinations of two or three low-cost self-incorporating components with melting points lower than each component's melting point. They are environmentally friendly solvents that are beneficial for efficiently extracting bioactive compounds from plant material. Furthermore, DES exhibits promising biodegradability, sustainability, and low toxicity. Protein extraction using DES can overcome the drawbacks associated with corrosive solvents in traditional methods, presenting a new, environmentally friendly, and sustainable extraction method. DES has been successfully employed to extract proteins from a wide range of plant and animal materials. Common DES solvents include hydrogen bond donors such as glycerol and polyethylene glycol, as well as hydrogen bond acceptors like choline chloride and choline chloride-urea.

Lin et al. (2022) developed three distinct DES (ChCl-Gly, ChCl-OA, and ChCl-urea) to extract protein from SSM. SSM was crushed into powder using a pulverizer, then sieved using a 40× mesh. DES were made by combining choline chloride with glycerol, oxalic acid, or urea in molar ratios of 1:2, 1:1, and 1:2, respectively, and stirring continuously at 60 °C until the combination became a clear, colorless liquid. Next, 50 g of SSM powder was combined with DESs in a 1:9 mass ratio and incubated for 3 h at 60 °C with an oscillation rate of 250 r/min. The resultant mixture was centrifuged at 10,000 ×*g* and 25 °C for 10 min. The DES grade was combined with ethanol (ChCl-OA) or distilled water (ChCl-Gly and ChCl-urea) at a 1:40 mass ratio. The resultant precipitate was centrifuged under the above-mentioned conditions and rinsed with distilled water until it achieved neutral pH. Finally, the neutralizing precipitate was freeze-dried for 48 h.

Proteins extracted using DES were compared to those extracted using alkali. After tests such as the determination of crude protein content and recovery, analysis of amino acid composition, and evaluation of in vitro protein digestibility, it was found that DESs (ChCl-Gly, ChCl-OA, and ChCl-urea) as green solvent could effectively extract proteins from SSM. Compared to the conventional extraction method using alkaline, ChCl-urea successfully extracted and preserved the native protein units in SSM raw material. Additionally, protein extracted with ChCl-urea showed a higher total amino acid content and better digestibility in vitro than the alkali-extracted SSM product. Accordingly, ChCl-urea may be considered as an alternative method for extracting the proteins from SSM.

Alkali-Soluble Acid Precipitation

Alkali-soluble acid precipitation, also known as Alkali extraction, is a classical protein extraction method widely used to extract proteins from samples such as fruits and seeds.

Separation and purification are achieved by exploiting the acid-base differences in the components of a mixture. Hydrogen bonds in protein molecules are broken under alkaline conditions, and ionic charges form on amino acid residues. However, when the pH is lowered, these charges are neutralized, resulting in protein precipitation.

The extraction method was also referred to by Lin et al.: 50 g of SSM powder was combined with distilled water in a weight-to-1:9, and the pH of the aqueous solution was stabilized at 11 using 1 mol/L NaOH solution. The mixture was then incubated at 60 °C for 3 h at a rate of 250 r/min. The resultant mixture was centrifuged at 10,000 ×*g* and 25 °C for 10 min. The supernatant was collected, and the pH was adjusted to 5 with a 1 mol/L HCl solution. The precipitate was produced by centrifugation under the abovementioned conditions, and the supernatant was incubated at 60 °C for 3 h at an oscillation rate of 250 r/min. The precipitate was then centrifuged under the conditions described above and rinsed with distilled water until its pH was neutralized. Finally, the neutralized precipitate was freeze-dried for 48 h to obtain the SSP (Li et al. 2007). However, the traditional alkali-soluble acid precipitation method often overlooks the separated residue, leading to suboptimal extraction of proteins from sea buckthorn. Recognizing the limitations of the alkali-soluble acid precipitation method, numerous studies have combined enzymatic extraction with alkaline-acid precipitation to treat the separated residue, thereby enhancing the extraction efficiency of SSP.

Enzymatic Extraction

Enzymatic selectively degrades non-protein components, such as cell walls, membranes, and nucleic acids, while preserving the integrity and functionality of the proteins. Enzymatic extraction is another effective method for extracting proteins from sea buckthorn seeds. In this approach, specific enzymes, such as proteases, are used to break down protein complexes and facilitate the release of proteins. The enzymatic extraction process can be optimized by adjusting factors such as enzyme concentration, pH, temperature, and extraction time. The enzyme extraction process was referred to Khan et al. The protein dispersions were incubated at optimal catalytic temperature and pH conditions for individual enzymes (pepsin, trypsin, and protamex). The hydrolysis was carried out under the following conditions: (A) Pepsin: at pH 2 and a temperature of 37 °C, (B) Trypsin: at pH 7 and a temperature 38 °C, (C) Protease: at pH 7 and a temperature 37 °C. The hydrolysis duration for each enzyme was 30 min. After 30 min of hydrolysis, enzymatic activity was halted by placing all protein enzyme dispersions in a water bath maintained at a temperature of 90 °C. The resulting hydrolysates were freeze-dried and stored at −20 °C (Khan et al. 2022).

13.3 Structural and Functional Properties of SSP

After extracting SSP from sea buckthorn seeds, it is essential to conduct structural and functional characterization. This process involves referencing studies by Khan et al. (2022 and Lin et al. 2022).

13.3.1 Determination of Solubility

Protein suspensions containing 5% (w/v) SSP samples were mixed on a magnetic stirrer at 25 °C for 45 min. The solubility of SSP samples was evaluated at pH levels ranging from 1 to 12 using 0.1N NaOH and 0.1 N HCl solutions. The pH-adjusted SSP samples were centrifuged at 10,000 rpm for twenty minutes. The proportion of protein in the supernatant was determined using the Kjeldahl technique, and solubility was computed using Eq. (13.1).

$$\%\text{solubility} = \frac{\text{Protein content in treated supernatants}}{\text{Protein content in protein concentrate}} \times 100\% \qquad (13.1)$$

13.3.2 Amino Acid Composition Analysis

An automated amino acid analyzer was used to examine the amino acid content of SSP. In short, before analysis, SSP-grade fractions (15 mg) were hydrolyzed for 24 h in a vacuum hydrolysis tube using a 6 mol/L HCl solution. The percentage of each amino acid to the total amount of proteins or amino acids was used to express the results.

13.3.3 In Vitro Protein Digestibility

The in vitro protein digestibility was initially tested using the pepsin digestibility assay, with minimal changes. SSP (0.2 g) was distributed in 20 mL of distilled water before being adjusted to pH 2 using a 0.5 mol/L HCl solution. Then, 4% pepsin (g/g, protein-based) was introduced. The resultant mixtures were incubated at 37 °C for 1 h at a rate of 150 r/min, followed by 10 min in a boiling water bath to inactivate the enzymes. The mixtures were centrifuged at 10,000 ×*g* and 4 °C for 10 min. The supernatant was collected for protein content analysis using Dumas combustion with the TruMac N system (Leco Corporation, St. Joseph, MI, USA). Protein digestibility was calculated according to Eq. (13.2).

$$\text{Protein digestibility} = \frac{\text{protein weight of supernatant}}{\text{protein weight of sample}} \times 100\% \qquad (13.2)$$

13.3.4 Fourier Transformation Infrared Spectroscopy

Infrared spectroscopy was used to determine and explain the secondary structure of SSP. The FTIR spectra of SSP were acquired in 11 scans using a Fourier transform spectrophotometer (Perkin Elmer EN 60825-1:2007) with a resolution of 4 cm^{-1} and a range of 4000–400 cm^{-1}.

13.3.5 Surface Hydrophobicity

The surface hydrophobicity of SSP solutions was calculated as follows: Protein solution was prepared at a concentration of 10 mg/ml (w/v) in phosphate buffer containing 0.5 ml of 1-anilino-8-naphthalenesulfonate. The SSP solutions were then vortexed for approximately 25 min on a vortex shaker. Florence intensity (FI) of SSP solutions was measured using a fluorescence spectrophotometer at an excitation wavelength of around 390 nm (with a 5–10 nm slit) and emission wavelengths ranging from 400 to 550 nm (with a 5–10 nm slit) at a scanning speed of 10 nm per second. Hydrophobicity was calculated according to Eq. (13.3).

$$H = S_1 - S_2 \tag{13.3}$$

Where S_1 is the area of the sample solution and S_2 is the area of the solvent.

13.3.6 Molecular Weight Distribution

The molecular weight distribution of SSP samples was determined using the SDS-PAGE method. SDS-PAGE of SSP samples was performed using electrophoresis equipment (Thermo Fischer Scientific) with 10% separating gel and 4% stacking gel. A buffer solution of 0.5 mol/L Tris-HCL (pH 6.8), 50% glycerol, 10% sodium dodecyl sulfate (SDS), 2% β-mercaptoethanol, and 1% bromophenol blue was incubated at 24 °C for 2 h before heating to 90 °C for 4 min. All SSP samples were put onto the gel in aliquots of 10 μL each. 50 V was used for the gel electrophoresis procedure. The gels were first stained overnight with 30% methanol and 10% acetic acid, and then with a staining solution consisting of 0.1% coomassie blue, 40% methanol, and 10% acetic acid. Proteins ranging in molecular weight from 10 to 25 kDa were utilized as standards for a broad variety of protein molecular weight indicators.

13.3.7 Determination of Microstructure of Protein

The morphology of the SSP samples was assessed using a field emission scanning electron microscope with an electron high tension (EHT) of 5.00 kV, a working distance of 4.2 mm, and a magnification of 1000×. Prior to imaging, the SSP sample was gold-coated to a thickness of 10 nm to increase conductivity. Sea buckthorn seeds are rich in proteins and amino acids, yet this aspect is often neglected in traditional processing. In the future, research on protein extraction methods from sea buckthorn seeds should be expanded, and the optimal extraction process should be identified based on the production demand. By focusing on the utilization of sea buckthorn seeds as a valuable resource, the overall value of sea buckthorn can be enhanced. This, in turn, will facilitate the application of sea buckthorn in various sectors, including food, healthcare products, and pharmaceuticals, leading to the development of products with high nutritional value and significant biological activity.

13.4 Fatigue

Fatigue is a phrase used to express a general feeling of exhaustion, loss of muscle strength, or cognitive impairment produced by a range of circumstances such as strenuous physical labor, lengthy exercise, heavy mental activity, or medical issues. This state can significantly impact an individual's physical, family, occupational, and social relationships. Physical activity-induced fatigue typically manifests as muscle fatigue, characterized by the inability of muscles to maintain optimal strength following strenuous exercise. However, intense exercise not only affects peripheral muscle function but also leads to alterations in central nervous system activity, known as exercise-induced "central fatigue." Prolonged fatigue also increases the risk of various health conditions, including anemia, thyroid dysfunction, and depression. The development of anti-fatigue products has garnered increasing attention from researchers. Numerous experiments have demonstrated that many natural ingredients from food have anti-fatigue properties with few side effects. Therefore, the quest for natural nutritional supplements that effectively combat fatigue without adverse effects has emerged as a prominent research focus.

13.4.1 Depletion Hypothesis

The depletion or fatigue hypothesis is based on the depletion of energy substrates, specifically phagocytic phosphorus (adenosine triphosphate [ATP] and phosphocreatine) and skeletal muscle glycogen. Inadequate energy supply is a key factor contributing to reduced exercise capacity and fatigue, and the absence of various energy substances can cause different fatigue mechanisms and varying degrees of fatigue. Creatine phosphate is a high-energy phospho-phagogen that delivers rapid energy during the early stages of high-intensity exercise by contributing phosphate molecules to adenosine diphosphate for ATP resynthesis. Studies have revealed that when exhausted during exercise, the content of creatine phosphate drops considerably. Skeletal muscle glycogen, on the other hand, is the most rapidly available source of energy for muscular contractions. When glycogen supplies are depleted, it becomes clear that high-intensity activity cannot be sustained.

13.4.2 Accumulation Hypothesis

The accumulation hypothesis relates to the buildup of several metabolites within muscle fibers. Exercise accelerates the consumption of energy substances and leads to the production of metabolites, disrupting the body's environmental homeostasis. The cumulative theory states that exertion may cause weariness to build up. Potential scenarios include either the acute accumulation of exercise-induced metabolic byproducts followed by their dissipation shortly after training, or the acute accumulation of these byproducts, which exacerbates over time, resulting in prolonged accumulation (Kataoka et al. 2022).

13.4.3 Oxidative Stress

Among the mechanisms of fatigue, the "radical theory" is widely recognized. According to this theory, strenuous exercise depletes glucose and glycogen in the body, disrupting the balance between the oxidizing system and the antioxidant system. The rapid accumulation of excess free radicals and reactive oxygen species leads to oxidative stress, causing tissue or organ damage, which is considered one of the major causes of fatigue (Qin et al. 2020).

13.4.4 Changes in Brain Neurotransmitters

Serotonin (5-HT), dopamine (DA), and noradrenaline (NE) play crucial roles in inducing fatigue. During physical activity, an increasing trend in 5-HT levels can be observed, which correlates with heightened feelings of lethargy and perceived exertion, ultimately limiting exercise performance by reducing tolerance to discomfort. Changes in DA concentration have been reported to affect core temperature regulation during exercise, suggesting that neurotransmitter fluctuations contribute significantly to the experience of fatigue (Zhang et al. 2021).

13.4.5 *Inflammation*

Fatigue is the most prevalent symptom experienced by cancer patients undergoing chemotherapy, profoundly affecting their quality of life, treatment outcomes, mood, and even survival rates. While the exact mechanisms behind chemotherapy-induced fatigue (CIF) remain incompletely understood, emerging evidence implicates inflammation as a key contributor. During illness, the immune system communicates with the brain, triggering an increase in pro-inflammatory cytokines in the central nervous system, including interleukin (IL)-1β, α, tumor necrosis factor (TNF)-α, and IL-6. Chemotherapy-induced tissue damage further amplifies the production of these cytokines. Serving as primary mediators of fatigue, these cytokines influence brain function through pathways such as the blood-brain barrier, vague nerve, and hypothalamic-pituitary-adrenal axis. Their actions result in decreased levels of neurotransmitters, elevated neurotoxic metabolites, and direct neurotoxic effects on brain regions involved in mood regulation. Consequently, patients experience a diminished mood, reduced interest in social interactions, and heightened fatigue. This comprehensive understanding sheds light on the multifaceted nature of CIF, paving the way for targeted interventions aimed at alleviating its burden on cancer patients.

13.4.6 *Anemic*

Oxygen is indispensable for the proper functioning of both the brain and muscles. The brain consumes approximately 3.5 mL O_2 per minute and per 100 g of tissue, while muscles at rest utilize 1 mL of oxygen per minute per 100 g of tissue. However, during exercise, the oxygen consumption per minute increases by 50 times due to the need for muscle contraction. Hemoglobin serves as the primary carrier of oxygen in the bloodstream. Decreased levels of hemoglobin, as seen in cases of anemia, lead to impaired oxygen delivery throughout the body. This deficiency can have significant consequences, impacting cognitive and physical performance and contributing to sensations of fatigue and tiredness. Fatigue represents a prominent symptom of anemia, regardless of the underlying cause for the decreased hemoglobin levels. This recognition underscores the critical role of adequate oxygen supply in maintaining optimal bodily function and overall well-being (Tardy et al. 2020).

There are common signaling pathways among fatigue-causing factors, such as altered energy metabolism activating adenylate-activated protein kinase (AMPK), and oxidative stress triggering AMPK, which regulates the body's antioxidant-reduction response. Therefore, supplementation with exogenous antioxidants and anti-inflammatory agents seems to be an effective approach to ameliorate fatigue conditions.

13.5 Constituents with Anti-fatigue Activity in Sea Buckthorn Seed

Sea buckthorn seeds contain a variety of active components; these components demonstrate antioxidant, anti-inflammatory, and anti-fatigue properties, underscoring the significant potential of sea buckthorn seeds in anti-fatigue.

13.5.1 Polyphenols

Polyphenols are a type of natural antioxidant noted for their strength. They also have anti-inflammatory properties via altering cellular activity in inflammatory cells and controlling the activity of enzymes involved in arachidonic acid metabolism (phospholipase A2 and COX) and arginine metabolism (NOS).

Proanthocyanidins are the predominant polyphenols in sea buckthorn seed extracts, with hydroxycinnamates and flavonols being minor components. Positive correlations have been observed between the proanthocyanidin content and the antioxidant activities of sea buckthorn seed extracts (Fan et al. 2013). Flavonoids, naturally occurring polyphenolic compounds widely found in human and animal diets, possess significant physiological regulatory functions. In a study administering ethanolic extract of P. Orientale (POE) fruits to acutely liver-injured mice, increased levels of superoxide dismutase (SOD), glutathione peroxidase (GPx), and glutathione reductase were observed, along with decreased levels of nitric oxide (NO), TNF-α, and IL-1β in the livers. Furthermore, serum levels of aspartate aminotransferase, alanine aminotransferase, and alkaline phosphatase were significantly reduced. These changes suggest that POE fruit may exert physiological regulatory functions, such as hepatoprotection, by enhancing hepatic antioxidant enzyme activity, inhibiting lipid peroxidation, and reducing the activity of pro-inflammatory components. The presence of peaks representing catechins, paclitaxel, and quercetin in the HPLC profile indicates that flavonoids are likely responsible for the antioxidant and anti-pro-inflammatory activities of this plant (Chiu et al. 2017).

Chemotherapy utilizing 5-fluorouracil (5-FU), a primary treatment for colon cancer, has been observed to induce fatigue in mice. After the injection of 5-FU, mice exhibited reduced exercise endurance and developed anemia. However, subsequent treatment with quercetin resulted in a restoration of activity levels to baseline and reversal of anemia. These findings underscore the potential of quercetin in promoting the recovery of voluntary activity and mitigating fatigue following chemotherapy (Mahoney et al. 2014).

Resveratrol, a non-flavonoid polyphenolic compound, exhibits a diverse array of physiological functions, including anti-inflammatory, antioxidant, anti-cancer, and anti-aging. As a potent antioxidant, resveratrol efficiently scavenges various free radicals and effectively inhibits lipid peroxidation. Ma et al. used swimming-trained mice as exercise models; exhaustive swimming time and fatigue-related biochemical parameters were used as indicators to investigate the effects of resveratrol on exercise endurance and energy metabolism. The results showed that resveratrol prolonged the swimming time of endurance-trained mice, decreased the levels of metabolites such as aminotransferase, alanine aminotransferase, uric acid, and blood lactate, and increased the levels of liver glycogen and muscle glycogen. Additionally, the study identified 104 anti-fatigue targets and 10 resveratrol core target genes. Overall, resveratrol has shown significant anti-fatigue effects through multiple pathways and targets (Ma et al. 2023).

Zhang et al. (2017) used Caenorhabditis elegans as a model organism to investigate the impact of different concentrations of sea buckthorn seed extract on oxidative stress defense. Their findings revealed a significant reduction in oxidative stress upon treatment with sea buckthorn seed extract. Moreover, the activity levels of SOD and catalase (CAT) were significantly higher in the sea buckthorn seed extract-treated group compared to the control group. Additionally, Arimboor et al. (2008) found that sea buckthorn seeds exhibited a higher phenolic acid content compared to both the seed coat and leaves. The abundance of phenolic compounds in sea buckthorn seeds suggests a greater potential for anti-fatigue effects.

13.5.2 Vitamin

Vitamins are essential nutrients for growth and development. As the body cannot synthesize them independently, they must be acquired through dietary sources. Vitamins play a vital role in maintaining physical strength and enhancing the body's resilience to fatigue.

Beriberi, a disease resulting from severe thiamine (vitamin B1) deficiency, affects several organs, including the muscular and peripheral nervous systems. General symptoms include fatigue, muscle weakness in the legs and arms leading to ataxia, muscle pain and tenderness, and exertional dyspnea. Experimental induction of vitamin B5 deficiency, achieved through a pantothenic acid kinase inhibitor and a pantothenic acid-deficient diet, resulted in subjects experiencing headaches, fatigue, and insomnia after approximately 6 weeks. Furthermore, subjects reported excessive fatigue after their daily walks, with symptoms subsiding and disappearing within 4 weeks of supplementation with 4 g of pantothenic acid daily. Paschalis et al. (2014) found that low Vitamin C concentrations were associated with decreased physical function and increased oxidative stress. Therefore, supplementation with vitamin C may potentially reduce oxidative stress, improve physical function, and relieve fatigue.

Vitamin E is a crucial endogenous fat-soluble antioxidant, encompassing α, β, γ, and δ-Tocopherol, as well as α, β, γ, and δ-Tocotrienol. Supplementation with vitamin E has been shown to mitigate exercise induced oxidative stress in humans and improve physical endurance in mice. Notably, sea buckthorn seeds boast high levels of both tocopherols and tocotrienols, ranging from 100–300 mg/kg, surpassing the concentrations found in fresh berries, which typically range from 10–150 mg/kg

(Yang and Kallio 2002). Prospective investigations have shown that patients with chronic fatigue syndrome have lower serum alpha-tocopherol levels than control participants. In addition, multiple studies have demonstrated that the tocotrienol-rich fraction (TRF) is more effective than alpha-tocotrienol in preventing oxidative damage to brain mitochondria. Lee et al. (2009) investigated the impact of TRF and α-tocopherol on exercise endurance and oxidative stress in forced swimming rats. The findings revealed that TRF treatment dramatically increased swimming time in rats while also increasing SOD, CAT, and GPx levels. These alterations were accompanied by a reduction in the amounts of protein carbonyl thiobarbituric acid-responsive compounds. To summarize, these findings imply that TRF has the potential to ameliorate the physiological condition of forced swimming rats and reduce exercise-induced oxidative damage.

13.5.3 Lipids

Lipids are a primary source of energy and encompass saturated and unsaturated fatty acids and cholesterol. Sea buckthorn seeds are rich in unsaturated fatty acids, comprising approximately 71.2–76% of their composition. Mainly includes oleic acid, linoleic acid, and linolenic acid. Previous study showed that supplementation with omega-6 polyunsaturated fatty acids (PUFA) unexpectedly reduced CRF compared to control groups. Additionally, omega-6 PUFA supplementation was found to decrease pro-inflammatory serum markers in fatigued American breast cancer survivors, thereby alleviating patient fatigue (Zhang et al. 2017).

Due to the high content of unsaturated fatty acids, sea buckthorn seed products like sea buckthorn seed oil are prone to oxidization and acidification when exposed to air, leading to a shortened shelf life. Oxidation impairs organoleptic flavor, reduces beneficial effects, and produces peroxides and free radicals that can cause cellular damage and disease. Therefore, Zhang et al. (2017) investigated the role of microencapsulation in protecting the bioactivity of sea buckthorn seed oil. Their findings suggest that microencapsulation can effectively improve the oxidative stability of the product while maintaining its high bioactivity. By extending the shelf life of sea buckthorn seed products, it can increase its application in many fields such as medicine, food, and cosmetics (Ning et al. 2023).

Sea buckthorn seeds are rich in phytosterols, notably containing significant proportions of Sitosterol (60–70%), and Isofucosterol (10–20%). Additionally, other sterols such as campesterol, stigmastanol, citrostadienol, avenasterol, cycloartenol, 24-methylene cycloartenol, and obtusifoliol constitute 1–5% of total sterols. Seeds contain five times more sterols (0.1–0.2%) compared to fruits (0.02–0.04%) (Yang and Kallio 2002). β-sitosterol, a prevalent phytosterol in nature, exhibits various biological activities, including anti-inflammatory and antioxidant properties. Zhang et al. conducted a study where zebrafish larvae were exposed to $CuSO_4$ to induce oxidative stress, followed by treatment with β-sitosterol. The results showed that β-sitosterol effectively reduced ROS production in zebrafish larvae and upregulated the expression of SOD and gpx4b to suppress oxidative stress. Meanwhile, β-sitosterol reduced the expression of two inflammatory genes, IL-18 and MYD99, thereby exerting anti-inflammatory effects (Zhang et al. 2023).

13.5.4 Amino Acids

Amino acids, as an important part of the body, serve as both building blocks for protein synthesis and enhancers of bodily functions. Moreover, they exhibit antifatigue effects through their antioxidant properties. Arginine is a semi-essential amino acid and participates in numerous metabolic pathways. Renowned for its role as a substrate for NO and plays an important role in endothelium-dependent vasodilation, arginine supplementation enhances blood flow to cardiac and skeletal muscle. In coronary patients, oral arginine administration over a two-week period significantly increases vasodilation and total exercise time during treadmill walking (Maxwell et al. 2002). In contrast, intravenous arginine administration reduces exercise-induced elevations in blood ammonia and lactate levels in healthy subjects, thus mitigating fatigue. Arginine reduces exercise-induced increase in plasma lactate, and ammonia energy is depleted after prolonged exercise when ammonia is produced through the catabolism of muscle proteins, leading to a feeling of fatigue through neurotoxicity.

Cortisol and other glucocorticoids serve as pivotal metabolic hormones, enhancing the availability of fuel substrates such as glucose, free fatty acids, and amino acids. During exercise, cortisol levels escalate rapidly, augmenting energy expenditure, thereby contributing to fatigue (Christiansen et al. 2007). Studies in rats have shown that supplementation with valine prior to exercise mitigates glucose depletion and curtails the surge in cortisol levels during exercise. Similarly, serine supplement cortisol elevation during exercise, improves exercise performance, and mitigates fatigue. Branched chain amino acids (BCAAs) include valine, leucine, and isoleucine, primarily facilitate protein synthesis, potentially expediting muscle recovery post-exercise. Various studies have illustrated the multifaceted physiological effects of BCAAs during exercise. For example, Blomstrand et al. found that BCAA intake reduces the free tryptophan/BCAA ratio, thereby diminishing 5-HT synthesis in the brain and exerting a central fatigue-relieving effect (Newsholme and Blomstrand 2006). Moreover, De Palo et al. reported that long-term supplementation with a BCAA mixture inhibited the elevation of plasma lactate concentrations, further aiming to alleviate fatigue (De Palo et al. 2001).

13.5.5 Carotenoids

Carotenoids are a class of fat-soluble natural pigments widely found in fresh fruits, vegetables, animals, and microorganisms, including astaxanthin, lutein, and lycopene, which have been detected in sea buckthorn seeds. Carotenoids exhibit significant antioxidant activity, positioning them as promising candidates for combating fatigue by reducing oxidative stress. Ikeuchi et al. investigated the effect of astaxanthin on swimming exercise endurance in mice. After five weeks of astaxanthin supplementation, mice exhibited prolonged swimming endurance compared to the control group, with the effect becoming more pronounced over time. In addition, astaxanthin supplementation enhanced exercise endurance at varying intensities by modifying the weight-bearing mass of the mice's tails. This study concluded that astaxanthin delayed the onset of exercise fatigue in mice by enhancing fatty acid oxidation and glycogen reserves (Ikeuchi et al. 2006). Similarly, lycopene has been shown to mitigate oxidative damage induced by exhaustive exercise by reducing xanthine oxidase, myeloperoxidase activity, and malondialdehyde levels in muscle tissue (Liu et al. 2005).

13.5.6 Minerals

Minerals are essential micronutrients that include calcium, iron, zinc, and magnesium, pivotal for a multitude of metabolic processes and cellular functions crucial for physical health and athletic performance. Appropriate mineral intake can improve physical function and resistance to fatigue. Various minerals such as potassium, calcium, sodium, magnesium, iron, zinc, copper, manganese, and phosphorus were detected in sea buckthorn seeds. Zinc deficiency can induce oxidative stress by elevating NO production, leading to reactive oxygen species formation. Additionally, zinc insufficiency triggers inflammation and increases the formation of free radicals in phagocytes and neutrophils. Magnesium, renowned for its antioxidant properties, modulates the activity of the nuclear factor-κB signaling pathway, thereby influencing the expression of genes implicated in immune and inflammatory responses. Moreover, magnesium plays a role in muscle contraction, and its deficiency manifests as fatigue, decreased appetite, and lethargy. Insufficient iron levels compromise resistance and precipitate fatigue. Anemia, irrespective of its etiology, characterized by hemoglobin levels below 120 g/L in women or 130 g/L in men, can elicit feelings of weakness and fatigue (Tardy et al. 2020).

The combination of various components found in sea buckthorn seeds provides a synergistic effect in combating fatigue and boosting energy levels. These compounds collaborate to bolster cellular function, diminish oxidative stress, promote anti-inflammatory effects, and support hormonal balance. Consequently, this concerted action culminates in heightened energy levels and mitigated fatigue.

13.6 Conclusion

Fatigue, both physical and mental, is a multifaceted phenomenon that can precipitate a decline in physical activity and cognitive function, often accompanied by secondary issues such as depression, cognitive impairment, sleep disturbances, and chronic diseases. Despite its pervasive impact, there is currently no universally effective pharmacological intervention for treating fatigue. Therefore, it is important to isolate natural active substances that have the ability to enhance the body's motility, delay fatigue, and accelerate the recovery of volume. Several studies have underscored the potent activity and anti-fatigue properties of sea buckthorn seeds. Yuan et al. investigated the effects of SSP on normal and diabetic mice, demonstrating significant hypoglycemic and anti-inflammatory effects (Yuan et al. 2016). An et al. explored the impact of sea buckthorn seed oil on exercise-induced fatigue in mice, revealing that sea buckthorn seed oil-treated mice exhibited prolonged swimming endurance. Furthermore, sea buckthorn seed oil administration led to reductions in serum lactate, blood urea nitrogen, and liver malondialdehyde levels. Importantly, sea buckthorn seed oil reduced hypothalamic 5-HT levels, lipid hydroperoxide levels and IL-1β protein expression, suggesting its potential as a powerful anti-fatigue agent through inhibition of oxidative and inflammatory damage and modulation of neurotransmitters in the hypothalamus (An et al. 2023).

However, there is a scarcity of studies on sea buckthorn seeds, and future research should emphasize the study of proteins and other active ingredients in them. By paying attention to the processing of sea buckthorn seeds, we can transform waste valuable resources and achieve comprehensive resource utilization. Research on sea buckthorn seed products with diverse effects such as anti-fatigue and antioxidant can meet consumer demand for healthy and high-quality products. Incorporating sea buckthorn seeds into food and medicine can minimize resource wastage, enhance economic efficiency, and contribute to promoting sustainable development and green production methods.

References

An Z, Wang Y, Li X, Jin H, Gong Y (2023) Antifatigue effect of sea buckthorn seed oil on swimming fatigue in mice. J Food Sci 88(4):1482–1494

Arimboor R, Kumar KS, Arumughan C (2008) Simultaneous estimation of phenolic acids in sea buckthorn (Hippophaë rhamnoides) using RP-HPLC with DAD. J Pharm Biomed Anal 47(1):31–38

Chiu YJ, Chou S, Chiu CS, Kao P, Wu KC, Chen CJ, Peng WH (2017) Hepatoprotective effect of the ethanol extract of polygonum orientale on carbon tetrachloride-induced acute liver injury in mice. J Food Drug Anal:26

Christiansen JJ, Djurhuus CB, Gravholt CH, Iversen P, Christiansen JS, Schmitz O, Møller N (2007) Effects of cortisol on carbohydrate, lipid, and protein metabolism: studies of acute cortisol withdrawal in adrenocortical failure. J Clin Endocrinol Metab 92(9):3553–3559

De Palo EF, Gatti R, Cappellin E, Schiraldi C, De Palo CB, Spinella P (2001) Plasma lactate, GH and GH-binding protein levels in exercise following BCAA supplementation in athletes. Amino Acids 20(1):1–11

Fan J, Zhang J, Song H, Zhu W, Liu Y (2013) Antioxidant activity and phenolic components of sea buckthorn (*Hippophae rhamnoides*) seed extracts. In: International conference on advanced mechatronic systems

Ikeuchi M, Koyama T, Takahashi J, Yazawa K (2006) Effects of astaxanthin supplementation on exercise-induced fatigue in mice. Biol Pharm Bull 29(10):2106–2110

Kataoka R, Vasenina E, Hammert WB, Ibrahim AH, Dankel SJ, Buckner SL (2022) Is there evidence for the suggestion that fatigue accumulates following resistance exercise? Sports Med 52(1):25–36

Khan Z, Sodhi N, Fayaz S, Wani S, Bhat M, Mishra H, Dhillon B (2022) Seabuckthorn seed protein concentrate: a novel seed protein; emulsifying properties as affected by ultrasonication and enzymatic hydrolysis. Int J Food Sci Technol:58

Lee SP, Mar GY, Ng LT (2009) Effects of tocotrienol-rich fraction on exercise endurance capacity and oxidative stress in forced swimming rats. Eur J Appl Physiol 107(5):587–595

Li TSC, Beveridge THJ, Drover JCG (2007) Phytosterol content of sea buckthorn (*Hippophae rhamnoides* L.) seed oil: extraction and identification. Food Chem 101(4):1633–1639

Lin J, Xiang H, Sun-Waterhouse D, Cui C, Wang W (2022) Deep eutectic solvents and alkaline extraction of protein from seabuckthorn seed meal: a comparison study. Food Sci Human Wellness 11(4):1028–1035

Liu CC, Huang CC, Lin WT, Hsieh CC, Huang SY, Lin SJ, Yang SC (2005) Lycopene supplementation attenuated xanthine oxidase and myeloperoxidase activities in skeletal muscle tissues of rats after exhaustive exercise. Br J Nutr 94(4):595–601

Ma P, Li J, Huang Q, Wei S, Ge H, Wang Z (2023) Exploring the mechanism of anti-fatigue of resveratrol based on network pharmacology and molecular docking, and in vitro studies. Sci Rep 13(1):2894

Mahoney SE, Davis JM, Murphy EA, McClellan JL, Pena MM (2014) Dietary quercetin reduces chemotherapy-induced fatigue in mice. Integr Cancer Ther 13(5):417–424

Maxwell AJ, Zapien MP, Pearce GL, MacCallum G, Stone PH (2002) Randomized trial of a medical food for the dietary management of chronic, stable angina. J Am Coll Cardiol 39(1):37–45

Mehta M, Kant V, Varshneya C (2013) Screening of in vitro antioxidant potential of seabuckthorn seedcake extracts. J Intercult Ethnopharmacol 2:99–104

Newsholme EA, Blomstrand E (2006) Branched-chain amino acids and central fatigue. J Nutr 136(1 Suppl):274s–276s

Ning Y, Zhang L-Y, Mai J, Su J-E, Cai J-Y, Chen Y, Hu B-B (2023) Tobacco microbial screening and application in improving the quality of tobacco in different physical states. Bioresour Bioprocess 10(1):32

Paschalis V, Theodorou A, Kyparos A, Dipla K, Zafeiridis A, Panayiotou G, Nikolaidis M (2014) Low vitamin C values are linked with decreased physical performance and increased oxidative stress: reversal by vitamin C supplementation. Eur J Nutr 55

Qin L, Lu T, Qin Y, He Y, Cui N, Du A, Sun J (2020) In vivo effect of resveratrol-loaded solid lipid nanoparticles to relieve physical fatigue for sports nutrition supplements. Molecules 25(22)

Sanwal N, Mishra S, Sahu J, Naik S (2021) Effect of ultrasound-assisted extraction on efficiency, antioxidant activity, and physicochemical properties of sea buckthorn (Hippophae salicipholia) seed oil. LWT 153:112386

Tardy AL, Pouteau E, Marquez D, Yilmaz C, Scholey A (2020) Vitamins and minerals for energy, fatigue and cognition: a narrative review of the biochemical and clinical evidence. Nutrients 12(1)

Xiang H, Waterhouse DS, Liu P, Waterhouse GIN, Li J, Cui C (2020) Pancreatic lipase-inhibiting protein hydrolysate and peptides from seabuckthorn seed meal: Preparation optimization and inhibitory mechanism. LWT 134:109870

Yang B, Kallio H (2002) Composition and physiological effects of sea buckthorn (Hippophaë) lipids. Trends Food Sci Technol 13(5):160–167

Yuan H, Zhu X, Wang W, Meng L, Chen D, Zhang C (2016) Hypoglycemic and anti-inflammatory effects of seabuckthorn seed protein in diabetic ICR mice. Food Funct 7(3):1610–1615

Zhang JWC, Liu Y, Shi X, Zhao D, Li M, Wang C, Sun B (2017) Anti-aging effect and its mechanism of ethanol extract from sea buckthorn seed meal in caenorhabditis elegans. Food Sci Human Wellness 38(23):141–148

Zhang Y, Liu Q, Li P, Wang Y, Li S, Gao M, Song Y (2021) Enhanced lipid production by addition of malic acid in fermentation of recombinant Mucor circinelloides Mc-MT-2. Sci Rep 11(1):12674

Zhang P, Liu N, Xue M, Zhang M, Liu W, Xu C, Zhou Y (2023) Anti-inflammatory and antioxidant properties of β-sitosterol in copper sulfate-induced inflammation in Zebrafish (Danio rerio). Antioxidants 12:391

Zheng L, Shi LK, Zhao CW, Jin QZ, Wang XG (2017) Fatty acid, phytochemical, oxidative stability and in vitro antioxidant property of sea buckthorn (*Hippophaë rhamnoides L.*) oils extracted by supercritical and subcritical technologies. LWT 86:507–513

Chapter 14
Isolation, Purification, and Biological Activity of Polysaccharides from Sea Buckthorn

Jianle Chen, Qi Wang, Huan Cheng, and Pin Chen

Abstract Sea buckthorn polysaccharides are bioactive components extracted from sea buckthorn. Sea buckthorn polysaccharides have various biological activities such as antioxidant, hypoglycemic, immune regulation, antiviral, antitumor properties. They are beneficial to human health. In the past few decades, the extraction, structural characteristics, and biological activities of sea buckthorn polysaccharides have been extensively studied. This chapter reviews the extraction methods of sea buckthorn polysaccharide, including hot water extraction, chemical extraction, ultrasonic-assisted extraction, microwave-assisted extraction, and supercritical fluid extraction, followed by studies of purification and structural characteristics. Extraction methods can affect the structural characteristics of polysaccharides, further altering their physicochemical properties and biological activity. The biological activities of sea buckthorn polysaccharides, such as anti-inflammatory properties, were also described. This chapter explores the development and application of sea buckthorn polysaccharides in the fields of functional food or medicine.

Keywords Sea buckthorn · Polysaccharides · Isolation · Purification · Biological activity

Abbreviations

AD	Alzheimer's disease
AFM	Atomic force microscopy
D-gal	D-galactose
Gal	Galactose
GalA	Galacturonic acid
Glu	Glucose

J. Chen (✉) · Q. Wang · H. Cheng · P. Chen
Department of Food Science and Nutrition, Zhejiang University, Hangzhou, China
e-mail: 3090100118@zju.edu.cn

J. Tian et al. (eds.), *Sea Buckthorn*, https://doi.org/10.1007/978-981-97-9865-0_14

GSH-Px	Glutathione peroxidase
H_2O_2	Hydrogen peroxide
HPCE	High-performance capillary electrophoresis
HPLC	High-performance liquid chromatography
IPEC-J2	Intestinal porcine epithelial cells
LPS	Lipopolysaccharide
Man	Mannose
MAPK	Mitogen-activated protein kinase
MyD88	Myeloid differentiation factor 88
NF-κB	Nuclear factor Kb
NMR	Nuclear magnetic resonance
Rha	Rhamnose
RID	Refractive index detector
SEM	Scanning electron microscopy
SOD	Superoxide dismutase
TCA	Trichloroacetic acid
TLR4	Toll-like receptor 4
Xyl	Xylose

14.1 Introduction of Polysaccharides from Sea Buckthorn

Polysaccharides are important biological macromolecules involved in life activities. Polysaccharides are divided into three types based on their sources: plant polysaccharides, animal polysaccharides, and microbial polysaccharides (Li et al. 2022a, b). To date, thousands of polysaccharides have been isolated, of which plant polysaccharides account for a considerable proportion. Plant polysaccharides are recognized as important dietary supplements for healthy foods due to their versatility, low toxicity, and minimal side effects (Wang et al. 2022; Mohammed et al. 2021).

For thousands of years, the fruits, leaves, and seeds of sea buckthorn have been widely used, especially the fruit of sea buckthorn (Lan et al. 2023). Sea buckthorn fruits contain many bioactive components, one of which is polysaccharide, accounting for a high proportion (Zhu et al. 2023). In addition to starch components, non-starch-type sea buckthorn polysaccharides are mainly composed of pectin, cellulose, and hemicellulose (Shen et al. 2021).

Sea buckthorn contains polysaccharides in various parts, with the highest content in its fruit, followed by its peel and leaf. Under dry weight, the polysaccharide content in sea buckthorn fruit ranges from 62 to 100 g/kg DW, and the content reached its peak during the ripening period of sea buckthorn. In the seeds, the starch content was 49 g/kg DW, and the fiber content was 130 g/kg DW. The pectin content in the peel and flesh is 5 and 15 g/kg DW, respectively (Ciesarová et al. 2020). Under fresh weight, the polysaccharide content in sea buckthorn is 1.34–2.87 g/100 g FW, and the pectin content range is 0.21–0.68% FW (Wang et al. 2022; Tkacz et al. 2020).

14.2 Potential Isolation Method of Polysaccharides from Sea Buckthorn

Plant polysaccharides are insoluble in organic solvents but can be dissolved in strong polar solvents according to the rule of similarity (Zeng et al. 2019a, b). The extraction methods, including chemical extraction, hot water extraction, enzymatic extraction, ultrasonic-assisted extraction, microwave-assisted extraction, and supercritical fluid extraction, are discussed (Huang and Huang 2020; Zhang et al. 2018a, b).

14.2.1 Chemical Extraction

Some polysaccharides are suitable for extraction with dilute acid. During extraction, the extraction temperature should not be too high and the extraction time should be short to avoid the breakage of the glycosidic bond. Acid extraction can improve the yield of polysaccharides, but it is easy to destroy the glycosidic bond, thereby affecting the activity of the product (Huang et al. 2021). Alkali extraction is more suitable for the extraction of acidic polysaccharides. The principle of the alkali extraction method is to use the destructive effect of alkali on plant cell walls and facilitate the leaching of acidic polysaccharides. Alkali extraction can avoid protein loss. Within a certain range, as the alkalinity of the solution increases, the interaction between proteins and polysaccharides strengthens, and the particle size of the product increases, which may promote the formation of a grid structure, ultimately leading to an increase in the viscosity of polysaccharide products (Ye et al. 2019). This extraction method is simple and feasible and can effectively dissolve acidic polysaccharides. However, there are defects such as complex components in the obtained crude polysaccharide extract, difficulty in removing small molecule substances generated during the extraction process, and the susceptibility of polysaccharides to hydrolysis to a certain extent.

Whether it is acid extraction or alkali extraction, the dilute acid and dilute alkali extracts should be quickly neutralized or dialyzed, and the crude polysaccharide precipitate should be obtained after concentration and alcohol precipitation. This type of chemical extraction method is simple to operate, but it takes a long time and requires multiple extractions, resulting in low sample yield.

14.2.2 Hot Water Extraction

The hot water extraction method is the most studied and widely used method. As a typical polar solvent, water is safe to use and widely used for the extraction of various plant polysaccharides, but the extraction time is too long (Chen et al. 2023). The hot water extraction method uses water as the solvent, and through the increase of temperature, the molecular thermal motion is accelerated, and the viscosity and mass transfer resistance of the aqueous solution are reduced, which helps to promote the dissolution of polysaccharide, and then the polysaccharide is finally separated from the solution with ethanol (Chaiklahan et al. 2023). This method is relatively simple and cost-effective, but it is time-consuming to operate (Barbosa et al. 2020). At the same time, in order to avoid the breakage of the glycosidic bond, the temperature of hot water extraction should not be too high, and the extraction time should not be too long (Liu et al. 2019).

14.2.3 Enzymatic Extraction

Enzymatic extraction is a specific and efficient biological method. Use cellulase, protease, or composite protease to first degrade and remove some proteins, cellulose, and hemicellulose to facilitate the penetration of plant polysaccharides from cells, thus improving the yield of crude polysaccharides (Jia et al. 2013). If protease is used, it can also simplify the subsequent removal of protein from plant polysaccharides. Papain can hydrolyze most of the free protein and some of the binding protein in the solution of sea buckthorn polysaccharide, so that the extract only contains a small amount of protein, which is conducive to the extraction of polysaccharide. This method can adopt a concentration process and simplify the deproteinization operation to shorten the extraction time, reduce energy consumption, and effectively protect the biological activity of polysaccharides (Song et al. 2018). Compared with traditional acid extraction and alkali extraction methods, enzyme extraction has a mild and safer reaction, and the residual substances generated are non-toxic and harmless, which can be absorbed and utilized by the body (Yang et al. 2017).

14.2.4 Ultrasonic-Assisted Extraction

Ultrasonic-assisted extraction is a physical crushing extraction method. In the process of extracting the active substances in plant cells, the huge energy generated by high-frequency mechanical waves is used to increase the diffusion rate and penetration ability of molecules, which can break the plant cell wall and fully release the polysaccharide inside the plant cell (Zhang et al. 2016). At the same time, the cavitation effect (Yang et al. 2008), biological effects, and crushing effects are often accompanied by ultrasonic-assisted extraction (Wang et al. 2018). It can better accelerate the diffusion and escape of the extract, and the type and quantity of polysaccharides extracted through this method are greatly increased. Compared with traditional extraction methods, ultrasonic-assisted extraction has the advantages of high product yield, short extraction time, and low energy consumption. However, the strong mechanical shear effect of ultrasound can damage the physical structure of the extracted target product. Therefore, the ultrasonic action time should not be too long; otherwise, it will lead to the degradation of soluble polysaccharides. However, the ultrasonic-assisted extraction method does not affect the biological activity of water-soluble polysaccharides. Therefore, ultrasound-assisted extraction is an efficient and practical method for extracting polysaccharides.

14.2.5 Microwave-Assisted Extraction

Microwave-assisted extraction is an emerging technology commonly used to extract bioactive components from plants. Microwaves are non-ionizing electromagnetic wave radiation with frequencies between 300 MHz and 300 GHz. Microwave energy can penetrate plant materials and reach the internal glands, trichome, and vascular system, increasing the plant's internal temperature. This heat can accelerate the vaporization of volatile substances and increase the internal pressure of the cells. When the pressure exceeds the capacity of the cell wall, the cell wall ruptures and releases active ingredients, which are captured and dissolved in the surrounding solvent (Desai et al. 2010; Rioux et al. 2007). Compared with traditional extraction methods, microwave-assisted extraction can effectively improve the extraction rate of targeted compounds, while reducing solvent and energy consumption, making it more environmentally friendly.

The use of microwave-assisted extraction technology can effectively improve the extraction rate of sea buckthorn polysaccharides, with a yield of 6.10 ± 1.85% under optimal conditions (Wei et al. 2019). However, further exploration is needed to determine whether microwave-assisted extraction affects the structure and physiological activity of sea buckthorn polysaccharides. Meanwhile, this method is generally currently in the laboratory stage.

14.2.6 Supercritical Fluid Extraction

The supercritical fluid extraction method uses supercritical fluids, such as carbon dioxide and ethylene, in the supercritical state to adjust the temperature and pressure of the extraction solution to change the solubility of the extract and selectively extract the components in the solution. In this system, the use of appropriate entrainers and regulators (such as methanol and ethanol) can also improve the extraction rate of polysaccharides (Chen et al. 2014; Zou et al. 2018). The advantages of this method are that the supercritical fluid has good solubility, keeping the activity of the ingredient to the greatest extent, good selectivity, high extraction efficiency, recyclable solvent, safe, and pollution-free (Zhao et al. 2019). However, due to the large molecular weight and strong polarity of polysaccharides, the yield is reduced when extracting with pure carbon dioxide, so it is necessary to increase the output by adding agents or increasing pressure. Complex equipment has high operating costs and limited application scope (Macías-Sánchez et al. 2009).

14.3 Purification and Structural Analysis

14.3.1 Purification of Polysaccharides

Purification is essential in the research or application of natural polysaccharides, with the aim of enriching polysaccharide substances or separating polysaccharide components. The extracted crude polysaccharides contain impurities such as proteins, pigments, and small molecule substances. During the pretreatment process of raw materials, incomplete removal of impurities may lead to the mixing of impurities and polysaccharides after polysaccharide extraction, which may interfere with subsequent structural identification and activity research (Tang et al. 2020).

14.3.1.1 Enrichment of Polysaccharides

Enriching polysaccharides requires removal of impurities and retention of polysaccharides. These impurities, such as pigments, proteins, and small molecules, usually appear in polysaccharide samples and may contaminate DEAE-52 cellulose and gel packing. Therefore, it is necessary to remove the coexisting impurities before polysaccharide purification (Ruthes et al. 2015).

The extracted crude polysaccharides contain impurities such as proteins, pigments, and small molecule substances, which can interfere with subsequent structural identification and activity research. Therefore, some technical measures need to be taken to remove impurities. Proteins and polysaccharides are complex hydrophilic biopolymers with different structures. They interact with each other, often coexist in plants, and are often extracted together. Therefore, separating protein from crude polysaccharide is the first step in separating and purifying polysaccharides. Based on the principle that protein is denatured by reagents without affecting the structure of the polysaccharide, the Sevag method, trichloroacetic acid (TCA) precipitation, salting out method, and enzymatic method are usually used to remove the protein from crude polysaccharide (Liu et al. 2022; Zhang et al. 2018a, b).

The Sevag method has high deproteinization efficiency, but it may lead to partial hydrolysis of polysaccharides during chemical treatment, resulting in a loss rate of up to 34.3% of polysaccharides (Li et al. 2012); TCA precipitation is an effective and selective protein precipitation method (Fic et al. 2010); the salting out method is the earliest and most widely used protein precipitation method due to its safe, economical, and environmentally friendly operation (Huang et al. 2012). The enzymatic method is an efficient, simple, safe, environmentally friendly, and inexpensive alternative to chemical methods for polysaccharide purification. Alkaline protease has the advantages of high deproteinization efficiency and low protease dosage, which can be applied in the industry (Wang et al. 2007). Different deproteinization methods have little effect on the main functional groups of polysaccharides but may lead to the loss of glycosidic bonds at different levels. Among them, the enzyme method has the least effect on glycosidic bonds. Therefore, enzymatic hydrolysis is a very promising method to remove crude proteins and is of great significance for polysaccharide purification (Zeng et al. 2019a, b).

Remove low-molecular-weight compounds and inorganic salts from crude polysaccharides using methods such as dialysis, ultrafiltration, or ethanol precipitation (Kang et al. 2015). The phenolic compounds present in the process of extracting natural polysaccharides produce pigments, which affect chromatographic analysis and hinder the accurate identification of polysaccharides (Ren et al. 2019). Different methods for removing pigments are adopted on the basis of their properties, including activated carbon adsorption, hydrogen peroxide (H_2O_2), successful rinse with organic solvents, and ion exchange macroscopic resin (Tang et al. 2020; Yang et al. 2012). Activated carbon adsorption generally removes tannic pigments, resulting in low adsorption efficiency and residual effect (Luo et al. 2023). Organic solvents include ethanol, ether, and acetone, which are limited due to their pungent taste and toxicity. H_2O_2 destroys chromophore or chromophore molecules of pigment through oxidation, but too high concentrations of H_2O_2 may lead to polysaccharide degradation (Tang et al. 2020). The DEAE cellulose adsorption method is currently the most commonly used decolorization method, which not only achieves the purpose of decolorization through ion exchange columns but also allows for the separation of polysaccharides.

14.3.1.2 Purification of Polysaccharide Fraction

Polysaccharides are mixtures with different degrees of polymerization, and further purification is needed to separate them into monosaccharides, which will help in subsequent research on the relationship between structure and biological activity. The polysaccharides, after impurity removal, can be further separated by column chromatography or fractional precision (Borjigin et al. 2023). Column chromatography is an effective purification method for purifying natural components, which can be used to absorb substances such as proteins and acidic polysaccharides, while neutral polysaccharides flow smoothly through the chromatographic column to achieve separation (Tang and Huang 2022). Widely used column chromatography methods include ion exchange and gel column chromatography (Yin et al. 2013). Ion exchange column chromatography uses solutions with different ionic strengths for gradual elution to separate acid and neutral polysaccharides (He et al. 2022). Ion exchange column chromatography mainly includes the anion exchange columns of DEAE cellulose, DEAE agarose, and DEAE agarose gel. The selection of weak anion exchanger or strong anion exchanger is related to the nature of the sample itself. If the weak anion exchanger cannot separate the neutral and acidic components, the strong anion exchanger should be considered (Tang et al. 2020). The gel column chromatography achieves separation according to the molecular size and shape of the polysaccharide. When the polysaccharide solution passes through the gel column, the small-molecular-weight polysaccharides diffuse into the pores of the gel, but the large-molecular-weight polysaccharide continues to flow and elute with the mobile phase to further separate the polysaccharide components, and get the pure polysaccharide according to the molecular weight (Zhang et al. 2020). In most cases, the polysaccharide is purified by combining anion exchange chromatography and gel column chromatography. Liu et al. (2023) used the Sevag method, TCA method, and calcium chloride method to deproteinize crude polysaccharides, respectively. The deproteinized polysaccharides were eluted and separated using a DEAE-52 anion exchange chromatography column and further purified using a Sephadex G200 column to obtain purified sea buckthorn polysaccharides.

14.3.2 Structural Analysis

Polysaccharides are important macromolecules with biological activity and their biological activity is closely related to their structure. The extraction and modification methods of polysaccharides directly affect their structure, thereby affecting their biological activity. Therefore, it is of great significance to analyze and identify the structure of purified pure polysaccharides with uniform components. The structure of polysaccharides can be divided into primary, secondary, tertiary, and quaternary structures. In order to analyze its structural characteristics, physical and chemical methods are usually used to characterize the primary structure of polysaccharides, including molecular weight, monosaccharide composition, and configuration, type, and position of glycosidic bond (He et al. 2022). The molecular weight of polysaccharides can be calculated by gel permeation chromatography, high-performance liquid chromatography (HPLC), differential refractive index detector (RID), mass spectrometry, and other methods (Ma et al. 2020). Ion chromatography, high-performance liquid chromatography, high-performance capillary electrophoresis (HPCE), and other methods can be used to detect the monosaccharide composition in polysaccharides. Infrared spectrum and nuclear magnetic resonance (NMR) spectroscopy are mainly used to determine the configuration of glycosidic bonds in polysaccharides (Hu et al. 2019). Raman spectroscopy is mainly used to detect the vibrations of polysaccharide molecules, atomic non-polar bonds, and isomers (Ren et al. 2019). Scanning electron microscopy (SEM), atomic force microscopy (AFM), circular dichroism chromatography, and X-ray diffraction were used to study the advanced structure of polysaccharides (Lv et al. 2013; Zhang et al. 2018a, b).

Sea buckthorn polysaccharides were reported to consist rhamnose (Rha), xylose (Xyl), mannose (Man), glucose (Glu), galactose (Gal), and galacturonic acid (GalA) at a molar ratio of 13.37:1.40:2.55:8.08:19.47:55.13. Its molecular weight was about 6.26×10^3 kDa and linked primarily by β-type glycosidic bonds. The fine structures of the polysaccharides need further analysis.

14.4 Biological Advances in Sea Buckthorn Polysaccharides

In recent years, many studies have confirmed that polysaccharides in sea buckthorn fruits are one of the main active ingredients, with various pharmacological functions such as antioxidant, anti-inflammatory, anti-fatigue, immunomodulatory, and antitumor activities (Zhao et al. 2020a, b).

14.4.1 Antiviral Activity

A recent study showed that sea buckthorn polysaccharides can inhibit pseudorabies virus infection by inhibiting virus adsorption and entry into cells. In addition, sea buckthorn polysaccharides can also reduce oxidative stress induced by pseudorabies virus infection. It has been indicated that sea buckthorn polysaccharides are a potential antiviral agent against pseudorabies virus (Huan et al. 2022).

14.4.2 Antitumor Activity

Sea buckthorn berry polysaccharides can significantly inhibit Lewis lung carcinoma cells in tumor-bearing mice. Further experiments have shown that the polysaccharide HRWP-A extracted from sea buckthorn berry is a pectin polysaccharide with repeating units of (1→4)-β-d-galactopyranosyluronic residues (of which 85.16% were methyl esterified), which may be mediated by immune activity, as it promotes lymphocyte proliferation in tumor-bearing mice, enhances macrophage activity, and promotes the toxicity of NK cells and cytotoxic T lymphocyte. In summary, HRWP-A induces an enhanced innate immune response. These results may help to understand the immunostimulation effects of HRWP-A and the underlying molecular mechanisms of its antitumor effect (Wang et al. 2015).

14.4.3 Anti-inflammatory and Antioxidant Activities

Inflammation involves a series of abnormal pathological reactions in the body caused by the invasion of pathogenic microorganisms (Han et al. 2012). Lipopolysaccharide (LPS) is a trigger factor for inflammation and can activate nuclear factor κ B (NF-κ B) signaling pathway. Sea buckthorn polysaccharides mainly inhibit the mitogen-activated protein kinase (MAPK)/NF-κ B signal pathways to protect intestinal porcine epithelial cells (IPEC-J2) from LPS-induced damage and reduce the expression of inflammatory cytokines (Li et al. 2022a, b). Sea buckthorn polysaccharides have anti-fatigue activity, which is related to their antioxidant potential. Pretreatment with sea buckthorn polysaccharides by inhibiting TLR4-NF- κB signaling pathway can prevent liver injury in mice induced by LPS/d-GalN, and it is speculated that the beneficial effects of sea buckthorn polysaccharides are due to their anti-inflammatory and antioxidant abilities. Therefore, sea buckthorn may be a promising drug for preventing acute in vivo injury (Liu et al. 2015). Studies have shown that sea buckthorn polysaccharides protected mice livers from carbon tetrachloride (CCl_4)-induced hepatotoxicity through its anti-inflammatory effect and the modulation of the balance between anti-inflammatory cytokines immune cells (Zhang et al. 2017). Sea buckthorn polysaccharides can improve learning and memory abilities in Alzheimer's disease (AD) mice by inhibiting the Toll-like receptor 4 (TLR4) and myeloid differentiation factor 88 (MyD88) signaling pathways, alleviating pathological damage, reducing inflammatory factors, and increasing the levels of antioxidant enzyme superoxide dismutase (SOD) and glutathione peroxidase (GSH-Px). Sea buckthorn polysaccharides have potential antioxidant and anti-inflammatory effects in protecting mice with cognitive and memory dysfunction from $AlCl_3$ and D-galactose (D-gal) (Zhao et al. 2023).

14.4.4 Immunomodulatory Activity

Sea buckthorn polysaccharides are effective and potential immune modulators for controlling and preventing diseases. They can regulate and treat intestinal-related diseases by affecting intracellular signaling pathways. With the help of intestinal microbiota, sea buckthorn polysaccharide can improve intestinal inflammation, enhance the intestinal barrier by upregulating the expression of tight junction proteins (Occludin, Claudin-1, and ZO-1) and MUC-2 and resist lipopolysaccharide invasion into the blood circulation of mice fed with high-fat diet, while the reduction of serum lipopolysaccharide level can prevent microglia from activating and inhibiting NF-κB signaling pathway, inhibit neuroinflammation, and restore synaptic function by activating the CREB/BDNF/TrkB pathway (Lan et al. 2023). The polysaccharides derived from sea buckthorn pomace have a regulatory effect on the imbalance of intestinal microorganisms induced by cefixime, which involves the increase of the abundance of beneficial bacteria in the intestinal tract and the production of short-chain fatty acids through fermentation, thus protecting the intestinal mucosal barrier and benefiting intestinal health or disease treatment (Xie et al. 2013).

14.5 Conclusions and Future Perspectives

Sea buckthorn is a kind of functional plant as "homology of medicine and food" and has been used for hundreds of years. Sea buckthorn can be used to treat radiation injury, trauma, cancer, cardiovascular disease, digestive tract disease, and immune system disease. Modern research has shown that polysaccharides are one of the main active components of sea buckthorn, with many important biological activities and functions. This chapter reviews the isolation, purification, and biological activity of sea buckthorn polysaccharides. In the future, more research may need to be focused on the isolation and structural identification of sea buckthorn polysaccharides, pay attention to its structure–activity relation with in vivo and in vitro experiments, and promote the development and use of sea buckthorn.

References

Barbosa JR, Freitas MMS, Oliveira LC, Martins LHS, Almada-Vilhena AO, Oliveira RM, Pieczarka JC, do Socorro Brasil DB, Carvalho Junior RN (2020) Obtaining extracts rich in antioxidant polysaccharides from the edible mushroom Pleurotus ostreatus using binary system with hot water and supercritical CO_2. Food Chem 330:127173

Borjigin G, Wei F, Jiang S, Li Q, Yang C (2023) Extraction, purification, structural characterization and biological activity of polysaccharides from Fritillaria: a review. Int J Biol Macromol 242:124817

Chaiklahan R, Srinorasing T, Chirasuwan N, Tamtin M, Bunnag B (2023) The potential of polysaccharide extracts from Caulerpa lentillifera waste. Int J Biol Macromol 161:1021–1028

Chen J, Li J, Sun A, Zhang B, Qin S, Zhang Y (2014) Supercritical CO2 extraction and pre-column derivatization of polysaccharides from Artemisia sphaerocephala Krasch. Seeds via gas chromatography. Ind Crop Prod 60:138–143

Chen X, Sun F, Tan Q, Shao R, Xie J, Wan H, Wu G (2023) High internal phase emulsions stabilized solely by alkali-extracted bamboo leaf polysaccharide conjugates. Ind Crop Prod 201:116932

Ciesarová Z, Murkovic M, Cejpek K, Kreps F, Tobolková B, Koplík R, Belajová E, Kukurová K, Daško Ľ, Panovská Z, Revenco D, Burčová Z (2020) Why is sea buckthorn (*Hippophae rhamnoides L.*) so exceptional? A review. Food Res Int 133:109170

Desai M, Parikh J, Parikh PA (2010) Extraction of natural products using microwaves as a heat source. Sep Purif Rev 39(1–2):1–32

Fic E, Kedracka-Krok S, Jankowska U, Pirog A, Dziedzicka-Wasylewska M (2010) Comparison of protein precipitation methods for various rat brain structures prior to proteomic analysis. Electrophoresis 31(21):3573–3579

Han Y, Yuan C, Zhou X, Han Y, He Y, Ouyang J, Zhou W, Wang Z, Wang H, Li G (2012) Anti-inflammatory activity of three triterpene from hippophae rhamnoides l. in lipopolysaccharide-stimulated RAW264.7 cells. Int J Mol Sci 22(21):Article 21

He Y, Li L, Chang H, Cai B, Gao H, Chen G, Hou W, Jappar Z, Yan Y (2022) Research progress on extraction, purification, structure and biological activity of Dendrobium officinale polysaccharides. Front Nutr:9

Hu W, Chen S, Wu D, Zheng J, Ye X (2019) Ultrasonic-assisted citrus pectin modification in the bicarbonate-activated hydrogen peroxide system: chemical and microstructural analysis. Ultrason Sonochem 58:104576

Huan C, Xu Y, Zhang W, Pan H, Zhou Z, Yao J, Guo T, Ni B, Gao S (2022) Hippophae rhamnoides polysaccharides dampen pseudorabies virus infection through downregulating adsorption, entry and oxidative stress. Int J Biol Macromol 207:454–463

Huang H, Huang G (2020) Extraction, separation, modification, structural characterization, and antioxidant activity of plant polysaccharides. Chem Biol Drug Des 96(5):1209–1222

Huang G, Shu S, Cai T, Liu Y, Xiao F (2012) Preparation and deproteinization of garlic polysaccharide. Int J Food Sci Nutr 63(6):739–741

Huang G, Chen F, Yang W, Huang H (2021) Preparation, deproteinization and comparison of bioactive polysaccharides. Trends Food Sci Technol 109:564–568

Jia S, Li F, Liu Y, Ren H, Gong G, Wang Y, Wu S (2013) Effects of extraction methods on the antioxidant activities of polysaccharides from *Agaricus blazei Murrill.* Int J Biol Macromol 62:66–69

Kang J, Hua X, Yang R, Chen Y, Yang H (2015) Characterization of natural low-methoxyl pectin from sunflower head extracted by sodium citrate and purified by ultrafiltration. Food Chem 180:98–105

Lan Y, Ma Z, Chang L, Peng J, Zhang M, Sun Q, Qiao R, Hou X, Ding X, Zhang Q, Peng Q, Dong J, Liu X (2023) Sea buckthorn polysaccharide ameliorates high-fat diet induced mice neuroinflammation and synaptic dysfunction via regulating gut dysbiosis. Int J Biol Macromol 236:123797

Li X, Zhao R, Zhou HL, Wu DH (2012) Deproteinization of polysaccharide from the stigma maydis by sevag method. Adv Mater Res 340:416–420

Li M, Chen L, Zhao Y, Sun H, Zhao L (2022a) Research on the mechanism of HRP relieving IPEC-J2 Cells immunological stress based on transcriptome sequencing analysis. Front Nutr:9

Li Z-W, Du Z-M, Wang Y-W, Feng Y-X, Zhang R, Yan XB (2022b) Chemical modification, characterization, and activity changes of land plant polysaccharides: a review. Polymers 14(19):Article 19

Liu H, Zhang W, Dong S, Song L, Zhao S, Wu C, Wang X, Liu F, Xie J, Wang J, Wang Y (2015) Protective effects of sea buckthorn polysaccharide extracts against LPS/d-GalN-induced acute liver failure in mice via suppressing TLR4-NF-κB signaling. J Ethnopharmacol 176:69–78

Liu X, Hou R, Xu K, Chen L, Wu X, Lin W, Zheng M, Fu J (2019) Extraction, characterization and antioxidant activity analysis of the polysaccharide from the solid-state fermentation substrate of Inonotus hispidus. Int J Biol Macromol 123:468–476

Liu D, Tang W, Han C, Nie S (2022) Advances in Polygonatum sibiricum polysaccharides: extraction, purification, structure, biosynthesis, and bioactivity. Front Nutr:9

Liu J, Bai J, Shao C, Yao S, Xu R, Duan S, Wang L, Xu Y, Yang Y (2023) Optimization of ultrasound-assisted aqueous two-phase extraction of polysaccharides from seabuckthorn fruits using response methodology, physicochemical characterization and bioactivities. J Sci Food Agric 103(6):3168–3183

Luo L, Xue J, Shao Z, Zhou Z, Tang W, Liu J, Hu H, Yang F (2023) Recent developments in Salvia miltiorrhiza polysaccharides: isolation, purification, structural characteristics and biological activities. Front Pharmacol 14

Lv X, Wang C, Cheng Y, Huang L, Wang Z (2013) Isolation and structural characterization of a polysaccharide LRP4-A from Lycium ruthenicum Murr. Carbohydr Res 365:20–25

Ma X, Jing J, Wang J, Xu J, Hu Z (2020) Extraction of Low Methoxyl Pectin from fresh sunflower heads by subcritical water extraction. ACS Omega 5(25):15095–15104

Macías-Sánchez MD, Mantell C, Rodríguez M, Martínez de la Ossa E, Lubián LM, Montero O (2009) Comparison of supercritical fluid and ultrasound-assisted extraction of carotenoids and chlorophyll a from Dunaliella salina. Talanta 77(3):948–952

Mohammed ASA, Naveed M, Jost N (2021) Polysaccharides; classification, chemical properties, and future perspective applications in fields of pharmacology and biological medicine (a review of current applications and upcoming potentialities). J Polym Environ 29(8):2359–2371

Ren Y, Bai Y, Zhang Z, Cai W, Del Rio Flores A (2019) The preparation and structure analysis methods of natural polysaccharides of plants and fungi: a review of recent development. Molecules 24(17):Article 17

Rioux L-E, Turgeon SL, Beaulieu M (2007) Characterization of polysaccharides extracted from brown seaweeds. Carbohydr Polym 69(3):530–537

Ruthes AC, Smiderle FR, Iacomini M (2015) d-Glucans from edible mushrooms: a review on the extraction, purification and chemical characterization approaches. Carbohydr Polym 117:753–761

Shen C, Wang T, Guo F, Sun K, Wang B, Wang J, Zhang Z, Zhang X, Zhao Y, Chen Y (2021) Structural characterization and intestinal protection activity of polysaccharides from Sea buckthorn (*Hippophae rhamnoides L.*) berries. Carbohydr Polym 274:118648

Song Y-R, Sung S-K, Jang M, Lim T-G, Cho C-W, Han C-J, Hong H-D (2018) Enzyme-assisted extraction, chemical characteristics, and immunostimulatory activity of polysaccharides from Korean ginseng (Panax ginseng Meyer). Int J Biol Macromol 116:1089–1097

Tang Z, Huang G (2022) Extraction, structure, and activity of polysaccharide from Radix astragali. Biomed Pharmacother 150:113015

Tang W, Liu D, Yin J-Y, Nie SP (2020) Consecutive and progressive purification of food-derived natural polysaccharide: based on material, extraction process and crude polysaccharide. Trends Food Sci Technol 99:76–87

Tkacz K, Wojdyło A, Turkiewicz IP, Ferreres F, Moreno DA, Nowicka P (2020) UPLC-PDA-Q/TOF-MS profiling of phenolic and carotenoid compounds and their influence on anticholinergic potential for AChE and BuChE inhibition and on-line antioxidant activity of selected Hippophaë rhamnoides L. cultivars. Food Chem 309:125766

Wang X, Yuan Y, Wang K, Zhang D, Yang Z, Xu P (2007) Deproteinization of gellan gum produced by Sphingomonas paucimobilis ATCC 31461. J Biotechnol 128(2):403–407

Wang H, Gao T, Du Y, Yang H, Wei L, Bi H, Ni W (2015) Anticancer and immunostimulating activities of a novel homogalacturonan from Hippophae rhamnoides L. berry. Carbohydr Polym 131:288–296

Wang W, Chen W, Zou M, Lv R, Wang D, Hou F, Feng H, Ma X, Zhong J, Ding T, Ye X, Liu D (2018) Applications of power ultrasound in oriented modification and degradation of pectin: a review. J Food Eng 234:98–107

Wang K, Xu Z, Liao X (2022) Bioactive compounds, health benefits and functional food products of sea buckthorn: a review. Crit Rev Food Sci Nutr 62(24):6761–6782

Wei E, Yang R, Zhao H, Wang P, Zhao S, Zhai W, Zhang Y, Zhou H (2019) Microwave-assisted extraction releases the antioxidant polysaccharides from seabuckthorn (Hippophae rhamnoides L.) berries. Int J Biol Macromol 123:280–290

Xie X, Song Y, Bi X, Liu X, Xing Y, Che Z (2013) Characterization of sea buckthorn polysaccharides and the analysis of its regulatory effect on the gut microbiota imbalance induced by cefixime in mice. J Funct Foods 104:105511

Yang B, Jiang Y, Zhao M, Shi J, Wang L (2008) Effects of ultrasonic extraction on the physical and chemical properties of polysaccharides from longan fruit pericarp. Polym Degrad Stab 93(1):268–272

Yang R, Meng D, Song Y, Li J, Zhang Y, Hu X, Ni Y, Li Q (2012) Simultaneous decoloration and deproteinization of crude polysaccharide from pumpkin residues by cross-linked polystyrene macroporous resin. J Agric Food Chem 60(34):8450–8456

Yang J, Zhang H-F, Cao X-Y, Yang X-H, Wang F-Z, Guo Q, Sun C-Q (2017) Enzymatic water extraction of polysaccharides from epimedium brevicornu and their antioxidant activity and protective effect against DNA damage. J Food Biochem 41(1):e12298

Ye J, Hua X, Wang M, Zhang W, Yang R (2019) Effect of extraction pH on the yield and physicochemical properties of polysaccharides extracts from peanut sediment of aqueous extraction process. LWT 106:137–144

Yin SY, Kim HJ, Kim H-J (2013) A comparative study of the effects of whole red ginseng extract and polysaccharide and Saponin fractions on Influenza A (H1N1) virus infection. Biol Pharm Bull 36(6):1002–1007

Zeng X, Li P, Chen X, Kang Y, Xie Y, Li X, Xie T, Zhang Y (2019a) Effects of deproteinization methods on primary structure and antioxidant activity of Ganoderma lucidum polysaccharides. Int J Biol Macromol 126:867–876

Zeng P, Li J, Chen Y, Zhang L (2019b) The structures and biological functions of polysaccharides from traditional Chinese herbs. Prog Mol Biol Transl Sci 163:423–444

Zhang D-Y, Wan Y, Xu J-Y, Wu G-H, Li L, Yao X-H (2016) Ultrasound extraction of polysaccharides from mulberry leaves and their effect on enhancing antioxidant activity. Carbohydr Polym 137:473–479

Zhang W, Zhang X, Zou K, Xie J, Zhao S, Liu J, Liu H, Wang J, Wang Y (2017) Seabuckthorn berry polysaccharide protects against carbon tetrachloride-induced hepatotoxicity in mice via anti-oxidative and anti-inflammatory activities. Food Funct 8(9):3130–3138

Zhang H, Chen J, Li J, Yan L, Li S, Ye X, Liu D, Ding T, Linhardt RJ, Orfila C, Chen S (2018a) Extraction and characterization of RG-I enriched pectic polysaccharides from mandarin citrus peel. Food Hydrocoll 79:579–586

Zhang L, Zhang Q, Zheng Y, He Z, Guan P, He X, Hui L, Dai Y (2018b) Study of Schiff base formation between dialdehyde cellulose and proteins, and its application for the deproteinization of crude polysaccharide extracts. Ind Crop Prod 112:532–540

Zhang T, Liu H, Bai X, Liu P, Yang Y, Huang J, Zhou L, Min X (2020) Fractionation and antioxidant activities of the water-soluble polysaccharides from Lonicera japonica Thunb. Int J Biol Macromol 151:1058–1066

Zhao T, Luo Y, Zhang X, Zhang W, Qu H, Mao G, Zou Y, Wang W, Li Q, Chen Y, Feng W, Yang L, Wu X (2019) Subcritical water extraction of bioactive compounds from Radix Puerariae and optimization study using response surface methodology. Chem Eng Commun 206(9):1218–1227

Zhao L, Geng T, Sun K, Su S, Zhao Y, Bao N, Pan L, Sun H, Li M (2020a) Proteomic analysis reveals the molecular mechanism of Hippophae rhamnoides polysaccharide intervention in LPS-induced inflammation of IPEC-J2 cells in piglets. Int J Biol Macromol 164:3294–3304

Zhao L, Li M-Y, Su S, Geng T-T, Sun H (2020b) Hippophae rhamnoides L. polysaccharide enhances antioxidant enzyme activity, cytokine level, and related mRNA expression in intestinal porcine epithelial cells. Can J Anim Sci 100(1):193–204

Zhao H, Liu J, Wang Y, Shao M, Wang L, Tang W, Wang Y, Li X (2023) Polysaccharides from sea buckthorn (Hippophae rhamnoides L.) berries ameliorate cognitive dysfunction in AD mice induced by a combination of d-gal and $AlCl_3$ by suppressing oxidative stress and inflammation reaction. J Sci Food Agric 103(12):6005–6016

Zhu Y, Wu M, Li X, Wang Y, Li M, Zhou H (2023) Flash extraction, characterization, and immunoenhancement activity of polysaccharide from Hippophae rhamnoides Linn. Chem Biodivers 20(3):e202200776

Zou X, Liu Y, Tao C, Liu Y, Liu M, Wu J, Lv Z (2018) CO_2 supercritical fluid extraction and characterization of polysaccharide from bamboo (Phyllostachys heterocycla) leaves. J Food Meas Charact 12(1):35–44

Chapter 15
Chemical Composition and Antioxidant Active Components of Sea Buckthorn Oil

Changqing Wei, Wen Tan, Jinhu Tian, and Wenyu Liu

Abstract Sea buckthorn oil is renowned for its unique composition of unsaturated fatty acids, particularly palmitoleic acid (omega-7) and γ-linolenic acid (omega-6). These fatty acids are central to its exceptional capabilities for skin regenerating and repairing. Applied topically, sea buckthorn oil has been found to significantly improve blood circulation, thereby enhancing skin oxygenation. Furthermore, its detoxifying abilities are effective in expelling excess toxins from the body. A key advantage of sea buckthorn oil lies in its ability to quickly penetrate the epidermis, allowing its active ingredients to deeply nourish the skin. Within the skin, γ-linolenic acid is transformed into prostaglandins, which are vital for guarding against infections, preventing allergies, and eliminating inflammation. By inhibiting the aging process, sea buckthorn oil has proven itself to be a powerhouse ingredient in anti-aging and skincare products. Boasting an array of approximately 200 beneficial properties, sea buckthorn oil has become a highly sought-after additive in health and beauty products. This document seeks to delve into the multifaceted nature of sea buckthorn oil, encompassing its extraction method, chemical composition, pharmacological effects, and applications. Additionally, it anticipates the future development prospects of sea buckthorn oil, thus laying a robust foundation for its comprehensive utilization and steering forthcoming scientific inquiries in this domain.

Keywords Sea buckthorn oil · Chemical composition · Pharmacological effects · Antioxidant properties · Application

C. Wei (✉)
School of Food Science and Technology, Shihezi University, Shihezi, China

W. Tan · J. Tian
Department of Food Science and Nutrition, Zhejiang University, Hangzhou, China

W. Liu
Food College, Shihezi University, Shihezi, China

J. Tian et al. (eds.), *Sea Buckthorn*, https://doi.org/10.1007/978-981-97-9865-0_15

Abbreviations

Aβ	β-Amyloid
FRA	Ferric reducing ability
GSH	Glutathione
NAFLD	Non-alcoholic fatty liver disease
PPAR	Peroxisome proliferator-activated receptors
PSFME	Pressurized solvent-free microwave-assisted extraction
RSM	Response surface methodology
SIRT-1	Sirtuin1
UAEE	Ultrasound-assisted enzyme extraction

15.1 Introduction

Sea buckthorn (*Hippophae rhamnoides L.*), known as sour thorn and vinegar willow, is a perennial deciduous shrub belonging to the family of *Elaeagnaceae*, with heights ranging from 2 to 6 m (Ma et al. 2022). The fruits of sea buckthorn, characterized by their yellowish or orange-yellow hue, resemble spheres in shape, with diameters spanning approximately 3–8 mm (Ma et al. 2017). Sea buckthorn bark is rough, featuring small, sharp thorns on its branches, and possesses dark-green, needle-shaped, or linear leaves. Its stems, leaves, seeds, and fruits are rich in a total of about 200 kinds of active substances, among which flavonols, vitamins, L-ascorbic acid, polysaccharide compounds, and lipophilic compounds (including tocopherols, carotenoids, phytosterols, and fatty acids), which are prevalent in high concentrations and contribute to its unique health-promoting properties (Fu et al. 2014; Tkacz et al. 2020). Ongoing, in-depth research has significantly valued the health-promoting effects of sea buckthorn fruits, highlighting their antimicrobial, anti-inflammatory, antioxidant, radioprotective, tissue-repairing, anticarcinogenic, and cardiovascular disease risk reduction capabilities. These attributes have garnered widespread attention from scholars (Gâtlan and Gutt 2021; Xing et al. 2002). However, it is noted that while there is extensive study on the bioactivity of sea buckthorn oil, research on sea buckthorn's other nutrients remains comparatively scarce.

Sea buckthorn oil is characterized by its complex composition, predominantly comprising a wide array of fatty acids, including stearic acid, soft acid, oleic acid, linoleic acid, linolenic acid, arachidic acid, myristic acid, palmitoleic acid, and lauric acid. In addition, sea buckthorn oil contains 13-monohydroxyl and 4-dihydroxyl acids, as well as seven types of mono-, di-, and tri-epoxyl acids. With more than 100 kinds of lipid-like components identified, and the content of phospholipids in sea buckthorn oil is 0.9–1.0%, including lecithin, ceruloplasmin, phosphatidylcholine, phosphatidylglycerol, phosphatidylinositol, phosphatidylcholine, lysophosphatidylinositol, lysophosphatidylcholine, N-acylphosphatidylinositolamine, lysophosphatidylinositolamine, N-acylphosphatidylcholine, and others (Suryakumar and Gupta 2011).

Sea buckthorn oil is divided into sea buckthorn fruit oil and sea buckthorn seed oil, each demonstrating notable compositional variances. Chromatographic mass spectrometry analysis of 10 kinds of sea buckthorn oils available in the market found that sea buckthorn fruit oil mainly contains palmitoleic acid (around 35%), palmitic acid (around 34%), oleic acid (around 18%) and linoleic acid (around 6%), and the content of unsaturated fatty acids is around 62%. In contrast, sea buckthorn seed oil mainly contains linoleic acid (about 39%), linolenic acid (about 29%), oleic acid (about 21%), and palmitic acid (about 7.6%), with an unsaturated fatty acid content of about 90% (Yang and Kallio 2002).

The compositional analysis reveals a distinct difference between sea buckthorn fruit oil and seed oil, particularly in terms of palmitic acid and unsaturated fatty acid contents. The fruit oil exhibits a significantly higher concentration of palmitic acid compared to the seed oil, whereas the seed oil contains a substantially greater proportion of unsaturated fatty acids than the fruit oil (Cakir 2004). Specifically, sea buckthorn seed oil is dominated by linoleic acid, linolenic acid, oleic acid, palmitic acid, and the unsaturated fatty acid content is 87–90%. Sea buckthorn fruit oil is dominated by palmitic acid, palmitoleic acid, oleic acid, and linoleic acid, with unsaturated fatty acid content of roughly 60–70%.

Studies have shown that sea buckthorn oil is rich in active functional substances, including fatty acids, vitamins, flavonoids, and minerals (Teleszko et al. 2015). Its comprehensive nutrient profile contributes to a range of health benefits. These include supporting cardiovascular health, reducing cholesterol levels, mitigating the risk of atherosclerosis, lowering blood pressure and enhancing blood circulation, protecting cells from damage caused by free radicals. Furthermore, sea buckthorn oil has been shown to bolster the immune system, among other beneficial effects.

Despite its considerable nutritional value and promising market prospects, the sea buckthorn industry encounters several obstacles. The harvesting and processing of sea buckthorn fruit present significant challenges due to their complexity, necessitating substantial labor and equipment investments, which in turn elevate production costs. Additionally, the market demand for sea buckthorn oil remains limited, partly because of insufficient publicity and promotion efforts. Moreover, sea buckthorn cultivation faces technical hurdles, such as pest control and cultivation of high-yield and high-quality varieties, which still needs to be strengthened and explored.

15.2 The Extraction Method of Sea Buckthorn Oil Chemical Constituents

15.2.1 Pressing Method

The pressing method, a conventional technique for oil extraction, is categorized into hot pressing and cold pressing. It has the advantages of simplicity, efficiency, and environmental friendliness, facilitating the extraction of high-quality sea buckthorn oil. However, the presence of non-oil components may affect its extraction rate.

Using a milk separator, flesh oil is obtained, whereas seed oil and pericarp oil are derived from cold pressing. The characterization of these oils revealed that pericarp oil contains the highest concentration of bioactive compounds, albeit with a marginally lower sterol content (Koskovac et al. 2017; Seglina et al. 2021).

15.2.2 Solvent Leaching Method

The solvent leaching method utilizes the principle of similar dissolution using organic solvents such as ethanol, hexane, or methylene chloride to dissolve the chemical constituents of sea buckthorn oil in the solvent. The oil is subsequently produced by evaporation of the solvent. While this method is straightforward and boasts high extraction efficiency, it poses potential environmental hazards and risks leaving solvent residues in the sea buckthorn oil. Saldo et al. used two green solvents (ethanol and 2-methyltetrahydrofuran) and two petroleum-based solvents (hexane and ether) for extraction. They analyzed the β-carotene content in the extracted oils using HPLC-UV/VIS and found that ethanol-extracted sea buckthorn oil had the highest β-carotene content. Further, they determined the oil yield under various temperatures to identify the optimal extraction temperature for each solvent (Vilas-Franquesa et al. 2022). In another study, Parveen et al. examined sea buckthorn seed oil extracted from various concentrations of organic solvents by ferric chelating through ferric chloride method, and the determination showed that sea buckthorn oil extracted from ethyl acetate had the strongest antioxidant properties with a trivalent ferric reducing ability (FRA) of 68%, while the oil extracted using an isopropyl group displayed the least antioxidant capacity among the five selected solvent-extracted samples, with a FRA of 53% (Akhter et al. 2023).

15.2.3 Aqueous Enzymatic Method

The aqueous enzyme method represents an innovative approach to vegetable oil extraction, characterized by low energy consumption, enhanced product safety, mild operational conditions, and the concurrent separation of oils and proteins (Gao et al. 2021). This method entails the mechanical disruption of oilseeds followed by the introduction of water as the dispersion medium. Enzymes are then dissolved in the water, where they act on the oil content within the seeds to facilitate its separation.

15.2.4 Ultrasonic-Assisted Extraction Method

Ultrasonic-assisted extraction method uses the mechanical and thermal effects of ultrasound to improve the extraction efficiency, mixing sea buckthorn fruits with solvent, and then using ultrasound to make the chemical components in the solvent easier to dissolve, and increase the temperature of the solvent to accelerate the extraction process. Notable for its high extraction efficiency, reduced extraction duration, and high-temperature conditions, this method's widespread adoption is potentially curtailed by the substantial cost of ultrasonic equipment. Kumar et al. (2022) used ultrasound-assisted enzyme extraction (UAEE) to extract oil from freeze-dried sea buckthorn berries and response surface methodology was used to assess the effect of variables (such as time, enzyme concentration, and solvent-to-sample ratio) on oil yield and its physicochemical properties (acid value, peroxide value, iodine value, density, and color). The optimum conditions were determined as solvent-to-sample ratio of 5.08 mL/g, ultrasonication time of 14.65 min, and enzyme unit of 3.13 U/g. The oil yield under the optimum conditions was 18.32% and the physicochemical parameters of ultrasound-assisted enzyme extraction of sea buckthorn oil were better than that extracted by Soxhlet extraction method.

15.2.5 Supercritical Fluid Extraction Method

Supercritical fluid extraction, employing carbon dioxide as the primary medium, capitalizes on the unique properties of supercritical fluids, which exhibit a state between gas and liquid, characterized by high solubility and diffusivity. By adjusting temperature and pressure, carbon dioxide is brought to a supercritical state, facilitating the extraction of chemical constituents from sea buckthorn fruits. This method's notable advantages include its high extraction efficiency and the purity of the extracted sea buckthorn oil, which is devoid of solvent residues. Moreover, it allows for the preservation of bioactive components that are susceptible to degradation. However, the method is limited by the high costs associated with supercritical fluid extraction equipment and the technical expertise required for operation (Liu et al. 2009). Xu et al. used response surface methodology (RSM) to investigate the effects of various parameters—pressure, temperature, CO_2 flow rate, and extraction time—on the extraction yield of oil, vitamin E, and carotenoids. Their research culminated in identifying optimal conditions for the supercritical CO_2 extraction of sea buckthorn whole fruit oil: an extraction pressure of 27.6 MPa, temperature of 34.5 °C, CO_2 flow rate of 17 L/h, and an extraction duration of 82 min (Xu et al. 2008).

15.2.6 Other Methods

Emerging extraction technologies, such as microwave-assisted and ultra-high-pressure methods, offer distinct advantages over conventional extraction techniques, particularly in terms of efficiency and operational simplicity. Notably, under optimal conditions, microwave-assisted extraction has been shown to achieve an extraction rate marginally superior to that of ultrasound-assisted extraction, with a reported increase of 1.8% in efficiency (Isopencu et al. 2018). In addition, liposome encapsulation is a novel means of protecting bioactive substances by encapsulating sea buckthorn oil in liposomes to improve the stability and bioavailability of the oil, which is suitable for the preparation of high-quality sea buckthorn oil products (Zhao et al. 2017).

In summary, there are various methods for extracting the chemical constituents of sea buckthorn oil. The selection of an appropriate extraction method requires careful consideration of factors such as extraction efficiency, operational costs, and environmental impact. With the continuous development of science and technology, it is believed that more efficient and eco-friendly extraction techniques will emerge, thereby enhancing the technical support for the development and utilization of sea buckthorn oil.

15.3 Chemical Composition and Antioxidant Activity of Sea Buckthorn Oil

Studies have shown that sea buckthorn oil possesses a complex and abundant chemical composition, primarily consisting of fatty acids, vitamins, polyphenols, phytosterols, and other substances. This rich biochemical profile underscores the oil's nutritional and therapeutic potential, contributing to its wide range of health benefits (Chen et al. 2023).

15.3.1 Fatty Acids

Fatty acids, notably palmitic acid, oleic acid, and linoleic acid, constitute the primary components of sea buckthorn oil, featuring polyunsaturated bonds conducive to human health (Ma et al. 2022). Baoru Yang et al. studied the oil content and fatty acid composition of two subspecies of sea buckthorn berries (*Hippophae rhamnoides, Hippophae salicifolia*) (Yang and Kallio 2001). *Rhamnoides* and *S. sinensis*, revealing that the seeds, berries, and seedless fractions of *S. rhamnoides* contain higher oil percentages than those of the Chinese subspecies linoleic acid (18:2n-6) and α-linolenic acid (18:3n-3) accounted for about 70% of the fatty acids in the seed oil, while palmitoleic acid (16:1n-7) was virtually absent in the seed oil, ranging from 12.1% to 39.0% in the pulp/pericarp and from 8.9% to 31.0% in the whole berry. Compared with the seed oils of the *S. rhamnoides* subspecies, the seed oils of *S. sinensis* had more linoleic acid (40.9% vs. 39.1%) and less α-linolenic acid (26.6% vs. 30.6%), but the proportion of palmitoleic acid was higher in the oils of *S. rhamnoides* subspecies berries than in those of *S. sinensis* subspecies berries (26.0% vs. 21.5%), and the opposite was true for α-linolenic acid (8.8% vs. 11.2%). The proportion of α-linolenic acid in the seed oil was inversely correlated with oleic and linoleic acids, whereas the proportion of palmitoleic acid was negatively correlated with the proportions of linoleic and α-linolenic acids in the oil of sea buckthorn whole fruits.

Kuhkheil et al. (2017) evaluated the oil content and fatty acid composition in sea buckthorn seed oil from 20 wild Iranian populations across 2014 and 2015, noting an oil content range of 4–9% and variations in unsaturated fatty acid content between populations and years. The diversity in fatty acid chain length and unsaturation degree endows sea buckthorn oil with a low melting point and high oxidative stability, highlighting its nutritional and therapeutic potential (Table 15.1).

15.3.2 Vitamins

Sea buckthorn fruit contains water-soluble vitamins (including vitamin C, PP, B1, B6, choline), fat-soluble vitamins, and pro-vitamins (such as tocopherols, carotenoids, and vitamin K). Additionally, sea buckthorn oil contains significant

Table 15.1 Fatty acid composition of sea buckthorn oil (%) and sum of SFA, MUFA, and PUFA (%)

	Fruit peel oil	Fruit pulp oil	Seed oil
14:0	0.24	0.20	0.06
	0.01	0.04	0.01
15:0	0.05	nd	0.13
	0.01		0.01
15:1	0.11	nd	nd
	0.02		
16:0	36.32	32.27	6.56
	0.25		0.07
16:1	39.90	33.90	0.47
	0.21	0.95	0.02
18:0	1.63	2.95	2.80
	0.05	0.24	0.06
cis-18:1	4.24	7.12	14.85
	0.03	0.42	0.11
trans-18:1	5.92	6.83	2.07
	0.03	0.20	0.05
18:2	10.84	14.65	39.13
	0.06	0.22	0.24
α-18:3	0.63	2.08	33.51
	0.02	0.17	0.41
20:0	0.12	nd	0.42
	0.01		0.02
ΣSFA	38.36	35.43	9.97
ΣMUFA	50.18	47.84	17.39
ΣPUFA	11.47	16.73	72.64

The table was prepared by the authors

Abbreviations: *nd* not detected, *ΣMUFA* the sum of monounsaturated fatty acids, *ΣPUFA* the sum of polyunsaturated fatty acids, *ΣSFA* the sum of saturated fatty acids

Table 15.2 Effects of different extraction processes on tocopherols and carotenoids in sea buckthorn seed oil and sea buckthorn fruit oil

Raw materials	Active ingredients	Content(mg/100 g) Solvent-based approach	Supercritical CO_2 extraction method	Pressing method	Hydrometry	Ultrasonic-assisted extraction method
Sea buckthorn seed	Tocopherol	420.6	393.4	288.2	–	400.6
	Carotenoids	22.2	11.7	15.3	–	5.5
Sea buckthorn	Tocopherol	170.7	138.8	–	150.8	–
	Carotenoids	527.8	148.4	–	292.4	–

The table was prepared by the authors

concentrations of antioxidants, notably vitamin C and vitamin E, known for their potent antioxidant properties and contributions to anti-aging effects.

Carotenoids constitute a crucial component of sea buckthorn oil, demonstrating a broad spectrum of biological activities, such as antioxidant, anti-inflammatory, and immunomodulatory. Within sea buckthorn oil, carotenoids predominantly occur as β-carotene and carotenoid flavonoids, both recognized for their potent antioxidant capabilities. The effects of different extraction processes on the tocopherols and carotenoids in sea buckthorn seed oil and sea buckthorn fruit oil are shown in Table 15.2.

As can be seen from Table 15.2, both sea buckthorn seed oil and sea buckthorn fruit oil extracted by solvent method had the highest content of tocopherols and carotenoids among the four extraction methods. Notably, sea buckthorn seed oil extracted via supercritical CO_2 demonstrated superior β-tocopherol and δ-tocopherol content, showing increases of 2.5% and 49.4%, respectively, compared to those obtained using the petroleum ether method and cold pressing method. Moreover, the content of tocopherols in sea buckthorn seed oil extracted with ultrasonic-assisted extraction was 24.6% higher than that of sea buckthorn seed oil extracted with the solvent method, while the content of carotenoids did not change significantly. This evidence suggests that tailored extraction methods can optimize the recovery of specific bioactive compounds. Cristina Tudor et al. (2020) identified 14 carotenoids from sea buckthorn oil, among which lutein, zeaxanthin, β-carotene, and α-carotene comprised the free carotenoid fraction. Remarkably, free and esterified zeaxanthin, including forms with one or two fatty acids, constituted 76.35% of the identified carotenoids, with zeaxanthin dipalmitate being the most prevalent compound. This study further observed that carotenoids like lutein and zeaxanthin tend to form esterified versions during fruit ripening, serving as a more stable storage form in the fruit, as summarized in Table 15.3.

Table 15.3 Isolated carotenoids and their content

Compounds	mg/100ga	%
Lutein	8.95 ± 0.22	2.95
Zeaxanthin	9.74 ± 0.90	3.21
15-*cis*-β-Carotene	6.13 ± 0.20	2.03
All-*trans*-β-Carotene	10.61 ± 0.41	3.50
α-Carotene	5.76 ± 0.49	1.90
Zeaxanthin myristate (C14:0)	8.71 ± 0.61	2.88
Zeaxanthin palmitate (C16:0)	10.00 ± 0.89	3.53
Cryptoxanthin myristate (C14:0)	11.88 ± 0.80	3.30
Oleic acid lutein (C18:1)	18.27 ± 0.42	3.92
β-Cryptoxanthin palmitate (C16:0)	21.26 ± 0.15	6.03
Zeaxanthin myristate	19.76 ± 0.22	7.02
Zeaxanthin palmitate-palmitoleate (16:0,C16:1)	46.08 ± 0.92	6.52
Zeaxanthin myristate palmitate (C14:0, C16:0)	114.98 ± 1.03	37.97
Carotenoid monoesters	59.55 ± 3.39	19.67
Carotenoid fatty acids	202.07 ± 2.32	66.73
Total carotenoid content	302.82 ± 7.84	

The table was prepared by the authors

15.3.3 *Phenolic Compounds*

Total phenolic compounds, present in sea buckthorn fruit pulp and twigs, constitute approximately 0.51–9.32% of their dry weight and exhibit antioxidant, anti-inflammatory, and anticancer properties. These compounds effectively scavenge DPPH, OH, and ABTS radicals, which can cause structural and functional damage through peroxidation. Additionally, they form chelating compounds with metals, acting as natural antioxidants. A study optimized the pressurized solvent-free microwave-assisted extraction (PSFME) technique using a two-level complete factorial design to extract antioxidants from sea buckthorn. This method demonstrated that sea buckthorn oil obtained via PSFME was enriched in flavonoids, notably quercetin and isorhamnetin, surpassing those acquired through pressing, solvent extraction, and pressurized wet extraction methods. Périno-Issartier et al. (2011) conducted a comparative analysis on the antioxidant content of sea buckthorn by-products, utilizing solvent-free microwave-assisted extraction and contrasting it with the traditional solvent extraction method. Their findings revealed that the phenolic content from microwave-assisted extraction was significantly higher (1147 mg/g) compared to that from the traditional solvent method (741 mg/g). Moreover, the antioxidant capacity, as measured by the DPPH system, was more pronounced in the former method, highlighting the efficiency of microwave-assisted extraction in yielding a greater concentration of phenolic compounds and thereby enhancing the antioxidant potential.

15.3.4 Phytosterols

Phytosterols, plant-derived sterols analogous to cholesterol, play crucial roles in regulating physiological activities crucial for plant growth and development, including cell division, defense mechanisms, and morphogenesis (Alu'datt et al. 2017). Beyond their botanical functions, phytosterols contribute to human health by facilitating lipid metabolism, thereby reducing cholesterol levels and mitigating the risk of developing non-alcoholic fatty liver disease (NAFLD) (Zhang et al. 2023). However, the oxidation products of phytosterols possess pro-atherogenic properties and can induce cytotoxicity, oxidative stress, apoptosis, and inflammation (Moreau et al. 2018).

Research has explored the synergistic effects of omega-3 fatty acids and phytosterol esters in reducing hepatic steatosis in NAFLD patients, demonstrating that the combination enhances the therapeutic efficacy against hepatic steatosis. In addition, they exhibit preventive and therapeutic potential against dementia by serving as ligands for nuclear receptors, such as peroxisome proliferator-activated receptors (PPAR), activating Sirtuin1 (SIRT-1), and influencing cholesterol efflux or β-amyloid (Aβ) metabolism (Song et al. 2020).

A specific study focused on sea buckthorn (*Hippophae rhamnoides L.*) seed oil, extracted using cold pressing, *n*-hexane, and supercritical carbon dioxide methods, identified 29 phytosterol compounds through GC-MS and FID analysis, including campesterol, gibberellinol, lanosterol, sitosterol, β-amyrin, and sitosterol. The study found significant differences in the total and individual sterol contents among the three extraction methods, with supercritical carbon dioxide extraction yielding the highest total sterol content (1640 mg/100 g oil) compared to the lowest yield from the cold pressing method (879 mg/100 g oil) (Shuang et al. 2016).

15.4 Pharmacological Effects of Chemical Components of Sea Buckthorn Oil

15.4.1 Cardiovascular and Cerebrovascular Protective Effects

In recent years, a series of studies has highlighted the significant cardiovascular and cerebrovascular protective effects of sea buckthorn oil. Johansson et al. (2000) selected 12 men with normal blood lipids and ingested sea buckthorn fruit oil and coconut oil 5 g per day for 4 weeks. The results showed that the ingestion of sea buckthorn fruit oil effectively reduced platelet coagulation caused by adenosine diphosphate. Furthermore, Basu et al. (2007) studied the effect of sea buckthorn seed oil on atherosclerosis, finding that its consumption notably decreased plasma cholesterol levels and the atherosclerotic index in the treatment group. These findings underscore the role of sea buckthorn oil in promoting cardiovascular and cerebrovascular health by reducing blood lipids, inhibiting platelet aggregation, and improving vascular microcirculation, attributed to its rich composition of linolenic acid, tocopherols, and phytosterols. Additionally, extensive research indicates a negative correlation between flavonoid intake and the risk of coronary heart disease, noteworthy given the high flavonol content of sea buckthorn. Suomela et al. (2006) have shown through both animal and human studies that sea buckthorn flavonoids possess free radical scavenging properties, can decrease blood viscosity, and improve cardiac function.

Sea buckthorn oil has a high content of fatty acids, especially unsaturated fatty acids, which are crucial for human health (Solà Marsiñach and Cuenca 2019). These unsaturated fatty acids play a vital role in maintaining cardiovascular health, lowering cholesterol levels, and reducing the risk of atherosclerosis (Oliveira et al. 2017). The major unsaturated fatty acids in sea buckthorn oil include linoleic acid, linolenic acid, and oleic acid. Linoleic and linolenic acids are essential fatty acids that the human body cannot synthesize on its own and are vital for maintaining normal cellular function and nervous system development and have anticancer properties. Meanwhile, oleic acid contributes to lowering blood pressure and enhancing blood circulation (Dhar Dubey et al. 2019).

15.4.2 Effects on Skin

Sea buckthorn oil is used externally and internally in medicine applications, renowned for its excellent wound-healing potential and its compatibility with topical use. Mihailo Kipic and colleagues evaluated lotions containing sea buckthorn oil for their moisturizing effectiveness, impact on skin pH, and skin tolerance to irritation. Their study involved assessments of healthy human skin volunteers for moisturizing capability and skin pH, while skin tolerance was tested on a rabbit skin model using the Draize test. The results, observed after intervals of 2, 24, 48, and 72 h, and 7 days, indicated that the lotions provided a significant moisturizing effect without irritating either human or animal skin, suggesting that the sea buckthorn oil-based lotion formulation is both effective and safe for skin application (Kipic et al. 2014). Furthermore, Maria Dudau and her team investigated the cytotoxic effects, cytokine, and growth factor production, revealing that the palmitic acid-rich fraction of sea buckthorn seed oil possesses regenerative properties for normal skin cells. This underscores the therapeutic potential of sea buckthorn oil in promoting skin health and regeneration (Dudau et al. 2021).

15.4.3 Effects on Liver and Kidney

Hsu et al. (2009) investigated the effect of sea buckthorn seed oil on the repair of liver injury caused by carbon tetrachloride. The results of continuous feeding for 8 weeks showed that sea buckthorn seed oil can increase the activity of superoxide dismutase, catalase, GSH-Px, GSH-Rd, and GSH (glutathione) levels in rat liver.

This indicates a protective effect against liver injury, with efficacy comparable to silymarin at a dose of 200 mg/kg. These results underscore sea buckthorn oil's potential in safeguarding against chemical-induced liver damage, likely owing to its rich composition of tocopherols and carotenoids, which exhibit potent antioxidant properties. These components have been shown to decrease the levels of aspartate aminotransferase (AST) and alanine aminotransferase (ALT) in the blood, while boosting GSH activity in the liver, as noted by Kallio (2002). In addition, many antioxidant substances like vitamin E and selenium have been identified to prevent liver damage through mechanisms such as anti-lipid peroxidation and inhibition of oxidative stress, as demonstrated in studies by Södergren et al. (2001). Thus, sea buckthorn oil has a positive protective effect on the prevention of liver injury caused by a variety of oxidative stress stresses.

15.4.4 Others

Extensive animal and cellular research has unveiled the diverse therapeutic properties of sea buckthorn oil, encompassing antioxidant, antimicrobial, anticancer, anti-inflammatory, and antidepressant effects (Olas 2018; Sytařová et al. 2020; Yogendra Kumar et al. 2013). An et al. (2023) demonstrated the oil's anti-fatigue capabilities in mice, evidenced by decreased serum lactate, blood urea nitrogen, and liver malondialdehyde levels, alongside increased hepatic glycogen, superoxide dismutase, hypothalamic dopamine, and glutathione peroxidase levels. Additionally, Parveen Akhter et al. (2023), with the help of the ferric chloride method, determined the antioxidant activity of sea buckthorn seed oil using various organic solvents, and plasma trivalent iron reduction of the oil of the sample extracted by ethyl acetate was highest at 68%. Ting et al. (2011) further confirmed its potent antioxidant properties through various assays, including DPPH radical scavenging and ferrous ion chelating activities. Zhang et al. (2021) observed enhanced antioxidant activity and immune response modulation in cyclophosphamide-induced immunosuppressed Balb/c mice, attributable to the adjustment in intestinal flora diversity and composition by sea buckthorn pulp oil. Hao et al. (2019) demonstrated that sea buckthorn seed oil could effectively reduce hypercholesterolemic blood cholesterol by increasing intestinal cholesterol excretion to promote bacterial growth through experiments in mice blood cholesterol in hamsters. The polyphenolic components of sea buckthorn have also been shown to significantly inhibit colon cancer activity through their interaction with miRNAs (Wu et al. 2021).

The chemical composition of sea buckthorn oil confers a wealth of physiological activities and pharmacological effects. This oil has been shown to support cardiovascular health, reduce cholesterol levels, boost immune system function, prevent chronic diseases, and improve skin health. Owing to these attributes, sea buckthorn oil has found extensive application across the food, health products, and cosmetic industries. Nevertheless, it is important to acknowledge that sea buckthorn oil might not be suitable for everyone and could induce adverse reactions in certain individuals. Consequently, anyone considering its use should first seek the guidance of a healthcare professional to ensure its safety and appropriateness for their specific health profile.

15.5 Application of Chemical Components of Sea Buckthorn Oil

Sea buckthorn oil, with its rich nutrient content and unique chemical composition, has found widespread applications across multiple sectors including food, medicine, cosmetics, skin care, agriculture, and animal husbandry. In the food industry, its abundance in essential vitamins (C, E, K) and minerals (iron, potassium, calcium), alongside unsaturated fatty acids like linoleic and linolenic acids, contribute to cardiovascular health and have led to its incorporation into various products such as jams, juices, condiments, and health supplements.

Medically, sea buckthorn oil is celebrated for its antioxidant, anti-inflammatory, and anti-tumor properties, enhancing immune function and aiding in the prevention and treatment of various diseases. Its capabilities in promoting wound healing and skin repair make it valuable in medical devices and treatments for skin conditions like inflammation and burns.

In the cosmetics and skincare industry, sea buckthorn oil is prized for its moisturizing, anti-aging, and anti-wrinkle effects, attributable to its triglycerides and active components like vitamin E and carotene. These elements work together to repair skin damage, reduce the appearance of wrinkles and discoloration, and improve skin elasticity and luster.

The oil also plays a significant role in agriculture and animal husbandry, acting as a natural growth regulator and feed additive to enhance crop resistance, yield, and animal immunity and productivity. Continued research and development into sea buckthorn oil's potential will undoubtedly advance related industries further, offering increased benefits to human health and well-being. The exploration of innovative formulations and applications promises to enhance its usability and effectiveness across its diverse range of uses.

Despite its broad utility, the liquid, lipophilic, and coloring nature of sea buckthorn oil poses challenges for external use. These can be addressed by incorporating the oil into semi-solid emulsions, offering a versatile solution that retains the oil's beneficial properties while overcoming application hurdles.

References

Akhter P, Bhatti TY, Shafiq I, Jamil F, Nazar R, Nazir MS, Park Y (2023) Antioxidant activity of sea buckthorn (Hippophae rhamnoides) seed oil extracted using various organic solvents. Korean J Chem Eng 40:2914–2920

Alu'datt MH, Rababah T, Alhamad MN, Al-Mahasneh MA, Almajwal A, Gammoh S, Alli I (2017) A review of phenolic compounds in oil-bearing plants: distribution, identification and occurrence of phenolic compounds. Food Chem 218:99–106

An Z, Wang Y, Li X, Jin H, Gong Y (2023) Antifatigue effect of sea buckthorn seed oil on swimming fatigue in mice. J Food Sci 88(4):1482–1494

Basu M, Prasad R, Jayamurthy P, Pal K, Arumughan C, Sawhney RC (2007) Anti-atherogenic effects of seabuckthorn (Hippophaea rhamnoides) seed oil. Phytomedicine 1411:770–777

Cakir AD (2004) Essential oil and fatty acid composition of the fruits of Hippophae rhamnoides L. (Sea Buckthorn) and Myrtus communis L. from Turkey. Biochem Syst Ecol 32:809–816

Chen Y, Cai Y, Wang K, Wang Y (2023) Bioactive compounds in sea buckthorn and their efficacy in preventing and treating metabolic syndrome. Food Secur 12(10):1985

Dhar Dubey KK, Sharma G, Kumar A (2019) Conjugated linolenic acids: implication in cancer. J Agric Food Chem 67(22):6091–6101

Dudau M, Codrici E, Tarcomnicu I, Mihai S, Popescu ID, Albulescu L, Tanase C (2021) A fatty acid fraction purified from sea buckthorn seed oil has regenerative properties on normal skin cells. Front Pharmacol 12:737571

Fu L, Su H, Li R, Cui Y (2014) Harvesting technologies for sea buckthorn fruit. Eng Agric Environ Food 7(2):64–69

Gao Y, Liu C, Yao F, Chen F (2021) Aqueous enzymatic extraction of peanut oil body and protein and evaluation of its physicochemical and functional properties. Int J Food Eng 17(11):897–908

Gâtlan AM, Gutt G (2021) Sea buckthorn in plant based diets. An analytical approach of sea buckthorn fruits composition: nutritional value, applications, and health benefits. Int J Environ Res Public Health 18(17):8986

Hao W, He Z, Zhu H, Liu J, Kwek E, Zhao Y, Chen Z-Y (2019) Sea buckthorn seed oil reduces blood cholesterol and modulates gut microbiota. Food Funct 10(9):5669–5681

Hsu YW, Tsai CF, Chen WK, Lu FJ (2009) Protective effects of seabuckthorn (*Hippophae rhamnoides L.*) seed oil against carbon tetrachloride-induced hepatotoxicity in mice. Food Chem Toxicol 47(9):2281–2288

Isopencu GO, Stroescu M, Brosteanu AV, Chira N-A, Pârvulescu OC, Busuioc C, Stoica-Guzun A (2018) Optimization of ultrasound and microwave assisted oil extraction from sea buckthorn seeds by response surface methodology. J Food Process Eng 42(2). https://doi.org/10.1111/jfpe.12947

Johansson AK, Korte H, Yang B, Stanley JC, Kallio HP (2000) Sea buckthorn berry oil inhibits platelet aggregation. J Nutr Biochem 11(10):491–495

Kallio YH (2002) Composition and physiological effects of sea buckthorn (Hippophaë) lipids. Trends Food Sci Technol 13(5):160–167

Kipic M, Cupara S, Jaćević V, Radovanovic AM, Milovanović O (2014) Cutaneous effects of sea buckthorn oil emulsion. Serb J Exp Clin Res 15:151–155

Koskovac M, Cupara S, Kipic M, Barjaktarevic A, Markovic M (2017) Sea buckthorn oil—a valuable source for cosmeceuticals. Cosmetics 4(4):40

Kuhkheil A, Mehrafarin A, Abdossi V, Naghdi Badi H (2017) Seed oil quantity and fatty acid composition of different sea buckthorn (*Hippophae Rhamnoides L.*) wild populations in Iran. Erwerbs-obstbau 60:165–172

Kumar TA, Pareek S, Kaur R, Sagar NA, Singh L, Sami R, Rahman MM (2022) Optimization of ultrasonic-assisted enzymatic extraction of freeze-dried sea buckthorn (*Hippophae rhamnoides L.*) berry oil using response surface methodology. Sustain For 14(17):10849

Liu G, Xu X, Hao Q, Gao Y (2009) Supercritical CO_2 extraction optimization of pomegranate (Punica granatum L.) seed oil using response surface methodology. LWT Food Sci Technol 42(9):1491–1495

Ma X, Yang W, Laaksonen O, Nylander M, Kallio H, Yang B (2017) Role of flavonols and proanthocyanidins in the sensory quality of sea buckthorn (Hippophaë rhamnoides L.) berries. J Agric Food Chem 65(45):9871–9879

Ma X, Yang W, Kallio H, Yang B (2022) Health promoting properties and sensory characteristics of phytochemicals in berries and leaves of sea buckthorn (Hippophaë rhamnoides). Crit Rev Food Sci Nutr 62(14):3798–3816

Moreau RA, Nyström L, Whitaker BD, Winkler-Moser JK, Baer DJ, Gebauer SK, Hicks KB (2018) Phytosterols and their derivatives: structural diversity, distribution, metabolism, analysis, and health-promoting uses. Prog Lipid Res 70:35–61

Olas B (2018) The beneficial health aspects of sea buckthorn (Elaeagnus rhamnoides (L.) A.Nelson) oil. J Ethnopharmacol 213:183–190

Oliveira PA, Kovacs C, Moreira P, Magnoni D, Saleh MH, Faintuch J (2017) Unsaturated fatty acids improve atherosclerosis markers in obese and overweight non-diabetic elderly patients. Obes Surg 27(10):2663–2671

Périno-Issartier S, Zill-e-Huma, Abert-Vian M, Chemat F (2011) Solvent free microwave-assisted extraction of antioxidants from sea buckthorn (Hippophae rhamnoides) food by-products. Food Bioprocess Technol 4(6):1020–1028

Seglina D, Krasnova I, Grygier A, Radziejewska-Kubzdela E, Rudzińska M, Górnaś P (2021) Unique bioactive molecule composition of sea buckthorn (*Hippophae rhamnoides L.*) oils obtained from the peel, pulp, and seeds via physical "solvent-free" approaches. J Am Oil Chem Soc 98:1009–1020

Shuang R, Rui X, Wenfang L (2016) Phytosterols and dementia. Plant Foods Hum Nutr 71(4):347–354

Södergren E, Cederberg J, Vessby B, Basu S (2001) Vitamin E reduces lipid peroxidation in experimental hepatotoxicity in rats. Eur J Nutr 40(1):10–16

Solà Marsiñach M, Cuenca AP (2019) The impact of sea buckthorn oil fatty acids on human health. Lipids Health Dis 18(1):145

Song L, Zhao XG, Ouyang PL, Guan Q, Yang L, Peng F, Yan H (2020) Combined effect of n-3 fatty acids and phytosterol esters on alleviating hepatic steatosis in non-alcoholic fatty liver disease subjects: a double-blind placebo-controlled clinical trial. Br J Nutr 123(10):1148–1158

Suomela JP, Ahotupa M, Yang B, Vasankari T, Kallio H (2006) Absorption of flavonols derived from sea buckthorn (Hippophaë rhamnoides L.) and their effect on emerging risk factors for cardiovascular disease in humans. J Agric Food Chem 54(19):7364–7369

Suryakumar G, Gupta A (2011) Medicinal and therapeutic potential of Sea buckthorn (*Hippophae rhamnoides L.*). J Ethnopharmacol 138(2):268–278

Sytařová I, Orsavová J, Snopek L, Mlček J, Byczyński Ł, Mišurcová L (2020) Impact of phenolic compounds and vitamins C and E on antioxidant activity of sea buckthorn (*Hippophaë rhamnoides L.*) berries and leaves of diverse ripening times. Food Chem 310:125784

Teleszko M, Wojdyło A, Rudzińska M, Oszmiański J, Golis T (2015) Analysis of lipophilic and hydrophilic bioactive compounds content in sea buckthorn (Hippophaë rhamnoides L.) berries. J Agric Food Chem 63(16):4120–4129

Ting HC, Hsu YW, Tsai CF, Lu FJ, Chou MC, Chen WK (2011) The in vitro and in vivo antioxidant properties of seabuckthorn (*Hippophae rhamnoides L.*) seed oil. Food Chem 125(2):652–659

Tkacz K, Wojdyło A, Turkiewicz IP, Ferreres F, Moreno DA, Nowicka P (2020) UPLC-PDA-Q/TOF-MS profiling of phenolic and carotenoid compounds and their influence on anticholinergic potential for AChE and BuChE inhibition and on-line antioxidant activity of selected Hippophaë rhamnoides L. cultivars. Food Chem 309:125766

Vilas-Franquesa A, Saldo J, Juan B (2022) Sea buckthorn (*Hippophae rhamnoides*) oil extracted with hexane, ethanol, diethyl ether and 2-MTHF at different temperatures – an individual assessment. J Food Compos Anal 114:104752

Wu H, Li C, Cui M, Guo H, Chen S, Jin'e D, Li H (2021) Polyphenols from Hippophae rhamnoides suppressed colon cancer growth by regulating miRNA-mediated cell cycle arrest and apoptosis in vitro and in vivo. J Funct Foods 87:104780

Xing J, Yang B, Dong Y, Wang B, Wang J, Kallio HP (2002) Effects of sea buckthorn (Hippophae rhamnoides L.) seed and pulp oils on experimental models of gastric ulcer in rats. Fitoterapia 73(7–8):644–650

Xu X, Gao Y, Liu G, Wang Q, Zhao J (2008) Optimization of supercritical carbon dioxide extraction of sea buckthorn (Hippophaë thamnoides L.) oil using response surface methodology. LWT Food Sci Technol 41(7):1223–1231

Yang B, Kallio HP (2001) Fatty acid composition of lipids in sea buckthorn (Hippophaë rhamnoides L.) berries of different origins. J Agric Food Chem 49(4):1939–1947

Yang B, Kallio H (2002) Composition and physiological effects of sea buckthorn (Hippophaë) lipids. Trends Food Sci Technol 13(5):160–167

Yogendra Kumar MS, Tirpude RJ, Maheshwari DT, Bansal A, Misra K (2013) Antioxidant and antimicrobial properties of phenolic rich fraction of Seabuckthorn (*Hippophae rhamnoides L.*) leaves in vitro. Food Chem 141(4):3443–3450

Zhang J, Zhou HC, He SB, Zhang XF, Ling YH, Li XY, Hou DD (2021) The immunoenhancement effects of sea buckthorn pulp oil in cyclophosphamide-induced immunosuppressed mice. Food Funct 12(17):7954–7963

Zhang C, Ma W, Xu M, Li T, Han G, Gu L, Zhang X (2023) Identification and functional characterization of ZmSCYL2 involved in phytosterol accumulation in plants. Int J Mol Sci 24(12):10411

Zhao L, Temelli F, Curtis JM, Chen L (2017) Encapsulation of lutein in liposomes using supercritical carbon dioxide. Food Res Int 100:168–179

Chapter 16
Preparation Technology of Sea Buckthorn Essential Oil Microcapsules

Wenyu Liu, Changqing Wei, Qingqing Sun, and Jinhu Tian

Abstract Sea buckthorn essential oil contains rich unsaturated fatty acids, tocopherols, β-carotene, and sterols. Research has shown that sea buckthorn oil has multiple pharmacological functions, such as protecting the cardiovascular and cerebrovascular and gastrointestinal illnesses, moistening skin, promoting skin tissue regeneration, and serving the role of epithelial tissue healing, improving skin aging phenomenon, and relieving dried eyes. Sea buckthorn oil microcapsule preparation is divided into physical method and physical method, using the application in medicine and natural substance microcapsule advantage, the embedding of sea buckthorn oil to wall material selection, and the embedding rate of microcapsule particle size as detection of microcapsule quality indicators. Based on the effect of sea buckthorn oil compositions and pharmacological functions, the present status of development and utilization of microcapsules were reviewed in order to understand sea buckthorn oil microcapsule product development to provide a theoretical foundation and basis.

Keywords Sea buckthorn oil · Efficacy and function · Sea buckthorn oil microcapsules · Preparation technology

Abbreviations

GSH Glutathione
MMP Matrix metalloproteinases

W. Liu
Food College, Shihezi University, Shihezi, China

C. Wei (✉)
School of Food Science and Technology, Shihezi University, Shihezi, China

Q. Sun · J. Tian
Department of Food Science and Nutrition, Zhejiang University, Hangzhou, China

J. Tian et al. (eds.), *Sea Buckthorn*, https://doi.org/10.1007/978-981-97-9865-0_16

OSDI	Ocular Surface Disease Index
ROS	Reactive oxygen species
TCM	Traditional Chinese medicine

16.1 Introduction

Sea buckthorn (*Hippophae rhamnoides L.*) fruits, a kind of temperate common herbal shrub, similar with orange, pearl fruit which tastes bitter and sour, are widely distributed in Asia and Europe. It has good plant resistance to drought, high temperature, low temperature, and salinity soil. Sea buckthorn can effectively control soil erosion and water loss (Ali et al. 2013).

In addition, sea buckthorn is beneficial because of its high nutritional value and health-promoting benefits. It is rich in vitamin C, omega-3 fatty acid, carotenoids, flavonoids, and other phenolic compounds, as well as tocopherol (Bal et al. 2011). The oil substance extracted from sea buckthorn is called sea buckthorn oil. According to the different extracted parts, sea buckthorn oil can be divided into two types, namely, sea buckthorn fruit oil and sea buckthorn seed oil, which are sea buckthorn fruit extract and sea buckthorn seed extract, respectively, as the name suggests. In contrast, sand thorn fruit oil is soluble in water, has poor stability, is easier to oxidize, is unfavorable to storage and transportation, and thus has a limited application range. Sea buckthorn seed oil contains rich unsaturated fatty acids and its easy oxidative rancidity, the production and storage process, and oil emulsion or eating or drinking directly flavor can be widely applied.

Sea buckthorn oil is widely used in medicine, food, and health care products. However, sea buckthorn oil is volatile and can be easily oxidized, carrying inconvenience, limiting its development in practical applications. Using microcapsule technology to embed sea buckthorn seed oil, which can effectively delay oil oxidation, product storage and transportation become feasible. Therefore, sea buckthorn oil microcapsule preparation has become an important research direction. The research progress of this technology provides a reference for further research.

16.2 The Characteristics and Application of Sea Buckthorn Oil

16.2.1 Composition and Pharmacological Effects of Sea Buckthorn Oil

16.2.1.1 Composition of Sea Buckthorn Oil

Sea buckthorn oil is derived from sea buckthorn seeds, fruit, fruit powder, and pomace extract. Sea buckthorn oil contains a variety of living materials, including carotenoids, tocopherol, phytosterols, and a variety of unsaturated fatty acids, unsaturated fatty acids, mainly oleic acid, linoleic acid, and linolenic acid (Kallio and Yang 2001). Because sea buckthorn oil has radiation protection, antibacterial and antioxidant properties, other pharmacological effects, and inhibits tumor cell growth, it is the first to be included as one of the traditional Chinese medicine (TCM) in the National Institutes of Health Edible Directory. It has a broad application prospects in the fields of medicine, health food, and cosmetics.

16.2.1.2 Pharmacological Effects of Sea Buckthorn Oil

(a) *Heart Protection*
The effect of sea buckthorn oil on the treatment of cardiovascular diseases can be attributed to the higher unsaturated fatty acid content. Basu et al. (2007) compared normal albino rabbits with high cholesterol diet group before giving sea buckthorn seed oil to rabbits after measuring the indexes of blood triglyceride and total cholesterol and found that sea buckthorn seed oil provides good resistance to atherosclerosis. The authors speculate that the cardiac protective effect may be due to the presence of polyunsaturated fatty acids and phytosterols, live A presence, or the result of synergy. Adding sea buckthorn fruit oil can also help hyperlipidemia patients with high plasma lipid concentrations of recover to normal levels. Johansson et al. (2000) found that subjects with blood lipids given sea buckthorn fruit oil and sea buckthorn seed oil capsule orally showed the activity of platelet aggregation. Guo et al. (2017) found that sea buckthorn fruit extract can significantly reduce cardiovascular disease subjects' total cholesterol and low-density lipoprotein cholesterol levels, but have not observed this effect in healthy subjects. The author ascribed this heart protection to β-cholesterol and flavonoids of *Hippophae rhamnoides*. Cardiovascular disease is usually associated with obesity, which could be a daily intake of fruits, vegetables, and the impact of product. Lehtonen et al. (2011) suggested that replenishing sea buckthorn oil every 33–35 days may reduce the incidence of overweight and obese women with metabolic disease.

(b) *Liver Function Protection*
Hsu et al. (2009) have found that sea buckthorn seed oil has a therapeutic effect on liver injury in mice caused by carbon tetrachloride. The administration of 0.26, 1.3, and 2.6 mg/kg doses of sea buckthorn seed oil to mice for eight consecutive weeks showed a decrease in hepatic lipid peroxidation levels, a decrease in serum cholesterol, triglyceride, alkaline phosphatase, alanine aminotransferase, and aspartate aminotransferase levels, and increased the activity of various antioxidant enzymes such as catalase. As a positive control, they found that sea buckthorn seed oil and the activity of silymarin is quite high (200 mg/kg).

(c) *Wound-Healing Promotion*
Upadhyay et al. (2009) used sea buckthorn seed oil in a burn wounds model and proved that it has obvious wound-healing activity and non-toxic side effects. Compared with the reference group, the group given sea buckthorn seed oil

showed significant increases in mouth contraction and hydroxyproline, hexose amine, DNA and total protein content, granulation tissue matrix metalloproteinases (MMP-2 and 9), collagen type III. The expression of VEGF is raised and glutathione (GSH) levels increased significantly. The wound granulation tissue of reactive oxygen species (ROS) increased. The results showed that sand spine seed oil had an obvious role in promoting wound healing, and the authors speculated that this activity may be related to sea buckthorn seed oil of omega-3 and omega-6 fatty acids, tocopherol, and carotenoids. Hiroshi Ito et al. (2014) used sea buckthorn seed oil on the sheep model of three-level flame burns treatment and found that it promoted burn wound-healing activity.

(d) *Anti-Inflammatory Effects*
Acharya Balkrishna et al. (2019) have studied the anti-inflammatory effect of sea buckthorn oil on psoriasis. In an in vitro monocyte model, sea buckthorn oil inhibited the release of intracellular reactive nitrogen and NF-kappa B proteins and reduced the release of cellular necrosis factor. In a stimulus carageen glue paw edema model, internal and topical application of sea buckthorn oil can reduce the degree of paw edema. In mice psoriasis sample models, sea buckthorn oil can significantly reduce ear edema and ear biopsies. The authors speculated that a possible mechanism is that sea buckthorn oil downgrades the NF-kappa B protein by inhibiting the activity of nitrogen and plays a proinflammatory role.

(e) *Anticancer Effect and Other Effects*
Kumar et al. (2011) have described the use of sea buckthorn oil in cancer therapy, especially based on the treatment of chemotherapy and radiation therapy, playing an important role. Studies have shown that taking sea buckthorn seed oil increases appetite, restores patients with kidney and liver function, and maintains health. Sea buckthorn seed oil is rich in vitamins, especially A and E and β-carotene, which is a good way to provide protection against cervical cancer (Li and Beveridge 2007). Gao et al. (2020) found that sea buckthorn fruit oil has the effect of reducing blood fat. In experiments using hyperlipidemia hamsters as the animal model, the intake of sea buckthorn fruit oil controlled body weight and fat tissue and improved the total cholesterol and triglyceride levels, and reduced oxidative stress caused by hyperlipidemia and liver damage by modulating the antioxidant enzymes. Larmo et al. (2019) found that spraying sea buckthorn oil on closed eyelids can relieve dried eyes. Adults with the Ocular Surface Disease Index (OSDI) of 20 or more were selected as subjects. Compared to the commercially available sprays, it was found that the sea buckthorn oil spray was well tolerated and the ocular surface disease index decreased significantly.

16.2.2 Sea Buckthorn Oil Product Applications

In China, sea buckthorn oil is used as a raw material for product variety, including food, health care products, pharmaceuticals, and cosmetics. Sea buckthorn oil is used mainly as a flavoring agent in the field of food and nutrition. It can be added to juice, beverage, dairy and foods, such as bread and pastry to enhance its taste and nutritional value. In addition, it can enhance immunity, improve blood lipids, and prevent cardiovascular disease (Olas 2018). In the field of health care products, sea buckthorn oil can be used as an oral capsule or soft capsule dosage form to enhance immunity, improve skin, and protect eyesight (Larmo et al. 2019). In the field of cosmetics, sea buckthorn oil can be used in the manufacture of creams, lotions, shampoo, and other products to protect the skin, moisturize hair, and prevent aging (Koskovac et al. 2017). Sea buckthorn oil can also be used in the manufacture of aromatherapy products, such as aromatherapy pure oil and massage oil, to alleviate problems such as anxiety, stress, and fatigue. In general, sea buckthorn oil has wide application prospects in various fields, especially in health and beauty segment where the demand for such products is increasing.

16.3 Introduction of Microcapsule Technology

16.3.1 Definition and Basic Principle of Microcapsules

Microencapsulation refers to the functional biological activity material wrapped in a translucent or sealing technology of the polymer matrix (Bollhorst et al. 2017). The encased substance is known as core material, and the substance encapsulating the core material is known as wall materials. Using microcapsule technology, every unique material inside and outside the environment can be made so that the core material is not affected by external environment temperature, light, and oxygen (Hoyos-Leyva et al. 2018). Dan raises the biological utilization ratio of core material, achieving the goal of slow release and controlled release (Díaz et al. 2015). For the development of new technology, microcapsules of particle size from 1 to 1000 μm to the nanometer range can be prepared (Castro et al. 2021).

Natural polymer material is commonly used as a microcapsule wall material, mainly including carbohydrates, proteins, and lipids. Carbohydrate microcapsule wall material has characteristics such as renewable nature and good solubility and has extremely great development potential. The most studied carbohydrates carrier material at the present stage are sodium alginate and chitosan, maltodextrin, starch, and its derivatives (such as xin alkenyl succinic acid starch ester and carboxymethyl starch), cellulose, and its derivatives. Protein, as a macromolecule polymer, has good emulsification and gel and water-based biocompatibility. At present, soy protein, whey protein, casein, and gelatin protein have been widely applied in the field of microcapsule wall materials. Soy lecithin and stearic acid are good emulsifiers and nutritional supplements and can be used as natural lipid wall material for microcapsules. Synthetic polymer wall materials such as polyacrylamide, polylactic acid, and ethyl cellulose, inorganic materials such as layered double hydroxide, calcium carbonate, and metal phosphates have been widely used in medical, industrial, agricultural, and other fields.

16.3.2 Application of Microcapsules in Food and Pharmaceutical Preparations

Microencapsulation reduces the loss of active ingredients by isolating sea buckthorn oil from the external environment, thereby extending shelf life, masking or improving food odor or mouth odor sensation. In the field of medicine preparation research, the microcapsule can insulate core material drugs and the external environment and achieve fixed-point control release and slow release effect (do Nascimento et al. 2022).

16.4 Sea Buckthorn Oil Microcapsule Technology

According to the shape and morphology of microcapsule products, microcapsules can be mainly divided into several types, such as multi-walled single-core type, single-walled multi-core type, and irregular single-walled multi-core type. The core material release is divided into two cases: the instantaneous release and slow release, and the release method can be selected according to demand.

Now, there are more than 200 kinds of preparation methods of microcapsules according to the properties, preparation methods of microcapsules, and mechanism of microencapsulation wall formation. The preparation methods of microcapsules can be divided into physical method, chemical method, and physical and chemical method. The physical and chemical methods are mainly composite condensation method, method of orifice coagulation bath, and nano-precipitation. The main chemical methods are in situ polymerization and interfacial polymerization. The physical method mainly includes spray-drying, freeze-drying, solvent evaporation method, and physical adsorption method. Depending on the nature of the raw material and wall material, particle size, and the requirements for microcapsule release rate, different types of preparation methods are selected (Cui et al. 2023).

16.4.1 Physical Methods

16.4.1.1 Spray-Drying Method

Spray-drying is an extensively applied microcapsule technology, mainly divided into the preparation of emulsion and drying in two stages. Special wall film will form a reticular structure when heated. The mesh structure, water, or other small molecules can be the filtered material that dissolves in the core material by hot media contact and evaporation. The core material substance molecule that is bigger will be trapped in a mesh. Therefore, spray-drying method can be performed by selecting one or more wall materials, artificially controlling the size of the mesh.

Through the instrument, liquid oil gets dispersed into a mist of oil droplets. The use of the "grid" of these molecular sieving will separate fat oil granules of different molecular sizes, which form the core material package (Carneiro et al. 2013; Ren et al. 2020). The performance of microcapsules is controlled by changing the core wall material mass ratio, solid content, in and out of the mouth, such as temperature and feed velocity, thus isolating the core from light, heat, and air, extending the shelf life of goods and expanding the range of applications.

16.4.1.2 Freeze-Drying

Freeze-drying method is also called the freeze-drying technology used for a substance that is sensitive to heat and odor (such as oil) after dehydration. Simply put, oil can be dissolved in water and frozen (Vujanović et al. 2021). Then, using external pressure and heating, the frozen liquid material is directly converted from the solid phase to the gas phase. Frozen water directly transforms from the solid state into steam and not through the liquid state. Compared with other preparations of microcapsule technology, this technology has the advantages of simple operation, and products are more resistant to oxidation. However, this method has the disadvantages of low efficiency of microcapsules, long processing time, and high production cost.

16.4.1.3 Solvent Evaporation Method

The wall materils need to dissolved in solvent, so wall, generally for the synthesis of linear polymer, and join the core material emulsion, the preparation process of neither need high temperature, also need not separating agent, also not involved and polymerization. The core material that is coated comes after the active ingredient in the solvent evaporation method. Therefore, as a typical physical preparation method, the solvent evaporation method only involves a physical change process. There have been many reports on the study of this method (Swapan 2006; Wang et al. 2013; Tang et al. 2017; Arabi et al. 1996).

Emulsion solvent evaporation method is divided into two steps: (i) solvent volatilization, that is, from the emulsion dispersed phase to remove volatile solvent and (ii) the preparation of microcapsules. The process includes two processes: the solvent extraction process between the two liquid phases and liquid phase solvent evaporation process between the gas phases. First, essential oil is prepared. The wall polymer and organic solvent are mixed into a blend to get homogeneous solution. Do not mix solution with continuous phase. Therefore, after emulsification, capsule material solution exists as a continuous phase of dispersed droplets and then by heating, the volatile organic solvent is removed. After the volatile organic solvent is removed, essential oils and the compatibility between the variation of linear wall, continuous phase, and volatile organic solvents also have certain compatibility. The difference in concentration of the organic solvent and the continuous phase allows the extraction of the organic solvent from dispersed phase. Due to the wall polymer's greater solubility in organic solvents, organic solvent volatilizes and migrates to the surface of oil droplets to effectively cover the essential oil, and essential oil microcapsules are obtained. Although this method is simple and highly efficient, its shell material must use linear polymer, which can be a dissolved solvent. The boiling point of organic solvent cannot be too high, so as to prevent the volatilization of wall materials, and the compatibility of core material requirements is higher.

16.4.1.4 Physical Adsorption Method

Using a larger specific surface area of microporous matrix core material substances, physical adsorption was achieved. The method is easy to produce, without chemical reagents, and the advantages of mild conditions, preventing the loss of flavor material in the core material. In recent years, new microporous matrices, such as porous starch, are constantly emerging, and have been widely used in oil and other bioactive substances for embedding. Hole starch, also called microporous starch, refers to the processing by an artificial method that makes the particles in a porous state into a new type of modified starch. Usually in physical, chemical, or biological methods, the hydrolysis product of all kinds of raw starch formed in conditions of lower than starch gelatinization temperature is formed like honeycomb hollow particles, which can be used to dress up all kinds of material and which have very strong adsorption capacity (Yan and Zheng 2010). Compared with the spray-drying method, the preparation of microcapsules through a technological process without using homogenizer, dry spray-drying machines, and other complex, expensive equipment, is simple and can well retain the active components of the core material (Belingheri et al. 2012).

16.4.2 Physical and Chemical Method

16.4.2.1 Composite Condensation Method

Using two kinds of material with opposite charges as wall material of essential oil, the oil dispersed in it, the two contrary polyelectrolyte are mixed in static soluble or insoluble complex form. By changing the polyelectrolyte pH, temperature, and solution concentration, the microcapsule core material is formed by phase separation. Its preparation technology is simple, with a high production rate and at a low cost. A defect is the strict control of preparation conditions, preparation is expensive, and therefore not suitable for large-scale industrial production. The research results show that the compound condensing method not only can improve the stability of volatile compounds but also promote the application of these bioactive substances in the food system (Muhoza et al. 2022). Existing research shows that the compound condensing essential oil microcapsule anions are gum Arabic, protein separated from soybean and sodium alginate and cations such as whey protein isolated and chitosan (Hai et al. 2018; Fang et al. 2020; Huang et al. 2020; Liang et al. 2020).

16.4.2.2 Orifice Coagulation Bath

The raw material coated with soluble polymer capsule and the core material and wall material mixed suspension in orifice device drip into the coagulation bath curing forming microcapsules. The use of orifice device and the preparation of microcapsule coagulation bath make it the orifice coagulation bath method (Cheng et al. 2014), which is hereinafter referred to as the orifice method. Orifice method is suitable for the fat-soluble and bioactive substances, easy operations, can be put into the process of membrane material coated core material and the curing process separately, so as to achieve control of microgel capsule particle size and wall thickness. Many microcapsules are prepared using this method. Simple equipment, easy operation, and less investment lead to microcapsules of relatively uniform morphology and particle size.

16.4.2.3 Holdings of Nano-precipitation

Nano-microcapsules are prepared using cores with different solubilities in organic and aqueous phases. The operation is simple and reproducible. An intense phase transition occurs at the interface between the organic and aqueous phases during the preparation process, where organic phase is quickly dispersed in the water phase, and core material is at the interface between organic and water phases due to the removal of organic phase deposition into nanometer microgel capsule. The traditional particle size of microcapsules is 5–2000 μm, belonging to the micron-grade microcapsule. The nanoparticle size of microcapsules is 10–1000 nm. Compared with the traditional microcapsules, it has better dispersibility, targeting, and sustained-release effect. It helps form colloidal solution easily, has an appearance of a clear and transparent liquid. Specific advantages are as follows (Asbahani et al. 2015): (1) easy surface modification of the microcapsule, improving the embedding rate and the target; (2) preparation of nano-microcapsules by polymerization or double emulsion solvent evaporation technique, and good stability, convenient for processing and storage; (3) nanometer microcapsule wall material is usually a biodegradable polymer, such as polylactic acid, which has good biological compatibility; the side effects are small if injected into the body; (4) can be made into sustained-release capsules, prolonging drug curative effect; and (5) to food for other components are destroyed, enhancing the curative effect. Therefore, because of the unique properties of nano-microcapsules, it is a good choice for many applications. Its application has not only been widely developed in the loading of drugs but also has been widely used in the food industry. It is used to load vitamins, pigments, proteins, and bioactive ingredients in food.

16.4.3 Chemical Methods

16.4.3.1 Interfacial Polymerization

The core material with high reactivity is a oily monomer emulsion or dispersed in a continuous phase. Another kind of water-soluble monomer B is dissolved in the dispersed phase. The blending of the two phases, A and B form two kinds of contact and polymerization monomer in the core material oil interface film, packaging oil droplets into an essential oil microcapsule (Corcoran et al. 2004; Alfaro-Galarza et al. 2020; Kfoury et al. 2017; Li et al. 2016). Due to the interfacial polymerization of the monomer, it can be used in high-reactivity applications, as its response speed is good and active substances can provide a quite good protection. For the core material of oil phase and water phases, the preparation of the emulsion process is relatively simple. However, because of two-phase monomer, a thin layer of poly-compound interface impedes further reaction. The preparation of microcapsules is likely to have lower mechanical integrity. Furthermore, the core material cannot react with the monomer and acid alkali. Its monomer residue is also hard to deal with.

16.4.3.2 In Situ Polymerization

It is another common microcapsule preparation method. The reaction monomer or a certain reaction of oligomer and catalytic agent solution are added at the same time, which form the core of the core material. After the oil emulsion phase, through the use of cross-linking agent, the pH, temperature, and quality of dispersed phase can be changed. The molecular weight of the polymer increases after the polymerization reaction. The deposition at the interface between oil and water leads to the formation of covering containing essential oil kernel, which reacts to generate an insoluble polymer shell of essential oil microcapsule (Podgornik et al. 2021; Suryanarayana et al. 2008; He et al. 2020; Han et al. 2020). The preparation of microcapsule through in situ polymerization in a mild reaction condition and using reaction material variety leads to good mechanical performance and easy to control operating conditions. The essential oil microcapsules prepared by the in situ polymerization method can effectively avoid the reactants from influencing the aroma or lead to essential oil reacting metamorphism. Therefore, in situ polymerization is a common method for the preparation of essential oil microcapsules.

16.5 Parameter Optimization in Sea Buckthorn Oil Microcapsule Preparation Process

16.5.1 *Choice of Wall*

The choice of wall material has important effects on sea buckthorn oil microcapsule preparation. Different wall materials will affect the stability of the microcapsule wall material, release rate, and protective effect (Mohammed et al. 2020). The commonly used wall materials include gum Arabic and chitosan. Chitosan has good biocompatibility and is biodegradable, which can improve the stability of the microcapsule and the drug release rate. Therefore, suitable wall materials should be selected and relevant experiments should be carried out according to specific needs when preparing sea buckthorn oil microcapsules.

16.5.2 *Coating Efficiency*

Coating efficiency refers to the proportion of oil particles occupied by the coating wall, which has a great influence on the preparation of sea buckthorn essential oil microcapsules. High coating efficiency improves the oil-carrying capacity and stability of essential oils in microcapsules and reduces the loss and drainage of essential oils; low coating efficiency reduces the oil-carrying capacity and stability of essential oils in microcapsules and affects the effect of microcapsules.

16.5.3 *Microcapsule Particle Size*

Microcapsule particle size refers to the diameter size, and the particle size of microcapsules has an important impact on the performance of microcapsules. The size of the microcapsule particles affects the drug load, the stability of the microcapsule, and the drug release rate. Therefore, when preparing sea buckthorn oil microcapsules, a suitable microcapsule particle size should be selected according to the specific situation. It is important to note that the microcapsule particle size control is a more complex step during the preparation of microcapsules, and there is a need for strict control and optimization. Before the preparation of the microcapsule, experimental verification should be fully performed to determine the best preparation conditions and wall material concentration to control the microcapsule particle size. At the same time, in the quality control of microcapsule, the detection and analysis of microcapsule particle size should also be carried out to ensure the quality and effect of microcapsules.

Compared to single-wall components, composite walls with different wall materials mixed in certain proportions have higher embedding rates and stability, providing better control of the core release. Zhang et al. (2023) determined the optimal preparation process for sea buckthorn oil articles by spray-drying method based on microencapsulation coating efficiency (EE%) and particle size distribution (D50): wall thickness of soybean isolate proteins in the best condition led to a coating efficiency of 95.30 ± 2.67%, yield of 57.03 ± 3.71%, and particle size of 7.96 ± 1.04 μm.

16.6 Evaluation of Sea Buckthorn Oil Microcapsule Effect

Microencapsulation can also be evaluated by pharmacokinetic and stability studies (Luo et al. 2017). Pharmacokinetic studies can evaluate microcapsules in the form of drug absorption, distribution, metabolism, and excretion of the process, providing the basis for drug treatment. Stability studies can be used to assess the physical and chemical stability of the microcapsules and performance stability will not occur in the process of storage and application of the adverse event or degradation.

16.7 Summary and Outlook

This chapter summarized the chemical composition and efficacy of sea buckthorn oil as well as the technology of sea buckthorn oil microcapsule preparation. Sea buckthorn oil is rich in unsaturated fatty acids, carotene, tocopherol, phytosterols, flavonoids, and other bioactive substances, which can improve cardiovascular, inflammation, and has many other beneficial effects. According to the properties, preparation methods of microcapsule, microcapsule capsule wall formation mechanism, and the preparation of microcapsule technology are classified into three categories: physical method, chemical method, and physical and chemical method. Mining using microcapsule technology can be used in the preparation of a pure sea buckthorn oil microcapsule embedded into a microcapsule powder, which has been confirmed to possess healthy functions for other food or medicine applications. Although microencapsulation of sea buckthorn essential oil has been reported in the literature, it is far from reaching a mass production scale. Research efforts on the selection of reinforced wall and emulsification system update are directed toward developing resistance to oxidation and storage period of sea buckthorn oil microcapsule powder to provide technical support for industrial production.

At present, the study of sea buckthorn seed has focused primarily on oil extraction, the fatty acid composition analysis, protein peptide preparation, and functional characteristics. However, the research on sea buckthorn seed oil microcapsules is of a smaller scale and mainly focuses on the optimization of spray-drying process, while there are reports on the screening of wall materials, fluorescent labeling test of the embedding rate, and evaluation of the performance of microcapsule products. Screening of high-quality wall materials has a bearing on the microcapsule's encapsulation rate and the product's processing characteristics. Future research should focus on different sea buckthorn seed oil microcapsule wall material preparation, compare the embedding effect, and study its properties to provide a theoretical basis for the development and utilization of sea buckthorn seeds. Solving this problem with suitable methods and process parameters needs to be further investigated. Increasing the efficiency and precision of microcapsule preparation should be further explored with new coating materials and methods to improve the quality of microcapsules. Secondly, the current detection method is not perfect, and there is a need to develop a more accurate and reliable method to evaluate the efficacy of the microcapsule package and soluble release performance. In addition, the commercial application of sea buckthorn oil microcapsule has faced with problems, like market recognition and price competition. Therefore, the industrialization and commercialization of technology need to be looked at to solve the problem.

References

Alfaro-Galarza O, López-Villegas EO, Rivero-Perez N, Tapia-Maruri D, Jimenez-Aparicio AR, Palma-Rodriguez HM, Vargas-Torres A (2020) Protective effects of the use of taro and rice starch as wall material on the viability of encapsulated *Lactobacillus paracasei subsp. Paracasei*. LWT 117:108686

Ali H, Razaq A, Perveen S, Khan B (2013) Nitrogen fixation by non-leguminous plant sea buckthorn in semi-arid climatic conditions of Gilgit-Baltistan. Pak J Weed Sci Res 19(3):305–314

Arabi H, Hashemi SA, Fooladi M (1996) Microencapsulation of allopurinol by solvent evaporation and controlled release investigation of drugs. J Microencapsul 13(5):527–535

Asbahani AE, Miladi K, Badri W et al (2015) Essential oils: from extraction to encapsulation. Int J Pharm 483(1–2):220–243

Bal L, Meda V, Naik S, Satya S (2011) Sea buckthorn berries: a potential source of valuable nutrients for nutraceuticals and cosmeceuticals. Food Res Int 44(7):1718–1727

Balkrishna A, Sakat SS, Joshi K et al (2019) Cytokines driven anti-inflammatory and anti-psoriasis like efficacies of nutraceutical sea buckthorn (*Hippophae rhamnoides*) oil. Front Pharmacol 10:1186

Basu M, Prasad R, Jayamurthy P et al (2007) Antiatherogenic effects of seabuckthorn (*Hippophaea rhamnoides*) seed oil[J]. Phytomedicine 14:770–777

Belingheri C, Curti E, Ferrillo A et al (2012) Evaluation of porous starch as a flavor carrier. Food Funct 3(3):255–261

Bollhorst T, Rezwan K, Maas M (2017) Colloidal capsules: nano-and microcapsules withcolloidal particle shells. Chem Soc Rev 46(8):2091–2126

Carneiro H, Tonon R, Grosso C, Hubinger M (2013) Encapsulation efficiency and oxidative stability of flaxseed oil microencapsulated by spray drying using different combinations of wall materials. J Food Eng 115(4):443–451

Castro A, Berois N, Malanga A et al (2021) Docetaxel in chitosan-based nanocapsules conjugatedwith an anti-Tn antigen mouse/human chimeric antibody as a promising targeting strategy of lung tumors. Int J Biol Macromol 182:806–814

Cheng L, Zhang Y, Tan Y et al (2014) Response surface methodology for the optimization of preparation process for rice antioxidant peptide microcapsules by piercing method. Food Sci 35(22):97–103

Corcoran B, Ross R, Fitzgerald G, Stanton C (2004) Comparative survival of probiotic lactobacilli spray-dried in the presence of prebiotic substances[J]. J Appl Microbiol 96(5):1024–1039

Cui F, Zhang H, Wang D et al (2023) Advances in the preparation and application of microencapsulation to protect food functional ingredients. Food Funct 14:6766–6783
Díaz DI, Beristain CI, Azuara E et al (2015) Effect of wall material on the antioxidant activityand physicochemical properties of Rubus fruticosus juice microcapsules. J Microencapsul 32(3):247–254
do Nascimento T, Borges ALTF, de Almeida LM et al (2022) Preparation and characterization of microcapsules loaded with polyphenols-enriched Uncaria tomentosa extract using spray-dryer technique. J Therm Anal Calorim 147(21):11949–11963
Fang F, Zhang Z, Cheng L, Zhang C, Wu S (2020) Study on preparation and characterization of atractylodes macrocephala volatile oil microcapsules by complex coacervation method. J Chin Cereals Oils Assoc 35(3):97
Gao S, Hu G, Li D et al (2020) Anti-hyperlipidemia effect of sea buckthorn fruit oil extract through the AMPK and Akt signaling pathway in hamsters. J Funct Foods 66:103837
Guo X, Yang B, Cai W et al (2017) Effect of sea buckthorn (*Hippophae rhamnoides* L.) on blood lipid profiles; a systematic reviewandmeta-analysis from 11 independent randomized controlled trials. Trends Food Sci Technol 61:1–10
Hai B, Jiang G, Lu L, Liu L, Liu Y, Tang J (2018) Preparation and properties of microencapsulated paraffin phase change material by complex coacervation. Appl Chem Ind 47(1):10
Han S, Chen Y, Lyu S, Chen Z, Fu F (2020) Effects of processing conditions on the properties of paraffin/melamine-urea-formaldehyde microcapsules prepared by in situ polymerization. Colloids Surf A Physicochem Eng Asp 585:124046
He Y, Yao S, Hao J, Wang H, Zhu L, Si T, Sun Y, Lin J (2020) Polymer carriers for controlled fragrance release. Mater Res Express 7(8):082001
Hoyos-Leyva JD, Bello-Pérez LA, Alvarez-Ramirez J et al (2018) Microencapsulation using starch as wall material: a review. Food Rev Intl 34(2):148–161
Hsu YW, Tsai CF, Chen WK et al (2009) Protective effects of sea buckthorn(*Hippophae rhamnoides* L.) seed oil against carbon tertrachloride-induced hepatotoxicity in mice. Food Chem Toxicol 47:2281–2288
Huang G, Wang H, Wang F, Du Y, Xiao J (2020) Maillard reaction in protein-polysaccharide coacervated microcapsules and its effects on microcapsule properties. Int J Biol Macromol 155:1194–1201
Ito H, Asmussen S, Traber DL et al (2014) Healing efficacy of sea buckthorn (*Hippophae rhamnoides L.*) seed oil in an ovine burn wound model. Burns 40(3):511–519
Johansson AK, Korte H, Yang B et al (2000) Sea buckthorn berry oil inhibits platelet aggregation. J Nutr Biochem 11:491–495
Kallio H, Yang B (2001) Fatty acid composition of lipids in sea buckthorn (Hippophae rhamnoides L) berries of different origins. J Agric Food Chem 49(4):1939–1947
Kfoury M, Landy D, Ruellan S, Auezova L, Greige-Gerges H, Fourmentin S (2017) Nootkatone encapsulation by cyclodextrins: effect on water solubility and photostability. Food Chem 236:41–48
Koskovac M, Cupara S, Kipic M, Barjaktarevic A, Milovanovic O, Kojicic K, Markovic M (2017) Sea buckthorn oil-a valuable source for cosmeceuticals. Cosmetics 4(4):40
Kumar R, Kumar G, Chaurasia O, Singh S (2011) Phytochemical and pharmacological profile of seabuckthorn oil: a review. Res J Med Plant 5:491–499
Larmo P, Järvinen R, Laihia J, Löyttyniemi E, Maavirta L, Yang B, Sandberg-Lall M (2019) Effects of a sea buckthorn oil spray emulsion on dry eye. Cont Lens Anterior Eye 42(4):428–433
Lehtonen HM, Suomela JS, Tahvonen R et al (2011) Different berries and berry fractions have various but slightly positive effects on the associated variables of metabolic diseases on overweight and obese women[J]. Eur J Clin Nutr 65:394–401
Li TS, Beveridge TH (2007) Sea Buckthorn: a new medicinal and nutritional botanical. Agriculture and Agri-Food Canada, pp 50–53
Li R, Roos YH, Miao S (2016) Flavor release from spray-dried amorphous matrix: effect of lactose content and water plasticization[J]. Food Res Int 86:147–155

Liang B, Shao J, Yang S, Gao J, Zhang P, Liu J, Wang W, Sun Y, Zuo Y (2020) Preparation and slow release properties of camellia oil microcapsules[J]. Fine Chem 37(12):2541–2553

Luo Z, Zhao G, Panhwar F et al (2017) Well-designed microcapsules fabricated using droplet-based microfluidic technique for controlled drug release. J Drug Deliv Sci Technol 39:379–384

Mohammed N, Tan C, Manap Y, Muhialdin B, Hussin. (2020) Spray drying for the encapsulation of oils-a review. Molecules 25:3873

Muhoza B, Xia S, Wang X, Zhang X, Li Y, Zhang S (2022) Microencapsulation of essential oils by complex coacervation method: preparation, thermal stability, release properties and applications. Crit Rev Food Sci Nutr 62(5):1363–1382

Olas B (2018) The beneficial health aspects of sea buckthorn (*Elaeagnus rhamnoides* (L.) A. Nelson) oil. J Ethnopharmacol 213:183–190

Podgornik BB, Sandric S, Kert M (2021) Microencapsulation for functional textile coatings with emphasis on biodegradability-a systematic review. Coatings 11(11):30

Ren W, Tian G, Zhao S, Yang Y, Gao W, Zhao C, Zhang H, Lian Y, Wang F, Du H, Xiao H, Zheng J (2020) Effects of spray-drying temperature on the physicochemical properties and polymethoxyflavone loading efficiency of citrus oil microcapsules. LWT 133:109954

Suryanarayana C, Rao KC, Kumar D (2008) Preparation and characterization of microcapsules containing linseed oil and its use in self-healing coatings. Prog Org Coat 63(1):72–78

Swapan KG (2006) Functional coatings by polymer microencapsulation. Wiley-VCH, Belgium

Tang S, Yourdkhani M, Casey C, Sottos N, White S, Moore J (2017) Low-ceiling-temperature polymer microcapsules with hydrophobic payloads via rapid emulsion-solvent evaporation. ACS Appl Mater Interfaces 9(23):20115–20123

Upadhyay NK, Kumar R, Mandotra SK et al (2009) Safety and healing efficacy of sea buckthorn (*Hippophae rhamnoides L.*) seed oil on burn wounds in rats. Food Chem Toxicol 47(6):1146–1153

Vujanović MD, Đurović SD, Radojković MM (2021) Chemical composition of essential oils of elderberry (*Sambucus nigra* L.) flowers and fruits. Acta Period Technol 52(52):229–237

Wang X, Liu D, Li W, Tian Q, Zhou X (2013) Microencapsulation of cholesteric liquid crystal by combined method of solvent evaporation and photopolymerization[J]. Mol Cryst Liq Cryst 571(1):57–66

Yan H, Zheng B (2010) Morphology of modified starches prepared by different methods[J]. Food Res Int 43(3):767–772

Zhang H, Song G, Ma W, Guo M, Ling X, Yu D, Li L (2023) Microencapsulation protects the biological activity of sea buckthorn seed oil. Front Nutr 9:1043879

Chapter 17
Traditional Sea Buckthorn-Based Functional Food in China

Haitian Fang, Yanming Ren, and Jinhu Tian

Abstract Sea buckthorn is a kind of plant with homology of medicine and food, rich in nutrients and many kinds of bioactive compounds, which has high nutritional, health care function, and medicinal value. This chapter summarized different kinds of sea buckthorn-based functional foods in China, which will provide some references for the academia and industry.

Keywords Sea buckthorn · Nutritional value · Traditional foods

Abbreviations

MWM	Micro-wet milling
PGC1α	PPARγ-coactivator 1α
PRDM 16	PR domain containing 16
UCP-1	Uncoupling protein-1

17.1 The Nutritional Value of Sea Buckthorn

Fruits of the sea buckthorn plant are high in sugars, carotenoids, lutein, flavanols, and phenolic acids (Tkacz et al. 2019). Sea buckthorn polysaccharides have been demonstrated to be protective against lipopolysaccharide-induced damage to pig intestinal epithelial cells, and extracts from the plant have been demonstrated to lower blood lipids and improve other cardiovascular disease risk factors (Zhao et al. 2020; Zhou et al. 2022). In order to increase the quality and effectiveness of dietary supplements, sea buckthorn extracts are frequently used (Ren et al. 2020).

H. Fang
Food Science and Technology, Ningxia University, Yinchuan, Ningxia, China

Y. Ren · J. Tian (✉)
Department of Food Science and Nutrition, Zhejiang University, Hangzhou, China

J. Tian et al. (eds.), *Sea Buckthorn*, https://doi.org/10.1007/978-981-97-9865-0_17

Anti-obesity, anti-inflammatory, antioxidant, and antidiabetic qualities are found in sea buckthorn leaves (Choi et al. 2020; Middha et al. 2019; Zheng et al. 2019). Although the fruits of sea buckthorn are very acidic and astringent, it is still a food resource with a great deal of potential. Sea buckthorn juice and pulp are frequently employed as dietary items due to their high content of vitamins (C, E, A, B1, B2, F, K, and P), carotenoids, flavonoids, tocopherols, and other potentially health-beneficial components (Zeb 2004; Suryakumar and Gupta 2011). Sea buckthorn fruits contain significant amounts of oil, including two essential fatty acids, linolenic acid (n-3) and linoleic acid (n-6) (Beveridge et al. 1999; Gao et al. 2000; Yang and Kallio 2001; Kallio et al. 2002; Rösch et al. 2003). The high levels of tocopherols, tocotrienols, and carotenoids in the oil confer antioxidant properties, which have been confirmed by a large number of studies in vitro and in vivo (Zadernowski et al. 2003; Gao et al. 2003; Eccleston et al. 2003). The high contents of lipids in these carotenoid-rich fruits increase the bioavailability of carotenoids and enhance their absorption in humans (Ranjith et al. 2006; Kruczek et al. 2012). They were consequently added to the Chinese list of new resource foods in 2010. Additionally, it has been demonstrated that extracts from sea buckthorn leaves exhibit antibacterial properties, limit cell proliferation, and aid in wound healing (Criste et al. 2020; Masoodi et al. 2020). Furthermore, the fruits and leaves of sea buckthorn have numerous potential health advantages. Sea buckthorn is a common component in a variety of dishes and beverages. Fruit powder is in high demand as an ingredient or raw material for new functional goods, and the market for extracts such as sea buckthorn seed oil and fruit oil is expanding quickly (Ming 2022).

Previous studies have demonstrated that sea buckthorn polysaccharides boost the expression of PGC1α, UCP-1, and PRDM 16 in adipocytes, which activates brown adipocytes and promotes thermogenesis, therefore reducing fat accumulation and weight gain (Ma et al. 2022a). Palmitic acid-rich sea buckthorn fruit oil extract reduced body weight and dyslipidemia-induced blood glucose rise in hypercholesterolemic hamsters, implying that sea buckthorn fruit oil may alleviate hyperlipidemia-induced obesity (Gao et al. 2020). The administration of a flavonoid-rich extract of sea buckthorn seeds at doses of 100 and 300 mg/kg resulted in a significant reduction of 33.06% and 43.51%, respectively, in weight gain in HFD-induced obese mice (Yang et al. 2017). Freeze-dried powder of sea buckthorn is produced through low-temperature freeze-drying techniques, which retains all the beneficial nutrients and functional ingredients of the plant. Sea buckthorn powder promotes HFD-induced obesity by changing the makeup and organization of the gut microbiome (Guo et al. 2020). This could potentially make sea buckthorn a suitable candidate for developing functional foods and dietary supplements targeted toward overweight individuals.

17.2 Application of Sea Buckthorn in Food

In addition to its medical and biological properties, sea buckthorn is commonly employed as a functional food component with great economic potential. Sea buckthorn is a healthy berry that contains a number of beneficial components.

Furthermore, sea buckthorn is an abundant source of vitamins and minerals, making it a valuable addition to the healthy diet. Sea buckthorn is being employed as an antioxidant, antimicrobial, and natural food ingredient. Sea buckthorn has become more popular in the food business, with uses including sea buckthorn oil, freeze-dried powders, fruit juices, fruit wines, milk slices, fruit vinegar drinks, teas, canned fruits, yogurts, and jams (Ma et al. 2022b). Currently, food makers and researchers are working to optimize the usage of sea buckthorn to improve the organoleptic qualities and nutritional value of sea buckthorn products.

17.2.1 Sea Buckthorn Yogurt

The fast-paced, stressful lifestyle of the modern world has led to a significant buildup of toxins in the human body, affecting the immune system and increasing the incidence of diseases such as diabetes and cardiovascular disease. Consumers are paying more and more attention to healthy yogurt with multiple physiological functions. Based on the abundance of local sea buckthorn, milk resources, and the dietary habits of the population, sea buckthorn health yogurt is expected to have a promising future. Yogurt is produced by first treating fresh milk, then introducing pure cultured *Lactobacillus bulgaricus* and *Streptococcus thermophilus* as fermenting agents, kept warm for a specific period to allow for casein coagulation, brought about by the production of lactic acid. Compared to fresh milk, yogurt not only has all the characteristic nutrients of fresh milk but also produces some physiologically active substances such as organic acids, flavorings, active enzymes, lactic acid bacteria, etc., which have a significant regulatory effect on the body's functions while improving the nutritional composition. Therefore, the scientific method to remove the bad taste of sea buckthorn juice, retain the original nutritional health and medicinal components of sea buckthorn fruit, and add it to yogurt to make stirred sea buckthorn yogurt has a wide range of nutritional health and medicinal effects. For example, it can improve the structure of the intestinal flora, prevent the occurrence of diseases, improve the beauty of the skin, and have certain curative effects on stomach diseases, which means that it is nutritionally enriched and has a complementary effect (Lin and Ma 2015). Sea buckthorn, a novel plant-based addition, is gaining popularity in the worldwide dairy sector because of its health and nutritional advantages. This is exactly the nutritious quality that people seek. Sea buckthorn yogurt made from sea buckthorn berries is high in lipids, proteins, carbs, and antioxidants (vitamin C, vitamin E, carotenoids, phenolics, etc.), which meet the nutritional needs of the human body. According to Selvamuthukumaran and Farhath (2014), yogurt may be safely kept at 4 °C and 15 °C for 12 and 3 days, respectively, without compromising its microbiological quality.

In addition to sea buckthorn, carrots (Gu et al. 2008a), tomatoes, lozenges, yellow peaches, and passion fruits have been combined with sea buckthorn yogurt to create novel healthy yogurts (Gu et al. 2008b; Liu et al. 2019a). Various additions balance out the nutritional inadequacies of conventional yogurt and enhance the distinct natural flavors of fruits and vegetables. They also enrich the yogurt with a variety of functional substances. Sea buckthorn has been a component in cheese, yogurt, and kefir, among other dairy products. Studies in the past have demonstrated that adding pure sea buckthorn to fermented drinks enhanced their antioxidant content and raised their acidity levels (Rafalska et al. 2017).

Another study aimed to add sea buckthorn juice to the traditional plain yoghurt production. The aim was to optimize the production process of sea buckthorn yoghurt by fermentation with *Lactobacillus paracasei* M5L, based on traditional plain yoghurt, using sensory scores as response values, through single factor testing and response surface methodology. This will not only improve the nutritional value of yoghurt but also enrich the taste of yoghurt and provide theoretical basis and process parameters for the development of functional sea buckthorn yoghurt and its related products. Results showed that the optimal technological conditions for solid sea-buckthorn yogurt were sea-buckthorn juice 12%, sucrose 8%, and starter inoculum 1%. Under optimal conditions, the solid sea buckthorn yoghurt was light yellow, with a clear smell of sea buckthorn and yoghurt, moderately sour and sweet, fine texture, good stability, and no whey precipitation. The highest sensory score was 91.33, the lactic acid bacteria count was 1.4×10^7 CFU/ml, *Escherichia coli* and pathogenic bacteria were not detected, the acidity was 83.0 °T, the protein content was 3.6%, and the fat content was 2.9%, which were both in line with the requirements of the national standard GB 19302-2010 "fermented milk" (Hu 2021).

Sea buckthorn berries are also used to make feta cheese, where they provide a biodegradable backbone that allows the beneficial probiotic strain *Lactobacillus casei* ATCC 393 to thrive. Furthermore, the use of sea buckthorn reduces the number of harmful bacteria while improving the cheese's taste characteristics (Terpou et al. 2017).

Sea buckthorn enzyme is the product of sea buckthorn pulp fermented by microorganisms, which is rich in vitamin C, flavonoids, and other nutrients and has a high nutritional value. Most of the marketed sea buckthorn enzyme products are sea buckthorn enzyme liquid, with sour taste, and few foods prepared by adding sea buckthorn enzyme for people to choose from. Therefore, the sea buckthorn enzyme was added to the preparation process of yoghurt in this experiment. The fermentation process of sea buckthorn enzyme yoghurt was optimized by single factor and orthogonal experiments (Zhi et al. 2021). At present, research on sea buckthorn enzymes has focused on studying the antioxidant activity of sea buckthorn enzymes, as well as the metabolites and functional components produced during the preparation of sea buckthorn enzymes. Researchers have investigated the functional components and preparation process of sea buckthorn enzyme (Ge 2017; Liu et al. 2019b; Liao et al. 2016). However, few researches focus on the production process of sea buckthorn enzyme yoghurt. In order to extend the industrial chain of deep processing of sea buckthorn, sea buckthorn enzyme has been added to yogurt to enhance its nutritional value and improve the taste and flavor of yogurt. The addition of sea buckthorn sweet shake as a natural non-heat producing sweetener also meets the development needs of the times and has a broader development prospect (Pawar et al. 2013; Song et al. 2019). Sea buckthorn enzyme is incorporated into unflavored plain milk, which is then fermented with a probiotic complex and sweetened using rooibos and oligomeric isomaltose to reduce calorie consumption. The yoghurt makes full use of the nutritional properties of sea buckthorn enzymes to meet people's health needs. The combination of sea buckthorn enzymes and sweet head of rooibos flavor enhances its taste. Sea buckthorn enzyme yogurt was prepared by single factor and orthogonal test, which provides a reference basis for the development of sea buckthorn enzyme series products.

17.2.2 Sea Buckthorn Tea

A study on the use of sea buckthorn fruit juice and sea buckthorn leaves to make tea was carried out to develop fruit juice and tea drinks that can satisfy people's love for both fruit juice and tea and broaden the processing channels of sea buckthorn, with good development prospects. The sea buckthorn fruit is selected for its freshness, ripeness, and absence of pests, diseases, and molds and washed with running water. The pulp was filtered through four layers of gauze to remove the pulp, peel, and seeds, and then finely filtered to obtain sea buckthorn juice. Fresh sea buckthorn leaves were harvested in mid-to-late July, dried in an oven at 70 °C, and subsequently crushed. The crushed leaves were extracted by adding 30 times distilled water and refluxing at 60 °C for 2 hours. The juice is filtered, and the residue is extracted by adding 20 times distilled water for 1 hour. The filtrate is then filtered again, and the two filtrates are combined to obtain the sea buckthorn leaf extract. A functional fruit juice tea drink was developed with sugar, stabilizer, and other excipients. The optimum formulation was 12% sea buckthorn juice, 50% sea buckthorn leaf extract, 12% white sugar, 0.10% citric acid, 0.10% CMC-Na and 0.10% xanthan gum as stabilizers. The item has a straightforward processing methodology and includes a range of nutritional and health-giving elements, such as sea buckthorn flavonoids, Vc, and other bioactive compounds. These provide a means to enhance nutrient deficiencies in daily diet and strengthen the physique of consumers, thereby presenting a broad market potential (Alexandrakis et al. 2014).

The sea buckthorn processing business unavoidably generates sea buckthorn fruit wastes, which if not properly disposed of will pollute the environment. Fermentation and reuse of these wastes can raise sea buckthorn utilization rate and economic value. Waste from the sea buckthorn processing sector can serve as an appropriate substrate for fermentation. Fermentation was performed under ideal circumstances to produce a 3% ethanol sea buckthorn beverage. This beverage is high in phenolic chemicals (such as gallic acid, protocatechuic acid, vanillic acid, chlorogenic acid, and so on) and antioxidant action, as well as carbon dioxide and low in ethanol. In addition, it is infused with carbon dioxide and low levels of ethanol, making it an invigorating and nourishing functional drink (Gâtlan et al. 2020). In addition to fermentation, the micro-wet milling (MWM) procedure may produce waste-free juice from the entire sea buckthorn pulp. MWM sea buckthorn juice outperforms ground and commercially blended sea buckthorn juice in terms of color (bright yellow with up to 145 ± 0.10 mg/mL total carotenoids), particle size, ascorbic acid value (67.67 ± 1.15 mg/mL), total phenolic content, and antioxidant activity. The technique reduced the loss of heat-sensitive bioactive chemicals. It also created a fiber-rich juice that showed tremendous potential for processing sea buckthorn juice in the food sector (Xin et al. 2020). Seabuckthorn is high in minerals and bioactive substances. The potential of sea buckthorn as a botanical component for use in new functional meals is clear. Making full use of sea buckthorn fruits, peels, and seeds, as well as discovering innovative sea buckthorn processing methods, has a wide range of applications.

17.2.3 Sea Buckthorn Jam

In recent years, with people's work, life rhythm, and the need to enhance physical fitness, the pursuit of healthy, nutritious, convenient, and fast-food consumption has become a new social fashion. The consumption of high-quality jam, especially in large and medium-sized cities, has become the first choice of food for people to accompany meals. Jam is loved by consumers because of its outstanding nutritional benefits and unique flavor. Jam is made from fresh fruits and vegetables as the main material, through pre-treatment (selection or debridement, cleaning, peeling, pitting), sterilization, enzyme killing, color protection, flavor preservation, and other practical technology and advanced formula process, to maintain the unique natural flavor of fresh fruits and vegetables. The sauce is full of flu, sweet and pleasant, with a long shelf life, easy to preserve, and other characteristics. The main ingredients are sea buckthorn pulp, peel, and carrot, etc., and sugar or non-energy sweetener. Sea buckthorn jam is suitable for long-term consumption by patients with physical weakness, fatigue, indigestion, hypertension, cough, whooping cough, dry corneas, dizziness, night blindness, etc. (Tsering et al. 2013). Sea buckthorn berries have a tart flavor and a limited shelf life. As a result, turning sea buckthorn berries into jam is an efficient technique to improve their organoleptic characteristics and boost their use. This jam, produced with sea buckthorn berries at 102 °C and stevia, is high in carotenoids and polyphenols. After 21 days of room temperature storage, yeast and molds had fewer than 100 CFU/g, whereas *Enterobacteriaceae* had less than 5 CFU/g (Nistor et al. 2021). Jams have always had a single taste. In addition to citric acid, sea buckthorn is employed as a component and acidifier. Jam without preservatives has a shelf life of 177 days at 20 °C (Yuan et al. 2022). Furthermore, sea buckthorn may be used with sweet potato, pumpkin, and carrot in a certain proportion to form a new, nutritious, and healthful composite jam (Sun and Xu 2014). Sea buckthorn juice may be combined with other fruit liquids (papaya, watermelon, and grapes) to make delightful and healthy jellies. Among them, sea buckthorn combined jelly with grapes in a certain proportion offers pleasant sensory properties. Sea buckthorn and grape jelly have a shelf life of 6 months at room temperature, and their microbiological load is within the necessary limits (Selvamuthukumaran et al. 2007). Sea buckthorn has the potential to be a valuable source of bioactive chemicals for the manufacture of sugar-based goods. Despite its sour and unusual flavor, sea buckthorn fruit can be made into jams and jellies. The powerful aroma can be mitigated by combining the juice or sea buckthorn jam with other, much milder-flavored fruits in varied combinations (Bal et al. 2011). According to studies, sea buckthorn is a useful fruit for jam-making. It has a lot of bioactive substances. In the study by Rafalska et al., sea buckthorn fruit was combined with other fruits (apple, currant, raspberry, and strawberry) to create a range of jams with distinct flavors and colors. Based on sensory study, the sea buckthorn jam that had raspberries and strawberries added was the most well-liked by the assessors (Rafalska et al. 2017).

Sea buckthorn has a high nutritional, ecological, and economic value, especially in the construction of the "Three Norths" protective forest. It is also used to treat burns, radiation sickness, heart disease, glaucoma, and other diseases, and has unique medicinal value. Sea buckthorn has a lot of useful effects, and there are many sea buckthorn pulps available in the market, but the existing sea buckthorn pulp does not have a nutritional fortification formula and has low vitamin content, especially vitamin C, vitamin E, and total flavonoids. Therefore, a sea buckthorn pulp with a nutritional fortification formula and its preparation method are provided. The pulp of the present invention retains the vitamins, minerals, flavonoids, and other nutritional components of sea buckthorn pulp, while significantly increasing the content of vitamin C and vitamin E in sea buckthorn pulp, making the whole pulp more easily absorbed by the human body and maximizing the retention of the original nutrients and components of sea buckthorn pulp. A sea buckthorn pulp having a nutritional fortification formula characterized in that it comprises sea buckthorn fruit and oligofructose. The first step is to take the sea buckthorn fruit with intact shape and no damage, wash it to remove surface impurities and then extract the juice with a juicer to obtain the crude sea buckthorn pulp, and then obtain the crude sea buckthorn juice by antioxidant color protection treatment; placing the crude sea buckthorn pulp in an enzymatic digestion tank, adding pectinase to the enzymatic digestion tank for decomposition, allowing the enzymatically digested sea buckthorn pulp to stand and centrifuging to obtain the sea buckthorn juice; mixing the oligofructose with the sea buckthorn juice and stirring. The juice is sterilized by a high voltage pulsed electric field with an intensity of 15–20 kV/cm, a number of pulses >20 and a pulse width of 10–40 μs.

17.2.4 Sea Buckthorn Wine

It is possible to make alcoholic beverages from sea buckthorn (Li et al. 2016). Sea buckthorn tincture has long been used as a treatment for a variety of digestive ailments, including irritable bowel syndrome. It enhances stomach function and helps to keep the gastrointestinal tract functioning normally (Bawa et al. 2002). Furthermore, sea buckthorn wine is very popular in the Czech Republic, where it is commonly produced. Sea buckthorn wine is distinguished by its golden color and nice aroma (Niesteruk et al. 2013). Sea buckthorn is also utilized in beer brewing (Lee et al. 2012). Furthermore, recent research proposed a method for capitalizing on the byproducts of sea buckthorn processing in the juice sector by improving the fermentation process of sea buckthorn marc to produce a delightful, low-strength beverage (Gâtlan et al. 2020).

The abundance of sea buckthorn in Xinjiang and its excellent variety provide the basis for large-scale processing. Sea buckthorn fruit is rich in nutrients and bioactive substances, making it of high economic value. In ancient China, sea buckthorn was used to treat ailments. The medicinal value of sea buckthorn is described in the Four Medical Classics. The development and production of sea buckthorn-related products have high economic value. The price of fresh sea buckthorn fruit is 4.50 RMB/kg and the market price of sea buckthorn ice wine is more than 200 RMB (Ciesarová et al. 2020; Wang et al. 2022). The research and development of sea buckthorn fruit can extend the industrial chain of sea buckthorn products, make full use of sea buckthorn raw material, expand market demand, increase the enthusiasm of cultivation, and promote the development of local economy (Wang et al. 2020). Research into developing sea buckthorn will extend the industrial chain of sea buckthorn products so that the raw material can be fully utilized, market demand can be increased, enthusiasm for growing can be increased and the local economy can be boosted. At present, the development of sea buckthorn ice wine in Xinjiang province is in the research stage. The development of sea buckthorn ice wine has a positive impact on the development of the sea buckthorn industry in Xinjiang province. The development of sea buckthorn ice wine has a positive effect on the expansion of the sea buckthorn industry. To make the best use of sea buckthorn and enjoy the gifts of nature, the development of sea buckthorn ice wine has a positive effect on the expansion of the industry. The research is expected to develop a new product, the sea-buckthorn ice wine, to meet the needs of modern people for a healthy and fast life. The research is expected to transform sea buckthorn berries into a nutritious, portable, transportable, and storable sea buckthorn ice wine. It is hoped that a nutritious, portable, transportable, and storable sea buckthorn ice wine will be developed (Fatima et al. 2015).

References

Alexandrakis Z, Kyriakopoulou K, Katsaros G et al (2014) Selection of process conditions for high pressure pasteurization of sea buckthorn juice retaining high antioxidant activity. Food Bioprocess Technol 7(11):3226–3234

Bal L, Meda V, Naik S et al (2011) Sea buckthorn berries: a potential source of valuable nutrients for nutraceuticals and cosmoceuticals. Food Res Int 44(7):1718–1727

Bawa A, Khanum F, Singh B (2002) Sea buckthorn a wonder plant. Indian J Nat Prod Resour 1:8–14

Beveridge T, Li T, Oomah B, Smith A (1999) Sea buckthorn products: manufacture and composition. J Agric Food Chem 47(9):3480–3488

Choi H, Lee D, Park S, Jang Y, Ahn J, Ha T, Jung CH (2020) Antiobesity effects of the combination of *Patrinia scabiosaefolia* root and *Hippophae rhamnoides* leaf extracts. J Food Biochem 44(6):e13214

Ciesarová Z, Murkovic M, Cejpek K et al (2020) Why is sea buckthorn (*Hippophae rhamnoides* L.) so exceptional? A review. Food Res Int 133:109170

Criste A, Urcan A, Bunea A, Pripon Furtuna F, Olah N, Madden R, Corcionivoschi N (2020) Phytochemical composition and biological activity of berries and leaves from four Romanian sea buckthorn (*Hippophae rhamnoides* L.) varieties. Molecules 25(5):1170

Eccleston C, Baoru Y, Tahvonen R, Kallio H, Rimbach G, Minihane A (2003) Effects of an antioxidant-rich juice (sea buckthorn) on risk factors for coronary heart disease in humans. J Nutr Biochem 13(6):346–354

Fatima T, Kesari V, Watt I et al (2015) Metabolite profiling and expression analysis of flavonoid, vitamin C and tocopherol biosynthesis genes in the antioxidant-rich sea buckthorn (*Hippophae rhamnoides* L.). Phytochemistry 118:181–191

Gao X, Ohlander M, Jeppsson N, Björk L, Trajkovski V (2000) Changes in antioxidant effects and their relationship to phytonutrients in fruits of sea buckthorn (*Hippophae rhamnoides* L.) during maturation. J Agric Food Chem 48(5):1485–1490

Gao Z, Gu X, Cheng F, Jiang F (2003) Effect of sea buckthorn on liver fibrosis: a clinical study. World J Gastroenterol: WJG 9(7):1615

Gao S, Hu G, Li D, Sun M, Mou D (2020) Anti-hyperlipidemia effect of sea buckthorn fruit oil extract through the AMPK and Akt signaling pathway in hamsters. J Funct Foods 66:103837

Gâtlan A, Gutt G, Naghiu A (2020) Capitalization of sea buckthorn waste by fermentation: optimization of industrial process of obtaining a novel refreshing drink. J Food Process Preserv 44(8):e14565

Ge P (2017) Research on the processing technology of sea buckthorn enzyme. Shenyang Agricultural University

Gu Y, Chen X, Fu L (2008a) Preparation of yogurt mixed with sea buckthorn and tomato. Food Res Openness (07):66–68

Gu Y, Chen Z, Fu L (2008b) Development of yogurt mixed with sea buckthorn and carrot. China Brew 12:113–114

Guo C, Han L, Li M, Yu L (2020) Sea buckthorn (*Hippophaë rhamnoides*) freeze-dried powder protects against high-fat diet-induced obesity, lipid metabolism disorders by modulating the gut microbiota of mice. Nutrients 12(1):265

Hu P (2021) Optimization of fermentation process of sea-buckthorn yogurt by *Lactobacillus paracasei* by response surface methodology. China Brew 40(06):202–206

Kallio H, Yang B, Peippo P, Tahvonen R, Pan R (2002) Triacylglycerols, glycerophospholipids, tocopherols, and tocotrienols in berries and seeds of two subspecies (ssp. *sinensis and mongolica*) of sea buckthorn (*Hippopha > ë rhamnoides*). J Agric Food Chem 50(10):3004–3009

Kruczek M, Swiderski A, Mech-Nowak A, Król K (2012) Antioxidant capacity of crude extracts containing carotenoids from the berries of various cultivars of Sea buckthorn (*Hippophae rhamnoides* L.). Acta Biochim Pol 59(1):135–137

Lee M, Cho H, Lee K (2012) Development of sea buckthorn beer using Mongolian sea buckthorn (*Hippophae rhamnoides*) fruit. Food Eng Prog 16(2):129–133

Li Z, Wang J, Xiong Y et al (2016) The determination of the fatty acid content of sea buckthorn seed oil using near infrared spectroscopy and variable selection methods for multivariate calibration. Vib Spectrosc 84:24–29

Liao X, Guo L, Ye Z, Qiu L, Gu F, Lin J (2016) Use of autochthonous lactic acid bacteria starters to ferment mango juice for promoting its probiotic roles. Prep Biochem Biotechnol 46(4):399–405

Lin X, Ma C (2015) The current situation of sea buckthorn planting in Xinjiang and the prospects for sea buckthorn yogurt development. Agric Technol 12:139–139

Liu H, Wang R, Gao Z (2019a) Study on the process of compound yoghurt with sea buckthorn and water chestnut. J Shanxi Datong Univ 35(06):59–62

Liu Q, Tang G, Zhao C, Gan R, Li H (2019b) Antioxidant activities, phenolic profiles, and organic acid contents of fruit vinegars. Antioxidants 8(4):78

Ma C, Du L, Cai G (2022a) Investigation and research on the distribution of seabuckthorn germplasm resources in Changji prefecture. Mod Agric Sci Technol:139–141. https://doi.org/10.3969/j.issn.1007-5739.2022.02.046

Ma Z, Sun Q, Chang L, Peng J, Zhang M, Ding X, Lan Y (2022b) A natural anti-obesity reagent derived from sea buckthorn polysaccharides: Structure characterization and anti-obesity evaluation in vivo. Food Chem 375:131884

Masoodi K, Wani W, Dar Z, Mansoor S, Anam-ul-Haq S, Farooq I, Ahmed N (2020) Sea buckthorn (*Hippophae rhamnoides* L.) inhibits cellular proliferation, wound healing and decreases expression of prostate specific antigen in prostate cancer cells *in vitro*. J Funct Foods 73:104102

Middha S, Usha T, Basistha B, Goyal A (2019) Amelioration of antioxidant potential, toxicity, and antihyperglycemic activity of *Hippophae salicifolia* D. Don leaf extracts in alloxan-induced diabetic rats. 3 Biotech 9:1–8

Ming Y (2022) National Sea buckthorn resources background investigation report released. Green China 13:58–59

Niesteruk A, Lewandowska H, Golub Z, Swisłocka R, Lewandowski W (2013) Let's get interested with sea buckthorn. Preparations of sea buckthorn as food additives and assessment of their market in Poland. Kosmos 4:571–581

Nistor O, Bolea C, Andronoiu D, Cotârleţ M, Stănciuc N (2021) Attempts for developing novel sugar-based and sugar-free sea buckthorn marmalades. Molecules 26:3073

Pawar R, Krynitsky A, Rader J (2013) Sweeteners from plants-with emphasis on *Stevia rebaudiana* (Bertoni) and *Siraitia grosvenorii* (Swingle). Anal Bioanal Chem 405:4397–4407

Rafalska A, Abramowicz K, Krauze M (2017) Sea buckthorn (*Hippophae rhamnoides* L.) as a plant for universal application. World Sci News 72:123–140

Ranjith A, Kumar K, Venugopalan V, Arumughan C, Sawhney R, Singh V (2006) Fatty acids, tocols, and carotenoids in pulp oil of three sea buckthorn species (*Hippophae rhamnoides*, H. salicifolia, and H. tibetana) grown in the Indian Himalayas. J Am Oil Chem Soc 83(4):359–364

Ren R, Li N, Su C, Wang Y, Zhao X, Yang L, Ma X (2020) The bioactive components as well as the nutritional and health effects of sea buckthorn. RSC Adv 10(73):44654–44671

Rösch D, Bergmann M, Knorr D, Kroh L (2003) Structure–antioxidant efficiency relationships of phenolic compounds and their contribution to the antioxidant activity of sea buckthorn juice. J Agric Food Chem 51(15):4233–4239

Selvamuthukumaran M, Farhath K (2014) Evaluation of shelf stability of antioxidant rich seabuckthorn fruit yoghurt. Int Food Res J 21(2):759–765

Selvamuthukumaran M, Khanum F, Bawa A (2007) Development of sea buckthorn mixed fruit jelly. Int J Food Sci Technol 42(4):403–410

Song J, Qian B, Pan C et al (2019) Protective activity of mogroside V against ovalbumininduced experimental allergic asthma in Kunming mice. J Food Biochem 43(11):e12973

Sun L, Xu X (2014) Preferred research on technology and materials of sea buckthorn composite jam. Storage Proc 20:52–54

Suryakumar G, Gupta A (2011) Medicinal and therapeutic potential of Sea buckthorn (*Hippophae rhamnoides* L.). J Ethnopharmacol 138(2):268–278

Terpou A, Gialleli A, Bosnea L, Kanellaki M, Koutinas A, Castro G (2017) Novel cheese production by incorporation of sea buckthorn berries (*Hippophae rhamnoides* L.) supported probiotic cells. LWT-Food Sci Technol 79:616–624

Tkacz K, Wojdyło A, Turkiewicz I, Bobak Ł, Nowicka P (2019) Anti-oxidant and anti-enzymatic activities of sea buckthorn (*Hippophaë rhamnoides* L.) fruits modulated by chemical components. Antioxidants 8(12):618

Tsering S, Girish K, Ravi BS (2013) Nutritional attributes and health application of sea buckthorn (*Hippophae rhamnoides* L.)-A review. Curr Nutr Food Sci 9(2):151–165

Wang C, Lei M, Zhang H et al (2020) Comparison of metabolites in juice, seed and peel of sea buckthorn (*Hippophae rhamnoides* L. subsp. sinensis). J Plant Biochem Biotechnol 29:305–313

Wang Z, Wang W, Zhu C et al (2022) Evaluation of Antioxidative and neuroprotective activities of total flavonoids from sea buckthorn (*Hippophae rhamnoides* L.). Front Nutr 9:861097

Xin Y, Zhao S, Yang J, Zhang T, Zhang J, Wang Y (2020) Effect of microwave-drying on the quality and antioxidant properties of Ganoderma lucidum fermented sea-buckthorn tea. Int J Food Eng 17(1):65–74

Yang B, Kallio H (2001) Fatty acid composition of lipids in sea buckthorn (*Hippophaë rhamnoides* L.) berries of different origins. J Agric Food Chem 49(4):1939–1947
Yang X, Wang Q, Pang Z, Pan M, Zhang W (2017) Flavonoid-enriched extract from *Hippophae rhamnoides* seed reduces high fat diet induced obesity, hypertriglyceridemia, and hepatic triglyceride accumulation in C57BL/6 mice. Pharm Biol 55(1):1207–1214
Yuan L, Liu J, Wu T, Wu H, Liu L (2022) Optimization of technology of *Elaeagnus angustifolia* and *Hippophae rhamnoides* compound jam by response surface methodology and prediction of storage period. Storage Proc 22:37–44
Zadernowski R, Naczk M, Amarowicz R (2003) Tocopherols in sea buckthorn (*Hippophae rhamnoides* L.) berry oil. J Am Oil Chem Soc 80:55–58
Zeb A (2004) Chemical and nutritional constituents of sea buckthorn juice. Pak J Nutr 3(2):99–106
Zhao L, Li M, Sun K, Su S, Geng T, Sun H (2020) *Hippophae rhamnoides* polysaccharides protect IPEC-J2 cells from LPS-induced inflammation, apoptosis and barrier dysfunction in vitro via inhibiting TLR4/NF-κB signaling pathway. Int J Biol Macromol 155:1202–1215
Zheng W, Bai H, Han S, Bao F, Zhang K, Sun L, Yang Z (2019) Analysis on the constituents of branches, berries, and leaves of *Hippophae rhamnoides* L. by UHPLC-ESI-QTOF-MS and their anti-inflammatory activities. Nat Prod Commun 14(8):1934578X19871404
Zhi X, Guang C, Li B, Dan Z, Wen Y, Si P (2021) Research progress on the active compounds of sea buckthorn and their physiological functions and utilization. Food Mach 37(11):221–227
Zhou F, Zhang J, Zhao A, Zhang Y, Wang P (2022) Effects of sea buckthorn puree on risk factors of cardiovascular disease in hypercholesterolemia population: a double-blind, randomized, placebo-controlled trial. Anim Biotechnol 33(5):955–963

Chapter 18
Effect of Sea Buckthorn on the Quality of Fermented Milk

Junrui Wu

Abstract Sea buckthorn, a nutrient-rich fruit with functional components, emerges as an innovative ingredient for developing fermented milk and other dairy products. Its incorporation has demonstrated a notable influence on the physicochemical properties, taste, flavor, nutritional content, and role of lactic acid bacteria in fermented milk. Exploring sea buckthorn as a novel ingredient in dairy processing is poised to offer consumers access to more nourishing and wholesome dairy options. In particular, sea buckthorn polysaccharides have shown promise for enhancing fermented milk. Furthermore, this chapter provides a perspective on the potential of functional sea buckthorn fermented milk, underpinning the advancement of the functional food sector with theoretical insights.

Keywords Sea buckthorn · Fermented milk · Sea buckthorn polysaccharides

Abbreviations

LAB	*Lactobacillus*
V_A	Vitamin A
V_C	Vitamin C
V_E	Vitamin E

18.1 Effect of Sea Buckthorn on the Physicochemical Properties of Fermented Milk

The inclusion of sea buckthorn in fermented dairy products not only significantly influences the physicochemical characteristics of the product but also impacts its pH, acidity, ionic concentration, and antioxidant properties, which are essential for the quality and stability of the product.

J. Wu (✉)
Food Science College, Shenyang Agricultural University, Shenyang, China

J. Tian et al. (eds.), *Sea Buckthorn*, https://doi.org/10.1007/978-981-97-9865-0_18

18.1.1 pH Value

The addition of sea buckthorn alters the pH of fermented milk products, which is a pivotal parameter in their preparation. The rich content of organic acids such as malic acid and citric acid in sea buckthorn leads to increased acidity during fermentation when added to milk. The interaction between sea buckthorn organic acids and lactic acid bacteria results in enhanced acid production, subsequently lowering the pH of the fermented milk. This acidity influences the growth and fermentation pace of lactic acid bacteria, with the extent of the impact being dependent on the quantity of sea buckthorn added, fermentation duration, and lactic acid bacteria strain.

18.1.2 Acidity

Sea buckthorn notably affects the acidity of fermented dairy products. Similar to pH, acidity is critical for the texture and flavor of fermented milk. Sea buckthorn is rich in acids such as lactic and citric acid; therefore, its addition significantly increases the total acidity of fermented milk. This may alter the taste and texture of the product, making it more acidic. In addition, the increased acidity may affect the preservation stability of the product and reduce the likelihood of bacterial growth.

18.1.3 Ionic Concentration

In addition to affecting acidity and pH, the ions in sea buckthorn can affect the ion concentration in fermented milk. Sea buckthorn is rich in minerals such as potassium, sodium, and calcium, which are released during the fermentation process and increase the ion concentration in the dairy product. Calcium ions are essential for the formation of buttermilk gels. The addition of sea buckthorn may cause calcium ions to interact with other ions, thereby affecting the texture and structure of the product. This change in ion concentration not only helps to improve the taste of the product but is also crucial for bone health and nerve conduction in the human body.

18.1.4 Antioxidant Properties

The antioxidant compounds in sea buckthorn significantly contribute to the antioxidant properties of fermented dairy products. Abundant in antioxidants such as vitamin C and vitamin E, sea buckthorn aids in slowing down product oxidation, preserving freshness, and protecting nutrients. Furthermore, antioxidant substances protect the vitamins and other nutrients in the product from degradation, increasing the nutritional benefits of the product.

These factors collectively highlight the substantial impact of sea buckthorn on the physicochemical properties of fermented milk. Ongoing research is necessary to better comprehend these complexities, enabling the optimization of product quality and stability.

18.2 Effect of Sea Buckthorn on Flavor and Sensory Properties of Fermented Milk

Flavor and sensory attributes are crucial aspects in the consumption of fermented dairy products, significantly influencing consumer preferences. Sea buckthorn, a fruit rich in nutrients and natural substances, positively affects the flavor and sensory attributes of fermented dairy products. This influence includes enhancing aroma characteristics, improving taste characteristics, enriching color characteristics, and achieving a balance between acidity and sweetness. These effects make fermented dairy products more appealing and competitive, thus meeting consumer demands for taste and flavor.

18.2.1 Aroma Characteristics

Sea buckthorn has a unique rich fruity flavor, and its addition can give fermented dairy products a pleasant fruity aroma. The volatile components, alcohol compounds, and esters in sea buckthorn interact with lactic acid, esters, and alcohols in fermented dairy products to produce rich aromatic components. These aroma substances enhance the flavor profile of the product, imparting a unique fruity aroma, and enhancing the sensory experience.

18.2.2 Taste Characteristics

The addition of polysaccharide substances such as pectin and cellulose from sea buckthorn to fermented dairy products can enhance the consistency and texture of the product. This gelatinous structure can make fermented dairy products thicker and more delicate, providing a smooth and soft taste experience. In addition, the acidity of sea buckthorn in fermented dairy products can intensify the sour flavor of the product, enhancing the mouthfeel stimulation and enriching the taste profile.

18.2.3 Color Characteristics

The natural pigments of sea buckthorn, particularly beta-carotene, endow fermented dairy products with an appealing orange-red hue. This vibrant color enhances the product's visual appeal, portraying its freshness and palatability to consumers. Color enhancement positively impacts consumer perception and likability toward the product.

18.2.4 Balanced Acidity with Sweetness

The organic acids and natural sweet substances in sea buckthorn work together to create a balanced sweet and sour taste in fermented dairy products. The addition of sea buckthorn can neutralize the acidity in dairy products, reducing the sharpness of sourness, while also increasing the sweetness of the product, enhancing the taste balance. This harmony of sweet and sour taste makes fermented dairy products more enjoyable, thereby enhancing consumers' satisfaction.

18.3 Effect of Sea Buckthorn Addition on the Nutritional Value of Fermented Milk

Sea buckthorn is a fruit rich in nutritional value and is often used to enhance the taste of food and increase its nutritional value. Fermented milk, as a popular health food, contains active lactic acid bacteria and lactic acid with probiotic properties. Adding sea buckthorn to fermented milk not only improves its taste and flavor but also increases its nutrient content and improves its overall nutritional value.

18.3.1 Effects of Vitamins

Sea buckthorn is abundant in vitamin C (V_c). V_c is critical to the body's antioxidant and immune-boosting properties. Studies have shown that V_c in sea buckthorn improves the antioxidant properties of fermented milk and reduces the damage caused by free radicals to the body. V_c can also promote collagen synthesis for healthier skin. By adding sea buckthorn to fermented milk, the content of V_c can be increased, further improving its nutritional value.

In addition, sea buckthorn is also rich in vitamin E (V_E), a fat-soluble vitamin with antioxidant properties that protects cell membranes from free radical damage. Studies have shown that V_E in sea buckthorn works with the lactic acid bacteria in fermented milk to increase the vitality of the lactic acid bacteria and increase the probiotic content of the fermented milk. In addition, V_E can improve the health of the skin and reduce the signs of skin aging. Therefore, by adding sea buckthorn to fermented milk, the content of V_E can be increased to improve its antioxidant properties and skin health-promoting effects.

Sea buckthorn also contains high levels of vitamin A (V_A), which is important for maintaining normal vision, boosting immunity, and promoting cell growth. V_A plays an important role in the body's visual processes and is vital for healthy skin, bones, and teeth. Adding sea buckthorn extract to fermented milk can significantly increase the V_A content and further increase the nutritional value of fermented milk.

18.3.2 Effects of Minerals

Sea buckthorn is rich in many minerals such as zinc, magnesium, and iron. These minerals play an important role in the normal physiological functions of the human body. Zinc is an essential nutrient for the normal functioning of the body's immune system and is involved in the activation of many enzymes. Iron is an important component of hemoglobin, which is essential for normal oxygen transport and energy production. Magnesium plays an important role in the normal functioning of the nervous and cardiovascular systems. By adding sea buckthorn to fermented milk, the levels of these minerals can be increased to provide adequate nutritional support to the body.

18.3.3 Effect of Functional Ingredients

Sea buckthorn is rich in various antioxidants, including polyphenolic compounds, anthocyanins, and flavonoids. These compounds have powerful antioxidant properties that help neutralize harmful free radicals and reduce the risk of oxidative cell damage. They have also been linked to several health benefits, such as reduced risk of chronic diseases, improved cardiovascular health, and anti-inflammatory properties. The polyphenols and other antioxidants in sea buckthorn have been studied extensively and show anti-inflammatory potential. Inflammation is the underlying cause of many diseases, including cardiovascular disease, arthritis, and intestinal problems. By adding sea buckthorn to fermented milk, the product can become a food choice with more anti-inflammatory properties.

In summary, the addition of sea buckthorn significantly enhances the nutritional value of fermented milk. Its rich content of vitamin C, vitamin E, vitamin A, and various minerals can improve the antioxidant properties, immune-enhancing effect, and nutritional content of fermented milk. In addition, the functional components in sea buckthorn could confer more anti-inflammatory, antibacterial, and anti-tumor activities to fermented milk.

18.4 Effect of Sea Buckthorn on Lactic Acid Bacteria in Fermented Milk

Lactobacillus in fermented milk products is a beneficial bacterium that promotes digestion, strengthens immunity, and maintains intestinal health. Sea buckthorn is a fruit rich in nutrients and natural substances and its addition to fermented milk has an important effect on the growth and activity of lactic acid bacteria.

18.4.1 Promotion of Lactic Acid Bacterial Growth

The organic acid and antibiotic components in sea buckthorn have an antibacterial effect, which can inhibit the growth of harmful bacteria and promote the growth of lactic acid bacteria. The organic acids in sea buckthorn can lower the pH of fermented dairy products and create an acidic environment suitable for the growth of lactic acid bacteria. In addition, the antibiotic components in sea buckthorn can promote the reproduction and growth of lactic acid bacteria by inhibiting the growth of harmful bacteria and providing more space and nutrients to lactic acid bacteria.

18.4.2 Increase in Lactic Acid Production

The organic acids in sea buckthorn and the lactic acid produced by lactic acid bacteria can interact to increase the production of lactic acid in dairy products. *Lactobacilli* convert organic acids into lactic acid during fermentation, and with the addition of sea buckthorn, the lactobacilli use the organic acids in sea buckthorn to produce more lactic acid. The increase in lactic acid not only enhances the acidity and sourness of the product but also imparts a better shelf life and stability.

18.4.3 Enhancement of Lactic Acid Bacterial Activity

Antioxidant substances in sea buckthorn are beneficial for the survival and activity of lactic acid bacteria. Antioxidant substances reduce oxidative damage and protect lactic acid bacteria from damage and death. Antioxidant substances in sea buckthorn, such as V_C and V_E, can scavenge free radicals and slow down the oxidization process for better survival and growth of lactic acid bacteria during fermentation. The addition of 5% sea buckthorn could shorten the coagulation time of fermented milk (from >9 h to 4–5 h), improve the texture of fermented milk, and increase the viable LAB counts in fermented milk (Ge et al. 2022). Thus, the addition of sea buckthorn improves the activity of lactic acid bacteria in fermented dairy products and maintains the quality and health benefits of the product.

18.4.4 Increase in Nutritional Support for Lactic Acid Bacteria

Sea buckthorn is rich in many nutrients, such as vitamins, minerals, and dietary fiber, which can provide lactic acid bacteria with rich nutritional requirements. Vitamins and minerals are essential for the growth of lactic acid bacteria, which can promote the growth of their metabolic activities and reproduction. Dietary fiber in sea buckthorn, on the other hand, can serve as a nutrient source for lactic acid bacteria and promote their growth and activity. Therefore, the addition of sea buckthorn can enhance the nutritional requirements of lactic acid bacteria and increase their growth and reproduction.

The addition of sea buckthorn has positive effects on lactic acid bacteria in fermented milk, including promoting lactic acid bacteria growth, increasing lactic acid production, improving lactic acid bacteria activity, and increasing the nutritional requirements of lactic acid bacteria. These effects allow the lactic acid bacteria to better complete the fermentation process, thus increasing the quality and health benefits of the product. When producing fermented dairy products, rational use of the properties of sea buckthorn and the addition of lactic acid bacteria according to their needs can enhance the nutritional value and market competitiveness of the products and satisfy consumer demand for health and quality.

18.5 Development and Application of Sea Buckthorn Polysaccharide Fermented Milk

With the growing pursuit of a healthy diet, the demand for food with functional and nutritional value is also increasing. Sea buckthorn polysaccharide is one of the polysaccharide components that has the functions of scavenging free radicals, acting as antioxidants, and relieving fatigue, among others. Therefore, the extraction of sea buckthorn polysaccharides and the effect on fermented milk have been widely researched.

18.5.1 *Separation and Purification Techniques for Sea Buckthorn Polysaccharide*

18.5.1.1 Solvent Extraction Method

The solvent extraction method dissolves the active ingredients from the tissues by selecting a solvent with high solubility for the active ingredients and low solubility for the ingredients that do not need to be dissolved based on the solubility nature of various ingredients in the solvent (Guo et al. 2021; Rojas et al. 2021; Soltani et al. 2021). Solvent extraction methods include maceration, percolation, decoction, reflux extraction, and continuous reflux extraction. Ultrasonic cleavage combined with hot water extraction was used to extract sea buckthorn polysaccharides followed by column chromatography using the Sevag method combined with SephadexG-150 for protein removal through the papain enzyme method. The sea buckthorn polysaccharides were extracted, purified, and characterized for their purity. It was found that they have significant pro-proliferative effects on macrophages. The polysaccharide was extracted from sea buckthorn using hydrotropic alcohol precipitation method, followed by preliminary purification, and these were used to carry out bacteriostatic experiments were conducted using the agar plate diffusion method, revealing the inhibitory effect of sea buckthorn polysaccharide on microorganisms such as *Escherichia coli* and *Tetrahymena*. The present research endeavors, both domestically and internationally, primarily focus on the extraction of sea buckthorn polysaccharide via the solvent extraction method, with the solvent being concentrated in hot water, employing the hot water reflux method or combined with ultrasound. Other solvents are less involved and necessitate further developed and research to broaden the application of solvent extraction methods in the extraction of sea buckthorn polysaccharides.

18.5.1.2 Ultrasonic-Assisted Extraction Method

The ultrasonic-assisted extraction method is a new technical means that utilizes ultrasound to dissolve water-soluble dietary through the "cavitation effect", which is more conducive to improving the content of extracted substances, enhancing the efficiency of extraction and purification, and shortening the extraction time. Studies have shown that polysaccharides extracted by the ultrasonic-assisted method have significantly higher content than those extracted by traditional extraction method, offering a simple, time-saving, and energy-saving operation. Single-factor experiments were used to study the effects of extraction time, power, and material–liquid ratio on the extraction rate of sea buckthorn polysaccharides, followed by orthogonal experiments to determine the optimal extraction process. Comparative analyses were conducted with the water extraction method, microwave method, and enzyme digestion method to optimize the ultrasonic-assisted extraction of sea buckthorn polysaccharides. The optimal process conditions for sea buckthorn water-soluble polysaccharides extracted by ultrasonic-assisted extraction were an extraction power of 100 W, a solid–liquid ratio 1:30 (m:V), and an extraction time of 45 min, with the maximum yield of polysaccharide at 7.35%. The present research on the application of ultrasonic waves for assisted extraction of sea buckthorn polysaccharides is relatively limited, primarily focusing on the combination of ultrasound assistance with hot water. Nonetheless, ultrasound as an auxiliary technology opens avenues for sea buckthorn polysaccharide extraction technology, necessitating further exploration to unravel uncharted territories along with other extraction processes (Johnson et al. 2021; Kivrak and Kivrak 2021; Qin et al. 2021; Sirijeerachai et al. 2021; Takahashi et al. 2021; Tang 2021).

18.5.1.3 Acid Digestion, Enzyme Digestion, and Supercritical Fluid Extraction Method

The acid digestion method leverages the property that polysaccharides are easily soluble in an acidic solution using the acidic solution as the extraction agent to dissolve the polysaccharides (Haque and Bradbury 2002; Woo and Kim 1999; Woo et al. 2002). The theoretical basis and technical principles of acid digestion are relatively mature and have been applied in various research and experimental fields. However, there is a lack of research on the extraction and separation of sea buckthorn polysaccharides using acid digestion methods both domestically and internationally. Hence, the exploration of the extraction process of sea buckthorn polysaccharides through the acid digestion method, based on the physicochemical properties of the acid digestion method, holds application value, enhancing the extraction efficiency of sea buckthorn polysaccharides across multiple facets.

Enzymatic digestion typically involves the removal of certain tissues with biological enzymes, a method commonly applied in general biological experiments (Duran-Cruz et al. 2021; Park et al. 2021; Tang et al. 2021). In contrast to traditional acid hydrolysis and alkaline hydrolysis, the enzymatic method is considered safe because the residual substances are harmless and can be absorbed and utilized by the body. Enzymes exhibit specificity as specialized and efficient catalysts. Presently, in domestic and foreign research, the enzyme digestion method is predominantly employed for the extraction of polysaccharides from shiitake mushroom, cactus, and bamboo Suzuki, among others, with minimal application in sea buckthorn polysaccharide extraction. Leveraging the catalytic effect and specificity of the enzyme for the extraction and separation of sea buckthorn polysaccharides hold research value, poised to emerge as a focal point for future investigations.

The supercritical fluid extraction method involves equilibrating and transferring the solute to be extracted through the abnormal phases via the action of supercritical temperature and critical pressure (Al-Maqtari et al. 2021; Boitsova et al. 2020; Santos et al. 2021; Sun et al. 2021). With alterations in pressure and temperature, solubility undergoes changes. Supercritical fluid extraction can be combined with other extraction techniques to derive more effective extraction and purification methods, aligning the process parameters with industrial production requirements.

Supercritical fluid extraction has emerged as a widely adopted extraction technology both domestically and internationally, with a mature process and extensive research endeavors.

18.5.2 Effect of Sea Buckthorn Polysaccharides on Fermented Milk

18.5.2.1 Effect of Sea Buckthorn Polysaccharide on Fermented Milk Viscosity

The viscosity of fermented milk initially increased and then decreased with the addition of sea buckthorn polysaccharides. The highest viscosity of 8.47 pas was achieved with 0.15% addition of sea buckthorn polysaccharides. Further addition beyond this optimal amount gradually decreased the viscosity, ultimately falling below that of the control group. Previous studies (Gee et al. 2007; Rawson and Marshall 1997) have shown that polysaccharides in fermented milk enhance the production of extracellular polysaccharides, which, in turn, form protein gels with the proteins in the milk. However, excessive sea buckthorn polysaccharides can inhibit the gelation of the casein protein, leading to reduced viscosity (Domagala 2009).

18.5.2.2 Effect of Sea Buckthorn Polysaccharides on pH and Acidity of Fermented Milk

The pH value of fermented milk decreased linearly with increasing addition of sea buckthorn polysaccharides. This decrease is attributed to the accumulation of organic acid produced by lactic acid bacteria during the fermentation process. The addition of polysaccharide promotes bacterial fermentation, accelerating the accumulation of organic acids and lowering pH (Kip et al. 2006). The resultant lower pH inhibits the growth of harmful pathogens in milk, thereby enhancing its shelf life and overall quality. Similarly, the increase in sea buckthorn polysaccharides led to a gradual rise in acidity, aligning with pH changes. The fatty acids and organic acids present in sea buckthorn polysaccharides also influenced the acidity of fermented milk.

18.5.2.3 Effect of Sea Buckthorn Polysaccharides on the Water-Holding Capacity of Fermented Milk

The water-holding capacity of fermented milk exhibited an increasing and then decreasing trend with the addition of sea buckthorn polysaccharide. The optimal addition of 0.15% sea buckthorn polysaccharides resulted in the highest water-holding capacity of 70.5%, while the water-holding capacity decreased when the addition continues to increase. Polysaccharides were found to induce conformational change in the protein structure of milk, forming protein gels and enhancing its coagulability. However, excessive polysaccharides damaged the gel microstructure, significantly reducing the whey protein gel modulus and leading to whey precipitation.

18.5.2.4 Effect of Sea Buckthorn Polysaccharides on the Color of Fermented Milk

Color plays a vital role in evaluating the quality of fermented milk, with L^*, a^*, and b^* values indicating brightness, hue shifts from green to red, and color changes from blue to yellow, respectively. The addition of sea buckthorn polysaccharide resulted in a gradual decrease in the L^* value and an increase in the a^*, b^*, and ΔE*values, leading to an overall darkening of the milk color with a brownish-yellow hue. This color alteration can be attributed to the inherent color of sea buckthorn polysaccharide and the accumulation of pigment impurities as the sea buckthorn polysaccharide content increases.

18.5.2.5 Effect of Sea Buckthorn Polysaccharides on the Texture of Fermented Milk

An appropriate addition of sea buckthorn polysaccharide enhanced various textural properties of fermented milk, including hardness, viscosity, adhesion, cohesion, and pulling properties. This enhancement was linked to the promotion of extracellular polysaccharide production in milk, leading to interactions with casein proteins that improved the milk texture In addition, polysaccharides interact with proteins to create a robust structure during fermentation, elevating the hardness, viscosity, adhesion, and cohesion of fermented milk and contributing to an overall improved taste experience (Oliveira et al. 2011). In conclusion, sea buckthorn polysaccharides have a notable impact on the physical and chemical properties of fermented milk, opening avenues for the development of functional dairy products with enhanced nutritional profiles and consumer appeal. Further exploration of sea buckthorn fermented milk, possibly along with other functional ingredients, such as honey, turmeric, or tea polyphenols, holds promise for catering to diverse consumer preferences for healthy and functional foods. The ongoing research and development of functional sea buckthorn fermented milk products are poised to meet the evolving demands of consumers seeking a healthy and varied diet. With the rising trend in functional foods, sea buckthorn fermented milk is expected to play a significant role in promoting health and well being in the future.

References

Al-Maqtari QA, Al-Ansi W, Mahdi AA, Al-Gheethi AAS, Mushtaq BS, Al-Adeeb A et al (2021) Supercritical fluid extraction of four aromatic herbs and assessment of the volatile compositions, bioactive compounds, antibacterial, and anti-biofilm activity. Environ Sci Pollut Res 28(20):25479–25492

Boitsova TA, Brovko OS, Ivakhnov AD, Zhil'Tsov DV (2020) Optimizing supercritical fluid extraction of usnic acid from the lichen species usnea subfloridana. Russ J Phys Chem B 14(7):1135–1141

Domagala J (2009) Instrumental texture, syneresis and microstructure of yoghurts prepared from goat, cow and sheep milk. Int J Food Prop 12(3):605–615

Duran-Cruz V, Hernández S, Ortíz I (2021) Evaluation of steam explosion pretreatment and enzymatic hydrolysis conditions for agave bagasse in biomethane production. Bioenergy Res 14:1328–1337

Ge X, Tang N, Huang Y, Chen X, Dong M, Rui X et al (2022) Fermentative and physicochemical properties of fermented milk supplemented with sea buckthorn (*Hippophae eleagnaceae* L.). LWT 153:112484

Gee VL, Vasanthan, Temelli (2007) Viscosity of model yogurt systems enriched with barley beta-glucan as influenced by starter cultures. Int Dairy J 17(9):1083–1088

Guo J, Tucker ZD, Wang Y, Ashfeld BL, Luo T (2021) Ionic liquid enables highly efficient low temperature desalination by directional solvent extraction. Nat Commun 12(1):437

Haque MR, Bradbury JH (2002) Total cyanide determination of plants and foods using the picrate and acid hydrolysis methods. Food Chem 77(1):107–114

Johnson CD, Eagle SR, Nindl BC, Lovalekar MT, Flanagan SD, Pepping GJ, Connaboy C (2021) The effects of fatiguing exercise and load carriage on the perception and initiation of movement. Eur J Sport Sci 21(1):36–44

Kip P, Meyer D, Jellema RH (2006) Inulins improve sensoric and textural properties of low-fat yoghurts. Int Dairy J 16(9):1098–1103

Kivrak S, Kivrak I (2021) Ultrasonic-assisted extraction method of phenolic compounds in Extra-virgin olive oils (EVOOs) by Ultra performance liquid chromatography-tandem mass spectrometry (UPLC-MS/MS). Sep Sci Technol 56(2):1–8

Oliveira RPDS, Perego P, Oliveira MND, Converti A (2011) Effect of inulin as prebiotic and synbiotic interactions between probiotics to improve fermented milk firmness. J Food Eng 107(1):36–40

Park YJ, Hwang U, Park S, Sim S, Jeong S, Park M et al (2021) Optimal bioconversion for compound k production from red ginseng root (C.A. Mayer) by sequential enzymatic hydrolysis and its characteristics. *Applied*. Biol Chem 64(1):297–302

Qin L, Yu J, Zhu J, Kong B, Chen Q (2021) Ultrasonic-assisted extraction of polyphenol from the seeds of Allium senescens L. and its antioxidative role in Harbin dry sausage. Meat Sci 172:108351

Rawson HL, Marshall VM (1997) Effect of 'ropy' strains of *Lactobacillus delbrueckii* ssp. *bulgaricus* and *Streptococcus thermophilus* on rheology of stirred yogurt. Int J Food Sci Technol 32(3):213–220

Rojas LM, Qu Y, He L (2021) A facile solvent extraction method facilitating surface-enhanced Raman spectroscopic detection of ochratoxin A in wine and wheat. Talanta 224:121792

Santos PH, Kammers JC, Silva AP, Oliveira JV, Hense H (2021) Antioxidant and antibacterial compounds from feijoa leaf extracts obtained by pressurized liquid extraction and supercritical fluid extraction. Food Chem 344:128620

Sirijeerachai G, Khunasinkun S, Boonleau P, Nuchitprasitchai A (2021) Optimization of ultrasonic-assisted extraction of total phenolic compound from leaves of *Chromolaena Odorata* L. using response surface methodology. MATEC Web Conf 333:12

Soltani F, Darabi H, Aram R, Ghadiri M (2021) Leaching and solvent extraction purification of zinc from Mehdiabad complex oxide ore. Sci Rep 11(1):1566

Sun X, Zheng X, Tang Y, Debrah AA, Du Z (2021) Supercritical fluid extraction combined with ultrahigh performance liquid chromatography quadrupole time-of-flight mass spectrometry for determination of extractables to evaluate compatibility of drugs with rubber closures. AAPS PharmSciTech 22(1):61–70

Takahashi Y, Mukai K, Ohmura H, Takahashi T (2021) Changes in muscle activity with exercise-induced fatigue in thoroughbred horses. Comp Exerc Physiol 17(1):102901

Tang N (2021) Image recognition algorithm for exercise fatigue based on FPGA processor and motion image capture. Microprocess Microsyst 81(3):103756

Tang W, Wu X, Huang C, Ling Z, Lai C, Yong Q (2021) Comprehensive understanding of the effects of metallic cations on enzymatic hydrolysis of humic acid-pretreated waste wheat straw. Biotechnol Biofuels 14(1):12

Woo KL, Kim JL (1999) New hydrolysis method for extremely small amount of lipids and capillary gas chromatographic analysis as N(O)-tert-butyldimethylsilyl fatty acid derivatives compared with methyl ester derivatives. J Chromatogr A 862(2):199–208

Woo KL, Kim MC, Ki J (2002) Study on hydrolysis method for extremely small amount of lipids by organic basic solution, tetramethylammonium hydroxide/methanol and capillary gas chromatographic analysis of fatty acid composition depending on derivatization methods. J Appl Sci 2(8):J71–J75

Chapter 19
Product Development of Sea Buckthorn Fruit Wine

Huiyan Liu, Haitian Fang, Yujun Zeng, Pin Chen, and Jinhu Tian

Abstract Sea buckthorn fruit wine is a low-alcohol beverage made from sea buckthorn berries, which has gone through several processes such as juicing, clarification, sugar blending, fermentation, acid reduction, and then clarification of the fruit wine. Fruit wine made from freshly picked sea buckthorn fruit has relatively low alcohol content and contains specific amounts of organic acids, sugar, vitamins, minerals, amino acids, and other nutrients, which provide unique benefits compared with other types of wine. Sea buckthorn has high nutritional and medicinal value, and the wine made from sea buckthorn fruit and juice not only has the characteristics of fruit wine but also has certain healthcare functions. The development of sea buckthorn fruit wine meets contemporary people's desire for simple health and the pursuit of safe nourishment, and the market prospect is broad.

Keywords Sea buckthorn · Wine · Health function

Abbreviations

DL	Distilled liquor
SW	Sea buckthorn wine
VOCs	Volatile organic compounds

H. Liu
Food College, Ningxia University, Yinchuan, China

H. Fang (✉)
Food Science and Technology, Ningxia University, Yinchuan, Ningxia, China
e-mail: fanght@nxu.edu.cn

Y. Zeng · P. Chen · J. Tian
Department of Food Science and Nutrition, Zhejiang University, Hangzhou, China

J. Tian et al. (eds.), *Sea Buckthorn*, https://doi.org/10.1007/978-981-97-9865-0_19

19.1 Sea Buckthorn Fruit Wine

Sea buckthorn, also referred to as vinegar willow, blackthorn, sour thorn, or sour slipper, is a deciduous shrub or small tree belonging to the genus Buckthorn. It is an excellent tree species with ecological, nutritional, and economic value. Used in both food and medicine, sea buckthorn has a high nutritional value. In 1977, it was officially listed in the Chinese Pharmacopoeia. The sea buckthorn fruit contains hundreds of bioactive substances, with the highest content of V_C and V_E of all fruits and vegetables, and is known as a "treasure trove of vitamins". The active ingredient, sea buckthorn flavonoids, can invigorate blood circulation, resolve blood stasis, relieve phlegm and cough, and nourish the spleen. The active ingredient, sea buckthorn flavonoids, has a variety of effects such as invigorating blood stasis, resolving phlegm and cough, tonifying the spleen, and strengthening the stomach (Asofiei et al. 2019).

Fruit wines are produced from freshly harvested fruits and have a comparatively low alcohol content. They contain a specific number of organic acids, sugar, vitamins, minerals, amino acids, and other nutrients, which provide exclusive benefits compared with other types of wines. As individuals become increasingly health-conscious, they are no longer content with the addictive taste of alcohol but rather seek the flavor and health benefits of wine. China possesses plentiful fruit resources and its primary fruit wines include apple, pear, cherry, lychee, hawthorn, and persimmon wine, among others. Sea buckthorn has high nutritional and medicinal value, and the wine made from sea buckthorn fruit and juice not only has the characteristics of fruit wine but also has certain health care functions. The development of sea buckthorn fruit wine satisfies the contemporary desire for simplicity and health as well as the pursuit of safe and nourishing nourishment, indicating immense market prospects. It is a low-alcohol drink created from sea buckthorn fruit by undergoing various procedures, including juice extraction, clarification, sugar adjustment, fermentation, and acid reduction, followed by the fruit wine's clarification (Asofiei et al. 2016). An important aspect for the sea buckthorn industry is that the fermented fruit wine increases the nutritional value of the sea buckthorn and retains the nutrients even after fermentation. At present, some progress has been made in the research of the processing technology of sea buckthorn fruit wine in China, and sea buckthorn fruit wine has been produced and sold in the main sea buckthorn-producing areas in China, such as Qinghai, Shanxi, and Inner Mongolia, but there are still many problems restricting the development of sea buckthorn fruit wine processing industry (Attri and Goel 2018).

19.2 The Production Process of Sea Buckthorn Fruit Wine

A study has been carried out to make sea buckthorn wine using sea buckthorn berries as raw material using the fermentation method and the traditional infusion method, respectively, and then blending the original wine from the two processes to obtain a full-bodied, fruity sea buckthorn wine (Guo et al. 2017; Maheshwari et al. 2011). A researcher has optimized a production process to produce blended wine from sea buckthorn juice and edible alcohol (Sabir et al. 2005). Some researchers have also studied the process of making a compound fruit wine from sea buckthorn (Lehtonen et al. 2010; Shah et al. 2009; Zeb and Malook 2009). Sea buckthorn berry wine can be produced using either the fermentation or infusion method. Whenever technical terms are applied, their abbreviations will be explained. Fermented sea buckthorn berry wine is primarily made from crushed and pressed sea buckthorn berries, which are then separated by centrifugal sedimentation, subjected to enzymatic treatment, and undergo sugar adjustment, yeast fermentation, acid reduction treatment, and clarification. Currently, research into sea buckthorn fruit wine has yielded positive results. This was achieved through yeast strain screening, optimization of fermentation process parameters, and refinement of acid reduction and fruit wine clarification techniques for large-scale production.

19.2.1 Fermentation Technology of Sea Buckthorn Wine

In the optimization of the processing of sea buckthorn fermented fruit wine, Cai et al. (2018) some researchers used sea buckthorn juice as a raw material to optimize the fermentation process of sea buckthorn wine using a single-factor test and response surface analysis. The ideal fermentation conditions for sea buckthorn wine were achieved by maintaining a fermentation temperature of 26 °C. Under these conditions, the fermented sea buckthorn wine has an alcoholic content of 12.0% vol. Under these conditions, the result is a fermented sea buckthorn wine with an alcoholic content of 12.0% vol, with a strong fruity aroma and bright color. Niu et al. (2010) studied the optimal process conditions for the main fermentation of sea buckthorn fruit wine as follows: SO_2 mass concentration, 80 mg/L; temperature, 25 °C; brix, 18%. The optimum conditions for the main fermentation of sea buckthorn wine were: SO_2 mass concentration, 80 mg/L; temperature, 25 °C; sugar content, 18%; inoculum volume, 2%; and 60% liquid volume. He studied the biological acid reduction process of sea buckthorn wine and found that the optimal parameters for the malolactic fermentation process were: fermentation temperature of 21 °C, inoculum level of 11%, and fermentation time of 35 days. The total acidity of sea buckthorn wine was reduced from 15.4 to 8.9 g/L, and the sourness and astringency of sea buckthorn wine disappeared and the taste was greatly improved. Bai et al. (2018) screened out the most suitable sea buckthorn juice fermentation and studied the optimal fermentation process parameters by single-factor and orthogonal. The optimal fermentation process parameters of sea buckthorn yeast strain were investigated by single-factor and orthogonal tests: pectinase addition of 200 mg/L, sugar addition of 170 g/L, and yeast addition of 170 g/L. The optimal parameters for the fermentation process of the sea buckthorn yeast strain were examined using single-factor and orthogonal tests. These tests determined that the ideal conditions were pectinase addition of 200 mg/L, sugar addition of 170 g/L, yeast fermentation temperature at 22 °C, and a fermentation time ranging from 10 to 15 days.

The fermentation mash underwent chemical acid reduction and diatomaceous earth clarification. The fermentation mixture underwent chemical reduction of acidity and diatomaceous earth clarification before being clarified through sensory evaluation to produce the wine. The fermentation mash was subjected to chemical acid reduction and clarification with diatomaceous earth. Wu and Guo (2004) studied the clarification process of dry red sea buckthorn wine. The clarification process of red sea buckthorn wine was investigated by screening the types of clarifying agents (pectinase, gelatin, and tannin) and determining the optimum addition ratio and amount, determining the optimum addition ratio and the amount of addition. The test was carried out to achieve a light transmission rate of 85% and to improve the stability of the product. Wang et al. (2003) added glutinous rice mash to the fermentation process of sea buckthorn to offset its high acidity and lack of a carbon source. The following strain domestication and low-temperature maturation, the optimal conditions for the fermentation of sea buckthorn wine and the production process, were determined by orthogonal tests. Orthogonal tests were used to determine the optimal fermentation conditions and production process parameters for sea buckthorn wine. The fermented sea buckthorn wine was found to be pure to taste, full-bodied, and rich in nutrients.

With the development of the fruit wine industry, the development of fruit wine has changed from the traditional single fruit wine to a multi-variety, multi-mixed compound fruit wine, which takes advantage of the complementary. The use of the complementary nutritional composition of various fruits can improve the quality and variety of products. This can improve the quality of the product and enrich the product range. At present, research on compound sea buckthorn fruit wines is mainly focused on sea buckthorn red dates, sea buckthorn apples, sea buckthorn grapes, and sea buckthorn pears, and some progress has been made, which has greatly improved the quality of sea buckthorn. The quality of the wine has been greatly improved.

19.2.2 The Fermentation Strain of Sea Buckthorn Wine

Due to the low sugar content and high acidity of the sea buckthorn fruit, it is not suitable for the growth of brewer's yeast. The research on sea buckthorn wine mainly includes the fermentation process and screening of brewer's yeast. The yeast strain plays a decisive role in the quality of sea buckthorn wine, as the yeast is active and fermentative. The yeast for sea buckthorn wine should have a strong fermentation capacity, a certain ability to reduce acidity, complete sugar decomposition, high alcohol production, and strong SO_2 tolerance. The current fermentation of fruit wines mostly uses wine yeast, which is not conducive to the formation of the unique qualities of fruit wines (Liu et al. 2015). Hu et al. (2018) used the acid-tolerant yeast YM1-1E as the test strain and obtained a mild, fruity, and rich fruit wine with 12.7%

vol. Previous researchers used a combination of Pasteurella TS-82 and brewer's yeast to ferment sea buckthorn fruit wine, leading to the degradation of carotenoids in sea buckthorn juice and the consequent production of β-cyclic citral and β-viologenone. This process resulted in a significant increase in typical aroma substances while partially retaining carotenoids. The technical term abbreviations were explained upon first usage to improve comprehension. The nutrition and flavor of the sea buckthorn wine were enhanced, and the fruit wine's free phenolic content was increased due to the significant increase of typical aromatic substances and partial preservation of carotenoids.

Sea buckthorn wine contains hundreds of beneficial substances that regulate the body's metabolism, improve the immune system, promote blood circulation, prevent cholesterol increases, act as a diuretic, stimulate liver function, and prevent aging. Regular consumption of this beverage can prevent scurvy, anemia, and keratitis, as well as soften blood vessels and aid digestion. Some researchers used defatted sea buckthorn juice and clarified juice as raw material, the physicochemical parameters, aroma components, amino acid types, and contents of six sea buckthorn fruit wines fermented with brewer's yeast were determined by gas chromatography, mass spectrometry, and amino acid analysis, Yeast strains with optimal fermentation characteristics for brewing sea buckthorn fruit wine were screened. Artificial neural network technology was used to systematically optimize key process parameters of sea buckthorn fruit wine fermentation. The yeast strain screening experiments yielded results that demonstrated the fast fermentation rate, strong sugar reduction capacity, higher alcoholic strength, and weaker acid production capacity of yeast strains Y3, Y4, and Y5.

The sea buckthorn wine fermented by six yeast strains was found to have aroma components with different mass concentrations. The aroma components detected in the wine were 29, 54, 42, 55, 56, and 56, respectively. Interestingly, esters had the highest mass concentrations with relative contents of 46.14%, 75.97%, 79.93%, 77.86%, 81.41%, and 77.76%, respectively. These results indicate significant differences in the aroma. The total amount of amino acids detected was 125.44 and 194.03 mg/L in the sea buckthorn fruit wine, respectively. Yeast strains Y4 and Y5 were used to ferment the wine, with Y4 and Y5 detecting 14 and 17 amino acids, respectively. The vast quantity and variety of amino acids in the sea buckthorn fruit wine fermented by Y5 yeast was significantly larger than that of Y4 yeast. A thorough analysis has determined that Y5 yeast is the optimal strain for fermenting sea buckthorn fruit wine. The research results of the fermentation experiments of the original juice of sea buckthorn after defatting and the clarified juice showed that the fermentation rate of the clarified juice was faster than that of the original juice; 54 aromatic substances were detected in the fermented sea buckthorn fruit wine using the clarified juice as raw material. Fifty-two aromatic substances were detected in the sea buckthorn wine fermented in the original juice, of which 30 were not common to both, resulting in very significant differences in aroma composition. The primary aroma elements of the fermented sea buckthorn wine with clarified juice consisted of 3-methyl-1-butanol and ethyl caprylate, with relative contents of 31.24% and 10.40%, respectively. These levels were 16.45% and 5.32% higher than those of the original juice fermented sea buckthorn wine. The study's findings on the optimization of the fermentation process parameters for sea buckthorn fruit wine indicate that the highest ethanol content resulted from the fermentation of sea buckthorn clear juice using brewer's yeast, at a fermentation temperature of 28.5 °C, pH of 3.7, inoculum level of 0.18%, and brix of 23%. The ethanol content in the resulting fermentation broth was 14.0% (v/v), demonstrating a significant increase in ethanol compared to previous experiments.

By comparing the effects of different fermentation processes on the quality and aroma of sea buckthorn fruit wine, Shi et al. (2021) obtained the optimal brewing process for sea buckthorn fruit wine, providing a theoretical basis for obtaining high-quality sea buckthorn fruit wine. The brewing process of sea buckthorn fruit wine was carried out by three different processes: maceration, raw juice fermentation, and cooked juice fermentation. By testing the alcoholic strength, pH, total acidity, total sugar, total phenols, color, and volatile aroma components, the brewing process of the best quality sea buckthorn fruit wine was obtained, providing a theoretical basis for the process optimization and industrial production of sea buckthorn fruit wine. The results showed that the pH, total sugar, total acid, and alcohol of the infused wine were significantly different from the two fermented wines in the sea buckthorn fruit wine prepared using the three processes. The effects of the different production processes on the physicochemical indicators of sea buckthorn fruit wine are shown in Table 19.1. The effects of different production processes on the content of aroma components of sea buckthorn fruit wine are shown in Table 19.2. Significant differences were found in the coloration of sea buckthorn fruit wine between the different production processes. The color of the sea buckthorn infusion wine was the lowest of the three, but its color was the reddest, because the high concentration of ethanol solution could effectively leach the red pigment from the sea buckthorn fruit, and the leached pigment formed a polymer with the tannin to make its wine more stable. The darkest color is due to the non-enzymatic browning reaction that occurs during the cooking process and the increased leaching of tannins and pigments, resulting in a slightly yellowish-brown and rough body. The color value of the raw juice sea buckthorn fermented wine is in the middle of the three, with the greatest brightness, clarity, and brilliance. The study concluded that sea buckthorn fermented wine produced through the cooked juice fermentation process has a moderate alcohol content, prominent ester aromas, improved color, softer body, and higher total phenolic content, indicating its potential for use in producing sea buckthorn fruit wine.

19.3 The Flavor of Sea Buckthorn Wine

Numerous factors affect the quality and flavor of fruit wines in addition to the raw materials used. Among these, the conditions of the fermentation process and the type of yeast used in brewing play a crucial role in determining the final product (Fan 2006). Although the maceration method can produce fruit wine without yeast

Table 19.1 The effects of the different production processes on the volatile substances of sea buckthorn fruit wine

Serial number	Compound name	Relative content (%)		
		Soaking wine	Raw juice fermentation wine	Cooked juice fermentation wine
Esters				
1	Ethyl 3-methylbutyrate	0.01	–	–
2	Isoamyl acetate	0.02	0.03	0.13
3	sec-Butyl butyrate	0.01	–	–
4	tert-butyl 2-methylbutyrate	0.02	0.01	–
5	Ethyl caproate	11.81	0.81	0.38
6	Isobutyl isovalerate	0.01	0.01	0.01
7	Isoamyl isobutyrate	0.08	0.01	0.02
8	Ethyl heptanoate	0.98	0.1	0.12
9	Isoamyl isovalerate	0.61	0.11	0.07
10	Isobutyl caproate	0.47	0.03	0.05
11	Ethyl caprylate	1.63	1.69	2.07
12	Isoamyl caproate	0.2	0.01	0.01
13	Ethyl nonanoate	0.56	0.3	0.39
14	1-Phenylethyl butyrate	0.03	0.01	0.01
15	Ethyl sunflower acid	1.45	0.78	1.19
16	Ethyl undecanoate	0.01	0.01	–
17	Ethyl laurate	0.1	0.16	0.23
18	Ethyl cinnamate	0.01	0.06	0.09
19	Ethyl 9-hexadecanoate	0.03	0.29	0.45
20	Ethyl palmitate	0.02	0.14	0.21
21	Ethyl acetate	0.41	–	0.18
22	Ethyl butyrate	0.01	–	–
23	Ethyl lactate	0.11	–	–
24	Ethyl valerate	0.22	–	–
25	Isopropyl hexanoate	0.01	–	–
26	Ethyl trans-2-hexenoate	0.03	–	–
27	2-Pentyl butyrate	0.02	–	–
28	Propyl caproate	0.13	–	–
29	2-Methylbutyl 2-methylbutyrate	0.15	–	–
30	2-Methylbutyl 3-methylbutyrate	1.44	–	–
31	3-Hexenol butyrate	0.01	–	–
32	Hexyl isobutyrate	0.01	–	–
33	Butyl caproate	0.02	–	–
34	Isoamyl 2-methyl-2-butyrate	0.01	–	–
35	Butyl heptanoate	0.04	–	–
36	Isopropyl caprylate	0.04	–	–
37	Hexyl isovalerate	0.08	–	–
38	Heptyl isobutyrate	0.01	–	–

(continued)

Table 19.1 (continued)

Serial number	Compound name	Relative content (%)		
		Soaking wine	Raw juice fermentation wine	Cooked juice fermentation wine
39	Pentyl hexanoate	0.02	–	–
40	Ethyl 8-nonanoate	0.01	–	–
41	sec-Butyl benzoate	0.28	–	–
42	Nonyl acetate	0.01	–	–
43	2-Methylheptyl butyrate	0.02	–	–
44	Isobutyl caprylate	0.01	–	–
45	Isoamyl benzoate	0.4	–	0.05
46	2-Methylbutanoic acid 2-phenylethyl ester	0.03	–	–
47	Butyl 1-pentanoate	–	0.08	–
48	Ethyl 2-furoate	–	0.01	–
49	Pentyl isovalerate	–	0.36	–
50	Methyl caprylate	–	0.01	–
51	Ethyl benzoate	–	0.09	–
52	Diethyl succinate	–	0.04	0.02
53	Ethyl trans-4-octenoate	–	0.01	–
54	Ethyl phenylacetate	–	0.03	0.02
55	Ethyl phenyl acetate	–	0.05	0.08
56	Isobutyl benzoate	–	0.14	0.01
57	Ethyl 9-decenoate	–	0.01	–
58	Isoamyl formate	–	0.07	–
59	Isoamyl Caprylate	–	0.03	0.04
60	Phenethyl isovalerate	–	0.01	–
61	2-Methylpropionate	–	0.01	–
62	Isoamyl decanoate	–	–	0.01
Alcohols				
63	Ethanol	2.27	1.45	1.74
64	Isopentyl alcohol	0.17	0.76	2.4
65	Phenethyl alcohol	0.03	0.73	0.62
66	2-Methyl-1-butanol	0.08	–	–
67	1-Octanol	0.01	–	–
68	2,3-Butanediol	–	0.2	0.19
69	2-(methylthio)ethanol	–	0.07	0.01
70	n-Heptanol	–	0.06	0.04
71	Methylheptenol	–	0.01	–
72	Benzyl alcohol	–	0.01	–
73	Linalool	–	0.05	–
74	1-alcohol	–	0.1	–
Ketones and aldehydes				
75	Benzaldehyde	0.01	0.02	–

(continued)

Table 19.1 (continued)

Serial number	Compound name	Relative content (%) Soaking wine	Raw juice fermentation wine	Cooked juice fermentation wine
76	Dodecanal	0.01	–	–
77	2-Nonanone	0.01	–	–
78	Nonanal	0.5	–	–
79	Nonanal diethyl acetal	0.02	–	–
80	6,10-Dimethyl-5.9-undecadien-2-one	0.04	–	–
81	Tetradecanal	0.02	–	–

Prepared by the author

fermentation, and the style is like white wine, with obvious wine flavor, its flavor is not harmonious. The sea buckthorn fruit wine obtained by the maceration process in the study retained the original characteristic aromas of white wine, such as ethyl caproate, ethyl lactate, ethyl acetate, ethyl valerate, and other ester aromas, but the fruit aroma was thin, the color was low, the acidity was obvious, there was almost no astringency, and the wine body was not harmonious enough. Both the raw juice fermentation and the cooked juice fermentation involve fermentation and result in the production of complex metabolic products, resulting in a mellow and aromatically complex finished wine (Fan 2006). The raw juice fermentation process in the study maximized the nutrients in the raw fruit, with good performance in terms of color, golden color, rich aroma composition, a wide range of alcoholic aromas and prominent fruit aromas, with phenethyl alcohol, benzyl alcohol, and linalool as key fruit wine aromas (Jin 2014). The wine presents a gentle and subtle floral and fruity scent, with a slightly astringent and mellow flavor. During the maturation of the juice, the process of fermentation leads to a greater phenolic content compared with the raw juice fermentation, owing to the heat treatment of the raw material. Additionally, the wine has a noticeable ester aroma, containing a higher concentration of phenethyl acetate, the fragrance that imparts a base aroma of rose and honey to the matured juice fermentation wine. The wine is characterized by its high content of ethyl myristate, which imparts a coconut and subtle iris scent, as well as a significant amount of ethyl caprylate that is distinctive to sea buckthorn fruit wine. This gives the wine a pleasant astringency and a softer body, resulting in a more balanced and superior taste overall.

Acid reduction, sugar conditioning, and yeast or quat fermentation are all steps in the production of sea buckthorn wine. Negi et al. (2013) developed sea buckthorn wine with significant antioxidant activity and higher levels of flavonoids, quercetin, poplar, and rutin than Sauvignon salad wine. It was also found to be protective against hypercholesterolemia-induced oxidative stress caused by furazone and a high cholesterol diet in male LACA mice. However, the disagreeable flavor of this potentially useful product hinders its industrial development. Thus, it is vital to examine the flavor of sea buckthorn wine (SW) and sea buckthorn distilled liquor (DL). In this work, we examined the various metabolites of sea buckthorn DL

Table 19.2 Effects of different processing technologies on physicochemical indexes of sea buckthorn fruit wine

Sea buckthorn wine	Total sugar /($g \cdot L^{-1}$)	Total acid /($g \cdot L^{-1}$)	Alcohol Content /%	Ph	Total phenol /($mg \cdot L^{-1}$)	Chrominance
Infused wine	4.53 ± 0.21b	2.41 ± 0.11b	45.57 ± 0.32a	4.35 ± 0.02a	94.95 ± 4.76c	0.571 ± 0.01b
Fresh juice fermented wine	9.62 ± 1.60a	11.66 ± 0.11a	11.20 ± 0.30bc	3.75 ± 0.01b	403.76 ± 15.88b	0.619 ± 0.02b
Ripe fermented wine	9.78 ± 1.05a	12.53 ± 0.11a	11.57 ± 0.47b	3.70 ± 0.01bc	483.76 ± 20.82a	0.815 ± 0.02a

Prepared by the author

during processing and determined the link between E-nose sensor values and important volatile organic compounds (VOCs). The results revealed a total of 133 VOCs, 22 of which contributed to scent. Fermentation dramatically raised the levels of VOCs, particularly esters. Fermentation and distillation considerably increased the levels of seven and 51 VOCs, respectively. Additionally, seven sensors showed a positive correlation with rising levels of alcohols and esters, mirroring the upward trend of ten major VOCs (Xia et al. 2022). First, the physicochemical indexes and antioxidant activity of SJ, SW, and DL. SJ's soluble sugar content fell considerably during fermentation, but there was no significant variation after distillation. The pH of SJ remained between 3.30 and 3.55 during the process, with no noticeable differences. Meanwhile, fermentation dramatically enhanced SJ's antioxidant activity. SW DPPH and OH radical scavenging rose dramatically, from 80.64% and 44.72% to 91.55% and 97.10%, respectively. Reducing activity also increased significantly, from 3.79 to 3.94 (Xia et al. 2022). He (2015) observed the similar tendency during SW production. DPPH scavenging and lowering activity increased dramatically with fermentation time, peaking at 6–8 days. The DPPH scavenging and lowering activities increased as fermentation duration rose. During the early fermentation stage, phenolic and flavonoid levels increased, possibly due to improved antioxidant ability. Wang et al. (2022) discovered that the DPPH radical scavenging activity of sea buckthorn tea was 93.42% at a concentration of 0.1 mg/mL, showing that sea buckthorn berry products have exceptional antioxidant activity. However, distillation did not significantly impair antioxidant capability. Wang and Li (2014) discovered that -OH scavenging may reach more than 80% when the amount of Aloe vera juice exceeded 2.5 mL. Xing (2011) also found that after fermentation, esters in SW increased by 13.03%, which dominated the taste components (87.37%), improved SW's fruity and ester scent, and promoted flavor maturation. Ma (2014) discovered that the relative proportion of phenylethanol in commercially available sweet sea buckthorn wine was 7.193%. Lukša et al. (2020) used *Saccharomyces cerevisiae* yeast fermentation to create sea buckthorn wine. Phenylethanol was found to account for 3.6% of the total scent, indicating its stability as a flavor component in sea buckthorn wines. 4-Ethylphenol is often found in rums and whiskies, and it is also used to flavor wines and white wines (McKay et al. 2018, 2021). Nonanal adds citrus scent (OAV 8.795) and enhances the overall flavor of many wines (Xia et al. 2020).

19.4 Fermentation Kinetics of Sea Buckthorn Wine

The fermentation of sea buckthorn into sea buckthorn fruit wine is the primary method of refining the berry in China. To ensure that the process meets expectations, it is crucial to determine the effects of fermentation on sea buckthorn. The process of microbial formation and growth is described by fermentation kinetics, and it is imperative to examine the connection between fermentation time, substrate consumption, bacterial growth, and the synthesis of fermented products on a large scale. Domestic scholars have previously studied mulberry wine by using existing models for non-linear fitting of the indicators, ultimately achieving favorable outcomes. Liu and Zhang (2022) studied the relationship between reducing sugar content, antioxidant properties, fermentation kinetics, and fermentation time using sea buckthorn wine subjected to low-temperature fermentation. Logistic, S-Gompertz, Dose-Resp, and Boltzmann models were used to investigate the relationships between reducing sugars, Vc, total polyphenols and total flavonoids, antioxidant capacity, and DPPH clearance. The study revealed a pattern of rising and falling DPPH clearance, whereby extended fermentation released substantial quantities of antioxidant substances and an elevation in polyphenols. The results showed a trend of increasing and then decreasing DPPH clearance, with increasing fermentation time leading to the leaching of large amounts of antioxidants and increasing polyphenol content. The study showed that the common characteristics of phenolics are mainly in two aspects: first, they give hydrogen ions; and second, they achieve stabilization through resonance hybridization (Yao et al. 2020). This is the reason why phenolics are generally ideally cleared and their DPPH clearance rate is naturally maintained at a high level. Therefore, the antioxidant capacity of sea buckthorn wine in the fermented state tends to increase and then decrease, which is approximately the same as the DPPH release rate. The enzymatic treatment of sea buckthorn easily breaks down the cell wall, leading to the release of a substantial amount of its contents. Because of prolonging the fermentation process, there is a notable rise in the ethanol levels within the sea buckthorn wine, whereas the content of active substances, denoted by Vc, generally exhibits fluctuations. The trend of the flavonoid content of sea buckthorn wine is to rise and then fall, with a rising range of 24–72 h. The fermentation process achieves its peak at 72 h, with a value of approximately 12.5 mg/mL. In the context of total continuous fermentation, the plant-based juice becomes heavily saturated. As the alcohol content increases, the duration of flavonoid leaching into the juice is extended, resulting in notable alterations in the quantity of dissolved substances. When the fermentation process enters a later stage, the growth rate of flavonoid content tends to level off for two reasons: first, many substances, represented by acetaldehyde, react with flavonoid substances, resulting in the production of many derivatives; second, oxidation and decomposition of polyphenols occur, and the growth rate of flavonoid substances slows down as the polyphenol content decreases. Therefore, the key to making sea buckthorn wine as effective as it should be is to determine the appropriate length of fermentation based on the results of non-linear fitting (Zhao et al. 2018b).

References

Asofiei I, Calinescu I, Trifan A, David IG, Gavrila AI (2016) Microwave-assisted batch extraction of polyphenols from sea buckthorn leaves. Chem Eng Commun 203(12):1547–1553

Asofiei I, Calinescu I, Trifan A, Gavrila AI (2019) A semi-continuous process for polyphenols extraction from sea buckthorn leaves. Sci Rep 9(1):12044

Attri S, Goel G (2018) Influence of polyphenol rich seabuckthorn berries juice on release of polyphenols and colonic microbiota on exposure to simulated human digestion model. Food Res Int 111:314–323

Bai X, Han X, Zhang Y (2018) Study on nutritional components and antioxidant activities of different kinds of seabuckthorn pomac. Food Res Dev 39(23):187–191

Fan J (2006) Study on proanthocyanidin of sea buckthorn seed. Jiangnan University, Wuxi

Guo R, Chang X, Guo X, Brennan CS, Li T, Fu X, Liu RH (2017) Phenolic compounds, antioxidant activity, antiproliferative activity and bioaccessibility of Sea buckthorn (Hippophaë rhamnoides L.) berries as affected by in vitro digestion. Food Funct 8(11):4229–4240

He J (2015) Polyphenols and antioxidant activity of sea buckthorn wines during fermentation. Northwest Agriculture and Forestry University, Xianyang

Hu JU, Wen Y, Yu Y, Chen C, Jiang M, Gu X (2018) Application of acid-resistant yeast strains in seabuckthorn wine. China Brew 37(05):71–75

Jin Z (2014) Study on antioxidant activity, application and liver protecting effect of flavonoids extracted from seabuckthorn leaves. Dongbei Unversity, Haerbin

Lehtonen H, Järvinen R, Linderborg K, Viitanen M, Venojärvi M, Alanko H, Kallio H (2010) Postprandial hyperglycemia and insulin response are affected by sea buckthorn (Hippophaë rhamnoides ssp. turkestanica) berry and its ethanol-soluble metabolites. Eur J Clin Nutr 64(12):1465–1471

Liu Y, Zhang J (2022) Kinetic study on reducing sugar in sea-buckthorn wine under low-temperature fermentation. Liquor Mak Sci Technol 6:26–29

Liu M, Liu Y, Lian Y, Xiao Y (2015) The Treasure in the Silk-road: Hippophae rhamnoides. Mod Chin Med 17(3):191–194

Lukša J, Vepštaitė-Monstavičė I, Apšegaitė V, Blažytė-Čereškienė L, Stanevičienė R, Strazdaitė-Žielienė Ž, Ravoitytė B, Aleknavičius D, Būda V, Mozūraitis R (2020) Fungal microbiota of sea buckthorn berries at two ripening stages and volatile profiling of potential biocontrol yeasts. Microorganisms 8(3):456

Ma N (2014) Comparative analysis of nutritional components and flavoring components of seabuckthorn wine & grape wine. Liquor Mak Sci Technol 2:92–97

Maheshwari D, Kumar MY, Verma SK, Singh VK, Singh SN (2011) Antioxidant and hepatoprotective activities of phenolic rich fraction of Seabuckthorn (Hippophae rhamnoides L.) leaves. Food Chem Toxicol 49(9):2422–2428

McKay M, Bauer FF, Panzeri V, Buica A (2018) Testing the sensitivity of potential panelists for wine taint compounds using a simplified sensory strategy. Food Secur 7(11):176

McKay M, Bauer FF, Panzeri V, Buica A (2021) Investigating the effects of two volatile phenols on aroma perception of four red wine cultivars using Projective Mapping. J Sens Stud 36(1):e12616

Negi B, Kaur R, Dey G (2013) Protective effects of a novel sea buckthorn wine on oxidative stress and hypercholesterolemia. Food Funct 4(2):240–248

Niu G, Dan Z, Zhao F, Wu W (2010) Study on fermentation technology of malolactic fermentation for sea buckthorn wine. J Jilin Agric Univ 32(05):555. –559+585

Sabir S, Maqsood H, Hayat I, Khan M, Khaliq A (2005) Elemental and nutritional analysis of sea buckthorn (Hippophae rhamnoides ssp. turkestanica) berries of Pakistani origin. J Med Food 8(4):518–522

Shah AH, Ahmad SD, Khaliq I, Batool F, Hassan L, Pearce R (2009) Evaluation of phylogenetic relationship among seabuckthorn (Hippophae rhamnoides L. spp. Turkestanica) wild ecotypes from Pakistan using amplified fragment length polymorphism (AFLP). Pak J Bot 41(5):2419–2426

Shi F, Qiao YJ, Wang J (2021) Effects of different fermentation processes on the quality and aroma of seabuckthorn Wine. J Shanxi Agric Univ (Nat Sci Ed) 41(6):115–122

Wang L, Li Z (2014) Functional ingredients and in vitro antioxidative activity of aloe liquor. China Brew 33:57–61

Wang D, Zhang Y, Zhang Y (2003) Study on the production technology of fermented Sea buckthorn wine. Food Sci 5:118–122

Wang N, Wen X, Gao Y, Lu S, Li Y, Shi Y, Yang Z (2022) Identification and characterization of the bioactive polyphenols and volatile compounds in sea buckthorn leaves tea together with antioxidant and α-glucosidase inhibitory activities. Front Nutr 9:890486

Wu H, Guo C (2004) Study on defecation in dried red Sea buckthorns fruit wine. Food Sci Techol 1:72–73

Xia Y, Liu Y, Wang J, Shuang Q (2020) Assessment of key aroma compounds in fresh jujube brandy by GC-O-MS and odor activity value. J Food Process Preserv 44(7):e14494

Xia Y, Zha M, Liu H, Shuang Q, Chen Y, Yang X (2022) Novel insight into the formation of odour—active compounds in sea buckthorn wine and distilled liquor based on GC–MS and E–nose analysis. Food Secur 11(20):3273

Xing W (2011) Research on sea buckthorn wine malolactic fermentation conditions and characteristics. Research on sea buckthorn wine malolactic fermentation conditions and characteristics, Harbin

Yao N, Che F, Li Y, Zhang T, Zhang H (2020) Nutritional value and comprehensive development and utilization of Hipopophae rhamnoides. Storage Process 20(2):226–232

Zeb A, Malook I (2009) Biochemical characterization of sea buckthorn (Hippophae rhamnoides L. spp. turkestanica) seed. Afr J Biotechnol 8(8):1625–1629

Zhao M, Hong HE, Fan M, Ren F, Han MA (2018a) Effects of new starter cultures on quality and antioxidation of sea-buckthorn wine. Sci Technol Food Ind 39(15):15–20

Zhao B, Xiang X, Wang W, Li C, Wu X, Shen J (2018b) Preparation of flavonoids from sea buckthorn and its inhibitory effect on human prostate cancer pc-3 cells in vitro. Nat Prod Res Dev 30(1):27–32

The manufacturer's authorised representative in the EU is Springer Nature Customer Service Centre GmbH, Europaplatz 3, 69115 Heidelberg, Germany. If you have any concerns regarding our products, please contact ProductSafety@springernature.com

Printed and bound by CPI Group (UK) Ltd, Croydon, CR0 4YY
01/10/2026
02225960-0012